T0234940

# Virtuality and Humanity

"From online meetings to immersive gaming, virtuality appears to be everywhere and nowhere. This ground-breaking book provides the first systemic account of the concept in all its dimensions and complexity. The thesis is deceptively simple: virtuality is not some revolutionary form of mediation brought on by digital technology; it is part and parcel of being human and therefore an integral component of our existential condition. Its demonstration, however, is multifaceted and intricate as Sam Lehman-Wilzig takes readers on an exciting and expertly guided journey of discovery, examining the ways that our human reality has always and already been virtual."

—David J. Gunkel, *Presidential Research, Artistry and Scholarship Professor, Department of Communication, Northern Illinois University, USA*

"Prof. Sam Lehman-Wilzig is a world leading scholar on virtuality. His latest book, *Virtuality and Humanity*, is a magnum opus. It is, without a doubt, the most wide-ranging and complete work on this topic available. As the title promises, Lehman-Wilzig vividly demonstrates that virtuality is an essential part of humanity. He points out that while time spent in the technological, virtual world is increasing apace, virtuality has in fact always been a part of human life. Drawing on his encyclopedic knowledge, Lehman-Wilzig takes his readers on a fascinating journey of the role played by virtuality throughout human history. The book is well written, highly thought-provoking and extremely relevant to our lives in the 21st century. I recommend it highly."

—Professor Yair Amichai-Hamburger (D.Phil., Oxford), *Director of the Research Centre for Internet Psychology, Reichman University, IDC, Herzlia, Israel*

Sam N. Lehman-Wilzig

# Virtuality and Humanity

Virtual Practice and Its Evolution from
Pre-History to the 21st Century

Sam N. Lehman-Wilzig ⓘD
School of Communication
Bar-Ilan University
Ramat Gan, Israel

ISBN 978-981-16-6525-7        ISBN 978-981-16-6526-4   (eBook)
https://doi.org/10.1007/978-981-16-6526-4

This Springer imprint is published by the registered company Springer Nature Singapore Pte Ltd.
The registered company address is: 152 Beach Road, #21-01/04 Gateway East, Singapore 189721,
Singapore

*To two people who had a major impact on my life, personally and professionally:*

*First, my wife Tami, whose constant prodding and support has pushed me to accomplish what she knew I was truly capable of achieving.*

*Second, to my late Harvard University professor, the noted sociologist Daniel Bell, whose teaching and scholarship opened my eyes to the vast potential of cross-disciplinary research.*

# Preface

Let us start with two short, related Rorschach tests. First, what do you see here?

## "virtualityinhumanity"

Second, how do you read this?

## *"virtualityvshumanity"*

I'll wait a few moments…

Did you see "virtuality **in**humanity"—or rather "virtuality **in** humanity"? Did you gloss "virtuality **vs [versus]** humanity" or perhaps "virtuality **us** humanity," or even "virtuality **is** humanity"?

This exercise is precisely the reason for writing this book. Too many people think of virtuality in negative, or at least threatening, terms. My basic argument is that there is little to fear and much to admire. Indeed, by our very nature humanity requires (yes, *requires*) virtuality, although that is not to suggest that virtuality IS humanity.

Like the air we breathe and electricity powering our lives—both hidden from sight although palpably enabling our world to function—so too with virtuality. The underlying thesis of this book is that the virtual experience is so embedded in the human condition that we are hardly aware of its omnipresence. Indeed, we are so oblivious to human virtualizing that it is only when the "unreal" becomes overly blatant or appears in new forms that we are wont to call it "virtuality."

One way of dealing with something that is too much a part of us to be properly appreciated is to remove ourselves from our embedded and embodied life condition and try for a moment to view the situation from the outside. To that end, let us try a mental exercise. Imagine an intelligent alien from another world who alights on our planet. Obviously, its psychology will be somewhat different than that of humans. In this particular case, there are two major (albeit related) differences between the alien and human beings.

First, the alien species cannot think about, imagine, or deal with anything that is not concretely real in the here and now—very similar to the way animals in our world consciously think (not including instinctual behavior), i.e., "what you see (smell, hear, etc.) is what you get." That includes anything that is symbolic or representational—beyond their ken. Second, whereas we can picture things differently from what we know to exist and even are capable of envisioning things that do not (and perhaps *cannot*) exist, the aliens can only comprehend things that "look" (sound, smell, etc.— or whatever perceptual senses they have) the same as what they come into contact in their daily experience.

Imagine the alien's puzzlement—even total bewilderment—as it walks around our world following and observing people and material phenomena. As humans look quite different from this alien species, the alien explorer has no choice but to kidnap several humans and steal their artifacts because the aliens back home would be incapable of imagining how we (human) "aliens" looked and acted— unless directly faced with real living flesh and blood humans. So upon returning to its home planet, the alien presents to its fellow beings the Earthlings and their "machinations," something that is not easy because the alien language does not have the relevant terms—and its mental framework is not geared for—many human concepts and patterns of behavior. Herewith is their report:

1.  *These Earth people exchange real objects of obvious value for this intrinsically worthless piece of colored paper that they call "money" or for the temporary use of a piece of plastic. Here's an example [shows them a dollar bill]. What's even stranger is that they make great efforts to stop other people from duplicating the same worthless paper!*

2.  *Many, many people go to a specific type of building (here's a good example from a place they call the Vatican) and mumble "prayers" to an unseen deity— once a week, every day or even a few times a day. I understand that in the past the number of people doing this was even larger but even so, the numbers today are huge.*

3.  *There are humans called "physicists" (there's one—the guy with the bushy hair) who talk of "particles" that they cannot see, and even stranger, they claim that such particles are sometimes corporeal and sometimes simply an immaterial "wave." Worse still, these physicists claim that the universe is mostly comprised of "dark matter" and "dark energy" that no one is able to see! They also have a problem coming up with a single explanation of all physical phenomena in the universe so that many have constructed a "theory" (your expressions of consternation are in place; I too have no idea what this means). This theory is about non-corporeal "strings" in eleven dimensions when they have never ever seen or otherwise provided any evidence for more than three dimensions! (They even call "time" a dimension; please do not ask me what they mean by that.) Oh yes, one final thing about these very strange*

*physicists: some of them, in fact the ones that seem to be considered the best, indulge themselves in something called a "thought experiment." They don't even bother trying to look at the physical world but just "think" about very weird situations, as if anyone could gain knowledge by "doing" nothing!*

4. *Their people sit alone for hours looking at squiggles on a paper object. They call it a "book"—here's what it looks like [shows them]. When I asked them what they're doing they answered that they are "reading" about another world or imaginary scenes, (re)creating scenes, events, and people in their minds. They call this "imagining"—don't ask...*

5. *Young people sit looking at a colored "screen" for hours, in obvious excitement as they manipulate pictures of non-living "creatures." Young and old people alike sit in front of another "screen" for hours looking at pictures of people and other objects that can only be seen unnaturally in two dimensions. I am about to show you this "television"; try not to get visually disoriented. Almost everyone watches this screen at some time of the day—and many do this for hours at a time—listening to sounds with no musical instruments anywhere to be seen, not to mention people's voices without a body in sight.*

6. *I continually saw very large numbers of people sitting, standing, and walking while holding a small object to their ear and talking to no one in the vicinity! Here's what this object looks like... They claim that the person they are talking to is elsewhere but he or she cannot be seen nor can they be touched—only heard.*

7. *On this strange planet there are people everywhere who feel loyal to (some are even willing to die for!) something called a "nation" or a "country" that they cannot see or touch. These "nations" can include millions of people (or more)—most of whom each individual member will never meet and almost all of whom they will never know. I am sorry, but it was impossible to bring you a concrete example of a "country."*

8. *When they are not thinking about the "nation," they spend many hours a day working within and for something they call a "company" or "organization"— again something that cannot be touched or directly seen—but which seems to demand and get their undivided loyalty. Many of these workers also send messages to people within the organization that they have never met in person because the latter live in another "nation"!*

9. *About a third of the time each day, all people fall into an unconscious state that they call "sleep." When asked afterwards what they were doing, they say they were "dreaming," which as best as I can understand from them means that they make up scenes in their head that have no connection to the real world, for no reason that they can adequately explain. Look over there—see that human lying down? Notice his eyelids fluttering? That's dreaming!*

10. *This does not happen only when they "sleep." Many people do this also when they are awake—just sit and think about things or scenes that are very different from what is in front of them. This is called "daydreaming" and seems to be quite similar to what is called "reading." Right now, none of the humans I*

brought along are doing that; obviously, they are too interested in us to have their "minds" go elsewhere (I'll return to this strange word a little later).

11. *Just like us, these people also eventually die. But when I asked why they have strange "ceremonies" when disposing of the body, most answer that they want to make sure the body reaches "heaven." Unfortunately, no one can really explain to me with any exactitude what or where this "heaven" is, but almost everyone tends to point up. Even stranger, they also mention something called "hell" which they also cannot describe because no one alive has ever seen it, but they are all sure that it is located somewhere "down" beneath the ground. I should mention that in another part of their world the answer to their body disposal ceremony is connected to my next point.*

12. *I decided to discuss with people from all over this world what makes a person alive. While some people used terms that I could understand (heat exchange, entropy, metabolism) others in a large part of the planet used a term that I cannot understand and find very hard to describe: "soul." This seems to be something they believe resides in their skull area but is separate from the organic matter of the brain (they used to believe that the "soul" was found in the chest cavity but now they "know" that this is no longer true). No one has ever seen it or even successfully measured it but most humans I talked to believe strongly that it exists. By the way, they claim that it is the "soul" that continues to live after the body dies and which goes to heaven. But when I asked why then they take such care in disposing of the body that doesn't go to heaven, they had no ready answer. In another part of their world, I received a different answer that I must say makes a little bit more sense to me: the "soul" returns in a new body, although not necessarily in the same type of creature. Obviously, I was unsuccessful in finding a "soul" to show you.*

13. *Many humans talk about things that they call "ghosts," "vampires," "saints," "angels," the "devil" (I do not know why there is only one of this last one) and several other creatures that people have never seen and cannot provide real evidence that they even exist! A few people claim to be able to talk to these unseen creatures and many others believe this and ask these "mediums" to send messages for them. Here too I obviously failed to bring you a physical example.*

14. *Almost everyone seems to do this next thing many times in their daily life: mathematics. Once again, these are squiggles on paper or on small electronic screens and it is claimed that they represent real things. If so, I have not been able to understand why one of these "numbers" is something called "zero" that is equal to nothing! And then there are other things called "irrational numbers" that are parts of something but that cannot ever be precisely measured. And if you think that's weird, then don't even try to understand something called an "imaginary number"—they tried to explain to me that it's a number that does not really exist! Not to mention "infinity": something that cannot be counted because it is far greater than all the atoms in the universe!! Another thing I obviously couldn't bring along to demonstrate to you.*

15. *It turns out that for many, many centuries the most intelligent humans have been doing something called "philosophy." I say "doing" but that's not really true—they just sit and think, without any connection to any object (other than occasionally writing down those squiggles on paper). And what are they thinking about? Many of them talk about non-things such as the "Ideal Form," "metaphysics" (meaning something that is above and separate from physical reality—whatever that means), and other things that do not seem to have any connection to anything that anyone can see, touch or hear; a sort of non-sense. Perhaps the strangest question of all that they ask is this: can people sense the world around them only through the organic matter in their brain or is there something there beyond physical matter (they call it "mind") that enables them to be "conscious"? I tried to explain that the answer is obvious—there cannot be anything beyond physical matter—but many people refused to be convinced.*

16. *As if all these non-physical phenomena were not enough, many people ingest real chemical substances (they call them "drugs") in order to influence their brain so that they can perceive other "quasi-sensory" things through their thoughts alone: "out of body experiences," "hallucinations," and other strange phenomena. Others don't even have to take drugs to do this: they sit quietly for hours with their eyes closed and claim to be able to transport themselves to elsewhere in a different mental state (that's a lie: I have sat watching them— they never move at all!).*

17. *You might ask whether each human learns these strange things from other humans. Not altogether—their strange abilities seem to be something that they are born with. How else to explain the fact, for example, that almost all little people ("children"), a short while after their birth, indulge in what is called by Earthlings "make believe"? These children will take objects, and act as if they were something (and somewhere) else completely. This would be very impressive for beings so young if it wasn't for the fact that it looks and sounds ridiculous. Even worse, their adult caretakers think this is an important stage in their development! But perhaps that should not be so surprising, as the adults go to large halls to watch other people (they call them "actors") doing the same make-believe activity. See that fellow in the long white gown over there? He's an example of how an actor dresses up, although if I would have tried to bring along all the different examples of such "dressing-up," my ship would not have gotten off the ground!*

18. *This strange activity called "acting" is just a small part of a much wider human phenomenon they call "the arts," much of which seems to me to be haphazard lines, sounds, and squiggles on paper or screen. They tried to explain the beauty of what they call non-representational art, fairy tales and myth narratives, science fiction, imaginary bestiaries, and synthesized sounds, but I couldn't find anything real by which I could evaluate these endeavors.*

19. *Finally, I thought at first that perhaps these strange human phenomena are caused by some temporary plague. However, I have also talked to people (called "Historians") who claim to have knowledge of other humans now dead, and they informed me that most (although not all) of these behavioral*

*patterns have been part of all previous humans when they too were alive. I can only conclude that the human race is inherently crazy and nothing of any worth can come out of interaction with them. We are worlds apart from them in more ways than one. I recommend sending all these unfortunate people back to their world. In fact, they have a place that seems to be appropriate for them all. They call it an "insane asylum."*

So much for the external perspective of exclusively material-minded aliens. As humans, we do "understand" what we are doing, even if it is not always easy to explain the rationale or function of our vast "virtual" behavior repertoire. In many instances, however, we do not even try—not so much because we believe that our beliefs and behavior are inexplicable but rather because they are so "normal" and (as the alien noted) so universal among human beings that we do not even notice that they could be perceived as "strange."

The astute reader might have noticed that there is (at least) one fundamental flaw with the above "mental exercise": It is extremely hard, and perhaps impossible, to conceive of any form of species, however alien, that is incapable of using abstraction, imagination, representation, incorporeal memory, communication-at-a-distance, and so on—in short, that cannot experience virtuality—and yet *managed to become technologically advanced*! Without the power to virtualize beyond the material world, all species must remain "animals" in the cognitively weak sense of the term.

The intention of the book before you is to answer the alien's befuddlement (what are these humans doing? why?) in order to lift a mirror before us so that we can appreciate how inextricably bound up "virtual" phenomena are, not only with what we consider to be "reality" but also with the very essence of our being, i.e., "humanity." Such an understanding, of course, has more than a purely "academic" purpose. In practical terms, it ought to enable us to reduce our apprehensions and fears regarding what contemporary society calls "virtuality," while accepting its many expressions with greater equanimity, tolerance, and even appreciation.

As such, the book is intended equally for two audiences: (1) scholars who teach and research STS (Science, Technology and Society) generally, and virtuality specifically; (2) the general, educated public interested in broad questions of human existence, i.e., humanity's history in our central fields of endeavor. Although I make no claim to be "on a par" with such luminaries as Steven Pinker, Jared Diamond, and Yuval Noah Harari, the present book does offer a similar, wide-ranging survey and analysis of human existence writ large—but on a subject that is generally thought to be rather narrow (incorrectly, as this book makes clear): virtuality.

The book is organized in the following order. An introductory chapter lays out its two main theses (strong and weak) and explains the study's interdisciplinary approach. This is followed by Part I with one chapter on the etymology of the term "virtuality" leading to a detailed taxonomy of thirty-five forms/types/levels of virtuality (listed in the open-access Appendix). A second, extended chapter analyzes the many ways our human brain "virtualizes" reality, perceptually and cognitively.

Part II constitutes the core of the book, with several chapters surveying numerous areas of life in their virtual incarnation from time immemorial: Religion and the

Supernatural; Mathematics, Philosophy, Physics and Astronomy/Cosmology; Music, Literature and the Arts; Economics (money, property, organization and ideology); Politics (community and nationhood, government, war); Communication (to/with/by the masses); and finally, Virtuality's Expansion in the Modern Era (the interplay of, and mutual influence between, nineteenth and twentieth-century media and contemporary developments in several of the fields covered in earlier chapters).

Part III starts with a chapter asking: "why do we virtualize so much?"—offering six different beneficial functions of our virtualization. A second chapter then describes and analyzes the integral relationship between virtuality and reality as two sides of the same coin, bringing examples from such varied fields as sex, gaming, education and training, shopping, museums, and the like.

Part IV concludes the book with two chapters. The first looks at several advanced, contemporary technologies—Neuroscience, Biotechnology and "compunications"—to suggest several new forms of virtualizing in the future (e.g., brain-to-machine-to-brain communication; lifelogging). The book's final chapter asks a social/philosophical question: Should there be limits to human virtualizing? No definitive answer is offered, but the book's general thrust does suggest how we (and future generations) can approach such a question with obvious revolutionary ramifications for human society.

Kfar Saba, Israel                                                                              Sam N. Lehman-Wilzig
October 2021

# Acknowledgments

Acknowledgments usually mention only those individuals who directly helped the author in researching, compiling, thinking about, and writing the book. However, I wish here to add all those people whose influence is indirect—but no less important in enabling me to produce what can be generously called an "ambitious undertaking."

The book's dedication page mentions the two most important of these. First, my wife Tami Lehman-Wilzig has been a major "prod" (as in the word "*prod*uctive") throughout my academic career to get the most out of myself. She might not have succeeded 100% in accomplishing this (all my "fault"), but the present book, years in the making, is certainly a product of such constant prodding and support.

Second, Prof. Daniel Bell—one of several incredibly bright and enlightening (for me) professors—I had the luck and pleasure learning from during my Ph.D. studies at Harvard University. He was not only an intellectual of the first order (and "mensch" to boot), but more germane, his incredibly wide-ranging scholarship provided the inspiration for my branching out during my later academic career into several, not obviously related, subject areas: politics, communications, technology and society, and legal philosophy. The present book is witness to that foundational concept and scholarly approach: In the end, all fields of human endeavor are related.

I wish to also thank a few additional, "indirect" influencers: Mr. Yaacov Aronson, my high school teacher whose approach to *analyzing* social science issues—and not merely remembering facts and events—was a major stimulus for me to enter political science in college and beyond; and my colleagues at Bar-Ilan University, especially Profs. Daniel Elazar (RIP), Bernie Susser, and Stuart Cohen, whose lively discussions and wide interests further stimulated my (perhaps overly) all-embracing research. The Bar-Ilan University graduate students in my seminar on virtuality that I taught for several years also helped me to sharpen many points included in this book.

I heartily thank those who helped me in different ways in the practical preparation of this book: Ms. Donna Susser, for her early research assistance; Dr. Isaac Herschkopf, for his astute comments regarding the chapter on Psychology; Prof. Ofer Feldman, for comments (substantive and editorial) and "also matchmaking" me with

Springer Nature. My gratitude also goes to Prof. David J. Gunkel and Prof. Yair-Amichai Hamburger for their encouraging words in this book's back cover, having read the manuscript under a very tight deadline. Last but certainly not least (quite the reverse!), my heartiest thanks to Ms. Juno Kawakami, my Editor at Springer, whose patience and immediate responses to my innumerable queries made completing this book project a lot easier and smoother than I had any reason to expect.

October 2021                                                    Sam N. Lehman-Wilzig

# Contents

# About the Author

Prof. Sam N. Lehman-Wilzig was born in New York City where he studied at CCNY (B.A. Political Science, 1971, summa cum laude) and then in Harvard University where he completed his Ph.D. in 1976 (School of Government). He taught (1977–2010) at the Department of Political Studies, Bar-Ilan University (BIU), Israel, and in BIU's recently established School of Communication (2010–2017). He also taught during sabbatical years at San Diego State University (1989–90); Brown University (2008–09); and the University of Maryland, College Park (2013).

He also served in the following capacities: Head of BIU's Division of Journalism and Mass Communication Studies (1991–1996); Chairman of the Israel Political Science Association (1997–1999); Founder of BIU's Communications Program within the Department of Political Studies (1994–2007); Chairman of the Political Studies Department (2004–2007); Chairman of BIU's School of Communication (2014–2016).

His fields of expertise are as follows: new media and communications; technology and society; and political communication. In these fields, he has published 42 academic articles (among them, the first-ever study of Artificial Intelligence and Law), 22 chapters, and two academic books in English, plus a popular textbook: *Handbook of Mass Communications* (in Hebrew; 1994). He has presented over 70 papers at international academic and university conferences, also appearing at numerous speaking venues such as weekend Scholar-in-Residence, public lectures on current events, and media appearances (newspaper, radio and TV). For more details, see: www.ProfSLW.com.

# Chapter 1
# Introduction

## 1.1 The Book's Underlying Purpose and Thesis

*Virtuality.* A word bandied about today by many critics, proponents, and neutral analysts of the Post-Modern/Post-Materialist/Digital/Information/Compunications[1] Era—from sundry disciplines: computers; anthropology; psychology; literature, cinema, and other arts; economics; physics; etc.—and especially the field of new media.

It is clear that something is afoot, and not just in the world of cyberspace. In popular culture, three cinematic hits—the trilogy *Matrix* (1999–2003, a fourth in 2021), *Avatar* (2009), and *Her* (2014)—indicate the public's curiosity regarding "virtuality." Dealing with Covid-19 (Zoom and social distancing) has only enhanced the feeling of virtuality taking over our lives. But how widespread is the phenomenon of virtuality itself? Is it in any way revolutionary—or perhaps merely an updated continuation of certain longstanding phenomena?

This book answers these questions and several others: how to define the term; what functions virtuality serves. It places virtuality in its "proper" social and historical contexts, a very wide scope indeed. As my main theses could be misunderstood or misconstrued, a bold-faced disclaimer follows immediately after presenting here the book's central arguments (those with a "strong" foundation of evidence are marked "S"; weaker ones, "W"):

1. Virtuality has been part of the human condition for as long as we can trace back our *Homo Sapiens'* ancestry (**S**), because...
2. Virtuality is intrinsic to humanity; indeed, it is one of the ways we can distinguish humans from other living creatures on Earth (**S**).
3. The course of human history can be described as a constant process (not necessarily linear) of widening the virtual experience—in number of life areas,

---

[1] Compunications: = Computer + Communications: computerized communications (cellphones using computer chips) or communicating computers (email from one's gadget). Increasing media convergence has rendered this a distinction without much difference.

© The Author(s), under exclusive license to Springer Nature Singapore Pte Ltd. 2021
S. N. Lehman-Wilzig, *Virtuality and Humanity*,
https://doi.org/10.1007/978-981-16-6526-4_1

depth/richness of the virtual experience, and the time we spend "virtualizing" (**W**).

4.  Overall, virtuality is not harmful but quite beneficial—otherwise humans would not widely engage in it, past and present (**S**).
5.  The modern era has seen a significant leap in virtuality, accelerating over the past 200 years, mostly for technological reasons (**S**).

*This is **not** to argue that all types of virtuality are constructive or beneficial,* nor that the end of this historical process will be a Matrix-like, 24/7 connection to a Virtual Reality machine. Of course, an interesting question (discussed in the final chapter) is how to determine the harm and/or benefit of virtuality expansion.

While the book's theses might raise some eyebrows, the overall thrust is deeply conservative. I am not arguing that a radical shift has recently occurred in human experience but rather that our virtual sensibility has very gradually evolved and expanded, quantitatively becoming a central component of human endeavor. That the past 200 years have seen a gradual acceleration in virtuality is also not a very radical idea, for it has gone hand in hand—influencing and being influenced by—scientific and technological developments that have been well documented by others.

What is new in the present book is that its basic thesis goes against the grain of most contemporary descriptive and prescriptive analyses regarding virtuality. First, many pundits and even serious researchers view this phenomenon as something quite modern, i.e., stemming from the influx of electronic, digital media and especially the computer revolution (Morse, 1998). Second, despite some differences of opinion regarding the effects of virtuality, many (and perhaps most) analysts decry the phenomenon as a whole or important aspects of it, pointing to effects detrimental to individuals and society.

The following are a small but representative sample of quotations (not taken out of context) from what can be loosely called the anti-virtuality/virtuality-skepticism camp, comprised of scholars and serious social pundits. In some cases, the term "virtuality" itself is not employed but rather "internet," "cyberspace" or some other virtual phenomena/media used as stand-ins.

Does the new virtuality signal a decisive step? That is, have we definitely reached the end of reality?. ...

[T]he internet creates a vast illusion that the physical, social world of interacting minds and hearts does not exist.... [There is a] connection between the rise of the Internet and the accelerating blur of truth and falsity in culture.... [T]he internet transvalues all experience into commercial experience." (Siegel, 2008, pp. 17, 25, 60)

People who skillfully manipulate today's fragmented modern media landscape can dissemble, distort, exaggerate, fake – essentially they can lie – to more people, more effectively, than ever before. (Manjoo, 2008, p. 15)

The speed and force of contemporary virtualization are so great that they exile beings and their attendant knowledge, alienate them from their identity, skills, and homeland. (Levy, 1998, p. 186)

It's an unreal universe, a soluble tissue of nothingness. While the Internet beckons brightly... [a] poor substitute it is, this virtual reality where frustration is legion and where – in the holy names of Education and Progress – important aspects of human interactions are relentlessly devalued. (Stoll, 1995, p. 4)

The deep meaning of personhood is being reduced by illusions of bits. Since people will be inexorably connecting to one another through computers from here on out, we must find an alternative. (Lanier, 2010, p. 4)

One of the extremely painful lessons... will be that the virtual is not an adequate substitute for the real. It will be painful because the notion of virtuality has become a psychological crutch for a culture that is recklessly destructive of real places, real experiences, real relationships with real people, and real notions of purposeful, decent behavior. (Kunstler, 2010)

Thanks to technology, people have never been more connected–or more alienated. ...what people mostly want from public space is to be alone with their personal networks. It... is more important to stay tethered to the people who define one's virtual identity, the identity that counts. (Turkle, 2007)

The cultural shift away from nature recreation appears to extend outside of the U.S. to at least Japan, and the decline appears to have begun 1981–1991. The root cause may be videophilia. (Pergams & Zaradi, 2008)

...exposure to violent video games is a causal risk factor for increased aggressive behavior, aggressive cognition, and aggressive affect and for decreased empathy and prosocial behavior. (Anderson et al., 2010)

Internet addiction is ...a deep 'craving.' And if we don't make the change in the way we classify Internet addiction, we won't be able to treat it in the proper way. (Scientist Redefines..., 2007)

Again, these are just a few of the more extreme expressions of what one could call "virtu-phobia." For the reader who wishes to delve more deeply into some additional, recent Cassandras of virtuality, especially of the internet/compunications/artificial intelligence variety, I recommend the following studies and extended thought essays penned by very serious thinkers and scholars: Sherry Turkle, *Alone Together* (2011) and *Reclaiming Conversation: The Power of Talk in a Digital Age* (2015); Evgeny Morozov, *The Net Delusion* (PublicAffairs, 2011); Saul Levmore & Martha C. Nussbaum, *The Offensive Internet* (2011); Susan Maushart, *The Winter of Our Disconnect* (2011); Elias Aboujaoude, *Virtually You: The Dangerous Powers of the E-Personality* (2011); Herbert L. Dreyfus, *On the Internet*, 2nd ed. (Routledge, 2008); Al Abelson, Ken Ledeen & Harry Lewis, *Blown to Bits* (2008); Nicholas Carr, *The Shallows* (2010); Andrew Keen, *The Internet is Not the Answer* (2015); Katajun Lindenberg et al. (eds.), *Internet Addiction in Adolescents* (Springer, 2020); three books by Jaron Lanier, a pioneer of virtual reality who turned into perhaps the most famous virtuality dystopian of all: *You are Not a Gadget: A Manifesto* (2011), *Dawn of the New Everything: Encounters with Reality and Virtual Reality* (2017), and *Ten Arguments for Deleting Your Social Media Accounts Right Now* (2018); and finally, John Brockman, *Is the Internet Changing the Way You Think?* (2011), encompassing interviews with dozens of pundits, some of whom are virtuality utopians, others are dystopians, and several somewhere in between.

This scholarly approach is not new: "As with earlier information revolutions, the most extreme manifestations of cyber-fear are articulated around metaphors of boundary-dissolving threats, intrusive alterities, and existential ambivalences created by the erosion of binary distinctions and hierarchies that are assumed to be constitutive principles of everyday life" (Sandywell, 2006, p. 40). All this can be seen within the larger critique of technology by such social critics as Foucault (Behrent, 2013) and Postman (1992).

Fiction authors by and large are of similar mind, as Chayka (2021) notes, listing several recent anti-internet/social media novels. The public seems to be ambivalent too, with more viewing the virtual world having greater negative aspects than positive—not only in the West. A Third World PEW survey found that "despite all the benefits of these new technologies, on balance people are more likely to say that the internet is a negative rather than a positive influence on morality, and they are divided about its effect on politics" (Internet Seen…, 2015). A non-academic, but serious blog post, summed up all the possible deleterious effects of the internet: (1) Addiction (2) Work Ethic (3) Cyber Crimes (4) Abandonment of Family (5) Physical Development (6) Cyber Bully (7) Privacy Disrupted (8) Cheating (9) Lazy (10) Bad Webs (11) Hesitant Toward Face to Face Communication (12) Lacking Creativity and Dependence (13) Wastage of Time (14) Insomnia (Nguyen, 2016).

The purpose of this book is not to "prove" that these beliefs, pronouncements, and analyses are incorrect. In fact, there are certainly some deleterious effects of the internet specifically, and virtuality more generally. Conversely, although probably a minority,[2] there are a few scholars and pundits who view virtuality in positive terms—from the techno-utopianism of Rheingold (1993) to Johnson (2005), a counter-intuitive admirer of an increasingly sophisticated, virtuality-oriented pop-culture—and occasional commentaries highlighting the positives in contemporary virtualities such as the internet (Gallagher, 2021). Overall, though, there are clearly more numerous, and certainly more vociferous, pundits who view virtual phenomena with great wariness, many of them with outright alarm.

However, there is one crucial point of wide (but not universal) consensus within both camps when writing about the internet, new media, and other manifestations of virtuality: the shared assumption that we are facing a relatively new phenomenon with little historical precedent. For instance, Hayles avers that "virtuality is clearly related to postmodernism" (1999, p. 78).

Nevertheless, at least two commentators do hold the opposite view. First, the person who coined the term "artificial reality" in 1973—Myron Krueger: "Humankind has always inhabited a conceptual universe that is every bit as important to it as the physical world" (Turner, 2002). Second, in his introduction to the edited book *Virtual Realities and Their Discontents*, Markley points out: "The blind spot of many critics of virtual technologies lies in… their casual assumption that we are living in revolutionary times in which technology intervenes in our subjectivity in

---

[2] The difficulty in finding pro-virtuality commentators might also be a result of the common tendency to publish books *against* something rather than *for*, although this does not explain the large difference between virtuality's limited number of supporters and much larger host of critics.

ways undreamt of before the late twentieth century" (1996, p. 9). Markley is technically wrong and substantively correct. The modern age *has* witnessed new forms of "re/presentational" technologies that could not have been dreamt of in the past; however, these are more (and mere?) *technical* improvements on virtualizing forms that have existed from time immemorial (as Markley astutely suggests), rather than *substantive* "improvements" on the essence of past virtualities.

It is the contemporary, general, underlying assumption and attendant idea that we are somehow undergoing a gigantic, "social experiment" with few historical precedents—and therefore must proceed with extreme caution—that the present book interrogates. To be sure, there are scholars who point out virtuality precedents here and there, but they almost always focus on one specific field of human endeavor as the harbinger of (or precedence for) contemporary virtuality e.g., Davis' *Visuality and Virtuality* (2017).

Perhaps the best example of this is Boellstorff, whose study's introductory comment excellently sums up the theme of the present book: "...our lives have been 'virtual' all along. It is in being virtual that we are human: since it is human 'nature' to experience life through the prism of culture, human being always has been virtual being. Culture is our 'killer app': we are virtually human" (2008, p. 5).[3] Very promising—but his book is an in-depth study of another narrow topic: virtual worlds (Second Life). In other words, even those researchers who do perceive virtuality as a longstanding, human phenomenon, use that insight as a springboard for a highly *focused* treatment of one or another virtual enterprise. They might have the right "recipe" for virtuality, but don't get around to cooking the meal. The proof is not in this or that "pudding" but in the entire repast.

This book is such a prepared banquet. It takes a wide-ranging, holistic approach, presenting numerous antecedents of virtuality throughout the past centuries, millennia and even tens of thousands of years. I will describe and analyze in some detail how virtuality can be found in numerous fields of human endeavor, and how/why at some point in the modern age it became a central locus of society. In other words, at a certain point *quantity = quality*. When human experience is surrounded by, and human thought immersed in, virtuality's many guises, one can say that it has become a *sine qua non* for humanity's social and intellectual life.

It goes without saying that this book makes no claim to a "complete" survey of all fields of human endeavor—something that would be difficult to do even in a multivolume encyclopedia. First, the treatment here will be largely (but not exclusively) Western-world oriented. My apologies to half the world's population in the East but trying to cover their long and venerable history of religions, philosophies, artistic accomplishments, and scientific/technological discoveries and inventions would have turned what already is quite a large book into an unwieldy tome. Readers more closely familiar with specific Far Eastern cultures can judge for themselves how close (or

---

[3] Among the few who are aware of the possibility of such a broad reading of historical virtuality, not all agree with such a perspective. For instance, Otto (2011, p. 6), argues that Boellstorff's definition of virtual reality "seems too broad." Interestingly, he places the start of virtual reality in the early nineteenth century Romantic era, providing a fascinating treatment of virtual entertainments during that period. But once again, we find here a very topic-specific treatment of the subject.

far) the relevant field-related chapters are to the Eastern experience. Nevertheless, occasionally I will offer Far Eastern cross-civilizational insights, as well as virtuality-related aspects from the Islamic and Jewish cultures and civilizations. In any case, half the present book's chapters deal with issues that are culturally global and universally human. For instance, human psychology and the ways/whys we virtualize are generally common to all human beings; similarly, the interaction between virtuality and reality is found everywhere (e.g., yin = body; yang = spirit); and so on. Thus, this book *does* relate indirectly to the East as part and parcel of humanity as a whole.

This book focuses on several central areas of life, emphasizing aspects within each that incorporate virtuality. My argument is not that most of our lives have always been taken up in virtuality; given the basic need for most humans through most historical periods simply to feed, clothe, and house their families, such a claim would be patently false. Rather, I argue that *despite* this fundamental, material need, humans have always found the time and energy to indulge in virtual activities—perhaps virtualizing in part *because* of the mind-numbing nature of feeding, clothing etc.

In any case, virtuality is not a phenomenon born of modern "compunication" (computer + communications) technologies but rather has existed from time immemorial. In our era, it is evident well beyond the restricted domain to which most analysts relate. In short, this book's new claim is that virtuality is not new. Quite the reverse: it has always been an integral component of human existence.

## 1.2 The Outline of the Book

In order to present the history of human virtuality, one must first define the term. Paradoxically, this turns out to be a very easy *and* a fiendishly difficult task. Easy because there are so many definitions to choose from; difficult because many of the definitions relate to widely disparate phenomena in different disciplines and fields of expertise. Part I starts with Chap. 2 attempting to bring some order to this definitional cacophony, first by presenting the term's historical evolution and then providing a spectrum of virtuality definitions and common denominators. Not every reader will accept every definition; "virtuality" is a slippery taxonomic concept.

Nevertheless, this varied list of meanings should not pose a serious problem regarding the description and analysis of virtuality in the central part of this book.[4] My underlying argument posits that it is the overall *quantity and breadth* of virtuality that renders it such an important part of the human experience, and not the qualitative essence of any specific form of virtuality—any one of which could be legitimately dismissed by different readers as not being "virtual" at all.

[4] The *Appendix: A Taxonomy of Virtuality*—listing by numbers 1–35, e.g. [V13] or [V34], the various types of Virtuality mentioned throughout the book—is freely available to the public online at https://link.springer.com/book/10.1007/978-981-16-6526-4.

Chapter 3 constitutes a bridge to, and the underlying foundation of, the book's second part. Here I analyze the human brain in light of perceptual and cognitive psychology. By describing how the brain senses the world (physical perception) and analyzes it (cognition), we understand why "virtuality" is not only attractive to humans but is inherent in the very nature of the brain's workings.

Part II of the book offers a survey of several central areas of human endeavor from the beginning of "human time" until the present day, from the standpoint of virtual activity found in each area. These chapters deal with the following major topic fields: religion (Chap. 4), mathematics, philosophy, physics and cosmology (Chap. 5), music, literature and the arts (Chap. 6), economics (Chap. 7), community and nationhood, government, war (Chap. 8), and finally mass communication until the early nineteenth century (Chap. 9). Clearly, distinctions between some fields are somewhat arbitrary, either because they were once considered to be unified (e.g., philosophy and physics) or largely overlapped (religion and the arts).

Indeed, virtuality involves a certain mindset or mental template that is transferable across disciplines. Thus, although specific chapters in Part II are devoted to one (or a few interrelated) general "field"(s), I shall note in passing their relationship to other, ostensibly unrelated, fields.

Each chapter's subject matter, therefore, is part of a vast tableau of human endeavor in which virtuality is found throughout. Part II does not include all relevant areas of life. Additional whole chapters could have been included on *sex* (pornography; fantasies)[5]; face-to-face/oral vs distance/textual *education* (Nguyen & Preston, 2006); games for problem-solving and simulation for *training* (Castronova, 2005, p. 252); medicine (mental-oriented psychology vs. pharmaceutical psychiatry; mind vs matter); and so on.[6] Again, the book's central thesis does not depend on any specific chapter, but rather on the complete picture *in toto*; readers are invited to consider virtuality aspects in other fields of life close to their heart or profession.

Chapter 10 (concluding Part II) focuses on the last 200 years when a quantum leap in the amount and intensity of human virtuality occurred. The chapter first analyzes the major factors and trends that brought about the tremendous acceleration of virtuality from the Renaissance through the present day. There exists a strong element of continuity between the pre-Industrial Age (post-medieval) and the modern era, despite differences in the extent of virtuality between them. Here I show the intricate connections of virtuality *between* these various fields of human endeavor i.e., how virtuality in specific areas of life generated greater virtuality in others through an ongoing, "virtuous" circle.

Part III asks a few cardinal questions emanating from Part II, including the most fundamental of all: *why* do human beings virtualize so much (Chap. 11)? Whereas Chap. 3 (brain psychology) offers one possible answer—mimicking/creating an external world similar to our own brain processes—the reasons for our extensive

---

[5] Interestingly, an Amazon search (2021) for "sex and virtuality" obtains *zero* non-fiction book results, and "virtual sex" only one book: a 12-page practical guide! Searching "sex and robots" delivers a plethora of books. Of course, sex with robots has a measure of virtuality [V6; V10].

[6] Some of these topics are touched on, especially in Part IV of the book.

virtualizing are far more varied. Chapter 12 serves as an "antidote" to Part II's exclusive focus on virtuality by showing how virtuality and corporeality/"reality" have always been, and continue to be, inextricably intertwined—the former fed by, feeding into, and overall strengthening, our concrete "real" life.

What of the longer-term future? Part IV raises questions and speculates about where the trend of increased virtuality might lead to. Chapter 13 forms a bridge to the future, surveying several "new" technologies with direct bearing on mid-to-long term virtuality: neuroscience, biotechnology (based mainly on genetics), and compunications. These three broad disciplines are already interrelated today and will become increasingly so in the not-too-distant-future.

However, the trend to broader virtuality will not necessarily be linear e.g., nanotechnology and 3D printing (additive manufacturing) will drive us back to the corporeal world of atoms and not bytes (a distinction first made by Dertouzos, 1997). Thus, the future might hold more of the same—only with greater intensity: irresistible technological forces pushing us towards increased virtuality, alongside the immovable object of our corporeal bodies pulling us back, or keeping us anchored, to the physical world that demands its non-virtual due.

This leads to Chap. 14 that concludes the book by asking whether there should be limits to the amount and type of human virtualizing. Humans will not and cannot completely abandon corporeal reality for virtuality but determining what should be its actual limits is fraught with many methodological and some philosophical difficulties. The chapter then ends by tying together several main arguments and theses, as well as offering an explanation as to how and why we moved from the Great Ape's non-virtual, mental world to *Homo Sapiens Sapiens'* widespread use of virtuality.

## 1.3 The Book's Epistemology and Inter-disciplinary Approach

This book employs two complementary approaches. Part II presents virtuality's many facets in *diachronic* fashion: each topic is surveyed in somewhat chronological fashion over human history. The ensuing two parts, however, offer a more in-depth, *synchronic* and *analytical* study in the modern and/or contemporary-to-future periods, cutting vertically across sundry areas of life, with greater emphasis on the *organic* interplay of their virtuality functions and characteristics.

Overall, then, this study is "about" virtuality, but not categorized as belonging to any specific discipline, and especially not exclusively to "new media/computer science." As such, the book's approach goes against the historical grain of the compunications discipline.

How so? Mass Communications studies emerged in the middle of the twentieth century from several disparate fields: politics, sociology, psychology, journalism, speech, electrical engineering, etc. Until the 1970s, most Communications scholars did their research within other related disciplines (e.g., Harold Innis from Economics; Marshall McLuhan from English Literature; Paul Lazarsfeld from Mathematics and later Sociology; Elihu Katz from Sociology; Claude Shannon from Mathematics and Engineering; and so on). By the latter part of the twentieth century, though, the field had coalesced sufficiently to be considered a bona fide discipline.

However, such a sharpened disciplinary approach also had its drawbacks, mainly a narrowing of focus (specialization). The present study attempts to return to the field's roots by freely drawing from other disciplines to understand a fundamentally, inter-disciplinary phenomenon—what Tilly (1984) called (somewhat bombastically): "world-historical research" that combines disciplines and spans the ages.

So much for methodology. Epistemologically, western philosophy and science offers two broad modes of thought: "essentialism" and "historicism." *Essentialism* attempts to uncover and analyze fundamental truths about the natural and social worlds (Rorty, 1979, pp. 361–65). Specific historical events or diverse social phenomena are viewed as part of an underlying foundation of immutable "laws." *Historicism*, on the other hand, perceives social and even natural phenomena as products of discrete historical contingencies. For instance, Newtonian physical determinism is essentialist (the same laws always hold sway everywhere), while Einsteinian relativity (everything depends on the relative position and speed of the object being investigated) and quantum mechanics (wave-particle duality) are the natural science's counterpart to social historicism (situational contingency).

This book argues from an essentialist perspective but with a heavy dose of historicism. Virtuality is a constant in human affairs because it is an integral outgrowth of human psychology. However, the extent to which—and ways through which—virtuality is expressed are historically contingent, as a result of technological capabilities and other social phenomena that encourage or discourage its full expression. Therefore, although virtuality is a constant in human society it does not necessarily express itself perpetually and consistently in the same fashion over different epochs and within sundry societies. Virtuality is manifested in many ways, but its appearance almost everywhere and anytime is itself a given.

A final word—on writing style. Although I am an academic, this book will not forego the word "I" for the more collective and neutral "we." In a sense, academics who use "we" are essentially "virtualizing" the author i.e., replacing the subjectivity of the individual writer for the more objective, "collective" wisdom of the group/audience.

Such a neutral style is not appropriate for this book because it is not based on mathematical or lab-experiment types of "proof," but rather makes its case through a broad-ranging survey and qualitative analysis of the human condition. I do not expect readers to agree with everything, nor even to accept that this or that piece of "evidence" actually supports the general argument. As this book advances a "personal argument," I use the first person singular when presenting a point or thesis, leaving the "we" for more general comments of a societal nature. The fact that you are absorbing my work through virtual, symbolic means (letters on a page or screen) should not obscure the fact that the evidence was collected, and the arguments are offered, by a real-life, fallible human.

# Part I
# Fundamentals of Virtuality: Definitions and Psychology

# Chapter 2
# Defining Virtuality

## 2.1 Introduction: Historical Etymology

The ubiquity of the word *virtual* in the contemporary Compunications Age has led to confusion, with different meanings in different disciplines and even among scholars and commentators within the same field. When meanings are widely divergent, analytical comprehension suffers. Thus, the first priority is to clear such confusion that even dictionary definitions illustrate[1]:

The Oxford English Dictionary: (1) *something possessed of force and power*; (2) *an essential nature of being, apart from external form or embodiment*; (3) *a virtual, as opposed to actual, thing, capacity*, etc., *i.e., a potentiality.*

The Cambridge Dictionary of American English: *almost, but not exactly or in every way.*

The Merriam-Webster Online Dictionary: (1) *being such in essence or effect though not formally recognized or admitted e.g., a virtual dictator*; (2) *of, relating to, or using virtual [computer] memory*; (3) *of, relating to, or being a hypothetical particle whose existence is inferred from indirect evidence, e.g. virtual photons*; (4) *being on or simulated on a computer or computer network.*

Finally, the American Heritage Dictionary of the English Language: (1) *Existing or resulting in essence or effect though not in actual fact, form, or name: "the virtual extinction of the buffalo."* (2) *Existing in the mind, especially as a product of the imagination. Used in literary criticism of a text.* (3) *Created, simulated, or carried on by means of a computer or computer network: "virtual conversations in a chatroom."*

From where did all these meanings emanate? The OED explains that the origin of virtual is Latin: the word "**vir**" denoted *male* and connoted *maleness*, what we today

---

A concise list of the 35 virtuality types discussed at length in this chapter is freely available at *Appendix: A Taxonomy of Virtuality* (in the book back matter). Suggestion: print it out and keep it by your side for quick reference as you read the book. This is also freely available to the public at https://link.springer.com/book/10.1007/978-981-16-6526-4.

[1] However, using "Virtuality," "Virtual," and other variations interchangeably is legitimate.

might refer to as *macho* traits: *strength, potency*, etc. The term "**virtus**" evolved out of this: the sum total of traits that a "real man" should have (*courage, decisiveness, stability, force, power*). Gradually, virtus came to mean *excellence, valuable, ability, capability*—and from there finally to *integrity, goodness, nobility of spirit*, and essentially any *positive trait*.

2000 years later, residues of these meanings are found in such words as "*virility*" on the original, physical side,[2] "*virtue/virtuous*" on the later, metaphysical side, and "*virtuosity*" somewhere between physical and metaphysical. This etymology underlies this book's theme: virtuality can be a positive phenomenon, and certainly isn't necessarily negative. Indeed, several *vir* words continue to have positive connotations: e.g., virtue, virility, and virtuous.

As to *virtual*, in medieval Latin from where English incorporated it, the term was *virtualis*, meaning *anything with physical abilities* (e.g., herbal medicines were virtual remedies), or the *potential* to emerge based on the intrinsic power within (the acorn is oak virtualis). Later, it lost some of its "physical" essence, and came to mean *anything that substituted for the physical trait itself* e.g., the wafer at Catholic mass replaced Christ's body virtually, but the Lord's strength was indeed ingested.

From here to modern denotations of the word *virtual*. First, "non-existent." Webster defines "virtual image" as "an image (as seen in a plane mirror) formed of virtual foci," from where rays of light seem to emanate but really do not (mirrors always display a reverse "mirror" image). From "illusory non-existence," it is but a small step to the connotation of dishonesty and illegitimacy.

A second, later meaning of virtual is something with the same influence as the real thing even if not physically present or without any physical presence. This is not the same as "virtual = potential," but it's understandable how that could lead to "virtual = incorporeal but still having the force of influence." This first appeared in 1959 with the term *virtual memory* within the computer, denoting something not existing physically but a program that makes it seem to exist.

To sum up: the term *virtual* has undergone a massive shift over time from some *thing* literally embodying physical reality to "something" not corporeal.[3] Second and related to this, a growing gap has evolved between the positive (and now neutral) *denotation* of virtual and the scholarly, mostly negative *connotation* of virtuality, as illustrated in the book's Introduction—ironic, given the original root's positive denotation.

Third, *virtualis* is not classically the polar opposite of *reality* but rather both form a dialectical relationship. The virtual is not the negation of the real; instead, it is something that has potential to actualize into something real. In short, reality and virtuality are bound together; some would even argue inextricably (the theme of Chap. 12).

---

[2] Also, "by virtue of..." i.e., on the basis of some force with the power to accomplish something.

[3] Ropolyi (2001, p. 4) differs, arguing that premodern virtuality had a "magic world view" whose fundamental feature was "indistinguishability of reality and virtuality." He suggests that virtuality underwent semantic transformation twice: from pre-modern/medieval times to the modern period, and today to a post-modern meaning that once again blurs reality and virtuality into one "hyperreality" (p. 9).

As noted above, though, one specific modern usage connotes virtuality as something akin to "illusion," even "fake(ry)." This might account in part for virtual(ity)'s negative image among broad swaths of commentators and scholars. Certainly, it is lexically confusing to have two contradictory, denotative and connotative meanings for the same word!

A fourth point is that we should not limit "virtuality" to the world of the computer, given that historically the term had a much wider context.

Of course, etymology is no "proof" that what we call virtual today is necessarily a "good" thing. Nevertheless, it is the contemporary world that has (mostly) changed our viewpoint regarding virtuality. Thus, historically analyzing virtuality in its sundry forms should be useful in attaining some perspective regarding virtu as something the ancients valued, sought, cherished, and even nurtured—and whether that might not be a better way of viewing the matter today as well.

There are two additional issues. First, different authors will define the term differently, leading to degrees of incommensurability among various analyses of the phenomenon. Second, most virtuality analysts discuss the phenomenon in one-dimensional terms (within one specific type/category), and consequently tend not to perceive virtuality commonalities across a spectrum of fields. These are the two factors underlying the confusion regarding virtuality's origins, multi-dimensional expression, and wide-ranging, social impact.

On the other hand, there is an obvious reverse pitfall: analyzing different meanings attributed to virtuality as a completely integrated whole. However, if different virtualities are not part of a unified picture, how then can the phenomenon be analyzed in any systematic fashion?

Two related metaphors can illustrate how to usefully approach this conundrum. First, the mosaic picture or jigsaw puzzle. "Virtuality" as defined by each (different) author will not form a fully integrated picture, but each can be seen as a piece of a larger mosaic/puzzle. Indeed, such a "*pixel*ized" picture can be viewed as a Cubist painting incorporating distinctly different perspectives. Although each angle is different, there does exist some correspondence between the various parts; likewise, there are different ways of explicating the phenomenon of virtuality.

Second, two sub-metaphors incorporating "intelligence": (1) The idea of General Intelligence ("g"—for general cognitive ability), the highest order factor that can be extracted from diverse cognitive tests using statistical factor analysis. Simply put, "g" is the product of several tests, each for a different mental ability, all positively correlated one with the other.[4]

(2) Howard Gardner's multiple intelligences approach (1983): human intelligence is not to be understood as one type of "analytical" cognition but rather as being comprised of different and separate types of intelligence: kinesthetic, linguistic, interpersonal, emotional, mathematical, etc.

Why use two metaphors: mosaic/puzzle and intelligence? Because they suggest two obverse perspectives on virtuality: objective and subjective, external and internal. The mosaic/puzzle metaphor represents virtuality's objective (external to the human

---

[4] I use "g" here as a metaphor and don't necessarily posit its scientific "factual truth."

being's) existence; general or multiple intelligence/s suggest a subjective (internal within the brain) component of virtuality. Chapter 3 will analyze both these sides, with different expressions of both described throughout this book. These two types of metaphors offer a way of considering how they exist independently and yet are still connected in subjective and objective ways. Just as the level of people's intelligence (whether cognitive, artistic, social etc.) determines how they perceive the mosaic picture, so too the quantity and quality of an individual's inclination or ability to mentally "virtualize" will affect how, and to what extent, that individual perceives and utilizes external forms of virtuality in the social environment.

## 2.2   The Virtuality Definition *Problematique*

Confusion reigns regarding virtuality's meaning, but the same situation exists with its alter-ego: "reality." Heim (1993, p. 117) relates how "reality" has undergone several metamorphoses: Plato's Ideal is the "really real"; Aristotle's reality comprises the individual substances we touch and feel; the medieval age added "superreal" messages to reality, giving them permanence, while materiality is less real i.e., terrestrial dross; Renaissance thinkers considered real only what could be counted and sensorily observed; in the modern period reality is based on atomic matter (equivalent to energy). To Heim's list one can add the contemporary undermining of commonsense reality: relativity theory, quantum mechanics' amorphousness, and post-modernism's hyper-subjectivity. In short, the ambiguity of virtuality can take comfort in the parallel, historical evolution and confusion regarding "reality"—the two sides of human life (Chap. 12).

The only previous scholar to attempt a somewhat comprehensive, virtuality catalogue is Ryan, who produced a "list of the connotations I collected in the course of my various readings on virtuality" (2001, pp. 27–28): *potential, counterfactual, possible, open, mental, abstract, general, incomplete, indeterminate, spectral, aura, irraditation, atemporal, deterritorialized, plural, blueprint, code, past and future, there, them, the other (as imagined), evanescent, intangible, ground, latent, absence, telepresence, mediated, electronic, information, cyberspace, that which does not count, presenting, passing as, role-playing, make-believe, fantasy and dreams, fiction, appearance, fake, simulation, illusion, falsity, copy, double image.*

Clearly, this list is not meant to be a formal taxonomy, but rather a kitchen-sink of all usages among modern writings. Indeed, several of these meanings seem to be duplicative e.g., "corporeal"/"inscribed in space"/"solid, tangible." This is not a criticism; Ryan is a literary critic, reporting what she found.

Although her collection is somewhat haphazard, I find two basic common denominators that can undergird a systematic taxonomy: *incorporeality* and *mis/non-representation*. Of Ryan's above forty terms, at least fifteen are incorporeally oriented: *mental, abstract, spectral, aura, irradiation, deterritorialized, blueprint/code, evanescent/intangible, latent, absence, telepresence, mediated, electronic, information, cyberspace*; another six are somewhat related to incorporeality: *potential, incomplete, atemporal, past/future, fantasy/dreams, fiction.* Regarding mis/non-representation, Ryan's list has ten: *presenting, passing as/role-playing, make-believe, fantasy and dreams, fiction, appearance, fake/simulation, illusion/falsity, copy/double, image;* another two are closely related: *general; indeterminate.* (Several virtuality concepts are repeated in both lists because they can incorporate both dimensions: incorporeality and mis/non-representation.)

My taxonomy (below) differs from Ryan's list: (1) It collapses the very closely related usages presented in her table into a few more discrete but broader virtuality elements. (2) It doesn't include a few definitions that aren't commonly used or that don't seem to have any connection with the roots "vir" or "virtus" e.g., "particular"/ "general"; "singular"/"plural." (3) It orders the list from those phenomena/elements that are "weakly virtual" i.e., closest to "reality" (either because they have residues of corporeality and/or a significant amount of realistic representation), all the way to "strongly virtual" phenomena, far removed from anything that exists in the world either corporeally and/or representationally. Neither corporeality nor representation took priority in forming the list, so that one could certainly differ about the placement of some definitions on this virtuality spectrum.

Moreover, these virtuality definitions aren't mutually exclusive and can be expressed in two or more ways e.g., a Cubist painting is both "two-dimensional" and "quasi-representational" (or "abstract"); and so on. Of necessity, the list below presents distinct definitions, but the reality of virtuality (an intentional oxymoron) is far messier—like a fuzzy cloud of negative charged electrons around the atomic nucleus, rather than a particle located at a single point.

## 2.3 The Taxonomy and Spectrum of Virtuality Meanings

1. *"duplicate"/"clone"*: a near/perfect imitation of a real object e.g., android, hologram, and biological clone, denoting a stand-in for the "real thing." Of course, this raises the question: what constitutes the difference between a real thing and its duplicate/clone? For instance, how do we relate to Roman statues created by Hellenistic sculptors who copied statues from the Age of Pericles? (Eco, 1983, p. 35). This is the weakest form of virtuality, involving a lack of originality but "virtually real."

2. *"action-at-a-distance"/"telepresence"*: manipulating or dealing with a physical or human entity far from the individual, without feeling remote. This is similar to the earlier and more general mediated incarnation called "presence" (Bracken & Skalski, 2010, p. 5). As Lee (2010) notes, both involve "a psychological state in which the virtuality of experience is unnoticed." However, another important element differentiates them: contemporary communication technologies enable us to be physically located in one place *and yet directly act upon a subject in another place,* such as remote surgery ("tele-surgery") that instantly translates a surgeon's movements into robotic actions on a patient

a great distance away (Marescaux et al., 2001). With the continued development of haptics that offers "virtual touch" (Gaudiosi, 2016), the progress of virtual reality technology, and vast improvements in bandwidth and transmission speed (5G), such examples of at-a-distance virtuality will continue to expand in the future.

Even more widespread is "social presence"—the virtual communication with other people who are out of sight e.g., computer chat rooms, phone conversations, and sophisticated, multiplayer computer games or artificial world play. Humans have a finely-honed ability to ignore and even temporarily forget that their fellow communicators are not physically present! Thus, although distant communication is wholly virtual, the feeling of "social presence" neutralizes the "loss" of nearness (Reeves & Nass, 1996).

As we'll see in Chap. 5, the sub-atomic world also displays this: "quantum entanglement." Thus, both at the micro-quantum level and the macro-human world levels, the action-reaction relationship is clear but without any physical connection.

3.  *"media(ted)"*/*"presence"*[5]: as opposed to face-to-face, any communication between people (or from a non-human source to a human) that has to go through some form of technology. By their very nature, media are "inter*media*ries" that add, subtract, or otherwise affect the original message.[6] Mediated virtuality is quite close to definition #5 below ("distortion"), but refers only to the media, whereas "distortion" can be caused by any (un)natural phenomenon e.g., a star's gravitation bends light, but its massive body is not a "medium" designed to transmit information.

What sets presence apart from telepresence or social presence is that the mediated message receiver doesn't interact with others but rather is alone with the mediated communication's contents. Anyone "really into" a good novel feels presence i.e., is "immersed" in the story. Thus, presence is passive; telepresence or social presence are interactive, involving a small degree of greater reality.

4.  *"incomplete"*: not having all the qualities or elements of the original. This definition is based on a common usage of "virtual" = "almost" e.g., "that theater production was virtually the same as the one I saw ten years ago with different actors." This is close to the next definition as well ("distortion") e.g., "that theater production was virtually the same as the one I saw ten years ago, but they changed the ending."

5.  *"distorted"*: appearing in different form than in its original incarnation. Virtuality-as-distortion can be caused by humans or occurs as a natural, physical phenomenon e.g., the mirror image is a *reverse* image of what exists in

---

[5] Not "mediated" involving human intercession between two parties in a dispute, but rather involving communication *media* exclusively.

[6] At the beginning of the eighteenth century, "virtual" was used in optics to describe the image reflection or refraction of an object (Woolley, 1992, p. 60).

our physical reality. From this perspective (pun intended), the mirror image can also serve as an example of our next virtual category.

6. *"substitution"/"similarity"*: something/someone that takes the place of, or acts as a stand-in, for the original. For instance, in eighteenth century Britain the concept of "virtual representation" was widely practiced. Members of parliament virtually represented citizens without voting rights; the rationale was that MPs were elected by "similar" voters.

Another form of substitution/similarity can be found in what we call stage "impersonations"—whether theatrical (all actors "impersonate" another person, whether fictional or historical) or in entertainment e.g., Elvis impersonators. This also includes hard-to-detect fakes and forgeries, indistinguishable from the original. However, this is somewhat complicated given the slipperiness of the term "original"—a point discussed below and in Chap. 7.

7. *"three-dimensional, representational object"*: a (relatively) precise rendition of an original object or human interaction. Examples: a hologram; 3D movie; lifelike sculpture; or "realistic theater" in which the actors mimic the behavior of real humans. These four examples have a common denominator of three-dimensional representation, but the first two differ significantly from the latter two regarding the corporeality denominator (#25 below). In any case, the verisimilitude of mimicry/representation falls somewhere in between completely accurate replication of the original and the following category.

8. *"two-dimensional, representational image"*: accurately portraying reality but missing a dimension e.g., actual human figures on a flat TV/movie screen, or a "realistic" painting on a canvas through which the viewer "must have visual substitutes for the things that are normally known by touch, movement of inference" (Langer, 1953, p. 73)—and one could add, sound as well. This she calls a "virtual scene" (p. 86) that demands memory of past experience and/or personal imagination to make sense of the two-dimensional artifact.

Moreover, a two-dimensional art(ifact) viewed in a picture frame or on a screen "has no continuity with the space in which we live; it is limited by the frame... or incongruous other things that cut it off.... The created virtual space is entirely self-contained and independent" of its surroundings (p. 72). As opposed to the earlier mirror image that isn't divorced from its surroundings (it reflects back whatever exists in front of it), screen images are not connected to the immediate environment.

Moreover, with two-dimensional representations, the viewer (above the age of toddler) cannot confuse the virtual with the real. Of course, we also consider a Rembrandt painting to be a "real" artifact in and of itself, thus serving as a (virtual) representation of something *and* a "real" object with its own intrinsic worth—another example of the real/virtual duality in objects of human interest.

9. *"description"*: oral or written depiction of a real situation, event, character e.g., a history book; a storyteller—"fiction as a virtual account of fact, or... a pretended speech act of assertion" (Ryan, 2001, p. 41). Once again, this had/has

a fixed, objective reality but the way it reaches the audience's cognizance is even further removed from that reality, given three levels of virtuality: (a) physical removal from the original source; (b) not seeing/hearing the phenomena in action but rather merely hearing *about* them; (c) the original "event" is filtered by the storyteller/author so that we cannot receive the "whole" story.

10.  *"simulation"*: replication of a dynamic scenario, a real-life interaction, and/or processes in the physical and social spheres—mechanically (a wind tunnel), mathematically (an econometric model), computer iteration (generational evolution of virtual beings), or even metaphorically (using a substitute population e.g., mice for humans in testing a new drug's efficacy). Here too the simulation is doubly "virtual": the act of mimicry itself, divorced from actual, real world actors/variables; the impossibility of including all the variables in the real-life situation.

11.  *"remembrance"/"associative memory"*: an inherent potential to bring up memories through a sensory immersion in the real. Walter Benjamin (1996, p. 218) described his experience in Paris, a city that he considered a virtual medium because for him personally its place and street names carried a multiplicity of histories. This is a *subjective* mentalization of virtuality, but no less important than the external, objective object or medium that stimulates the virtualizing process.

Past memory can also influence what we perceive as "real" today. As Baudrillard noted: "when the real is no longer what it used to be, nostalgia assumes its full meaning" (1994, p. 6). In other words, memory is not only virtual because it distorts our perception of the past, but consequently of the present as well.

12.  *"simulacrum"*: Baudrillard (1994) uses "simulation" and "simulacrum" interchangeably, but they are intrinsically different. He suggests that a simulacrum substitutes artificial signs-of-the-real for the real, thereby neutralizing the essential meaning of the original e.g., Disney's Epcot Center that provides such a strong feeling of the original that most visitors believe that they have "truly experienced" the foreign culture. This differs from virtual reality (VR) that belongs in definition #1 (duplication) or #7 (3D representation)—much closer to the original "real."

Baudrillard would not agree to place the simulacrum within our taxonomy here, because for him the contemporary age has completely and indistinguishably collapsed the real and the virtual one into the other. For him, the problem lies in the difficulty for modern persons to perceive a real referent, having been overfed a diet of virtual images through our slavish consumption of mass media and other types of virtual stimuli such as Disney World.

Nevertheless, we are still far from a situation where the average person cannot distinguish real referents from their virtual incarnations. For now, Baudrillard's virtuality/reality collapse theory is at best a warning of what might be, and not a realistic description of what is.

13. *"artificial"/"synthetic"*: All simulacra are artificial, but not everything artificial is a simulacrum, if we mean by "artificial" something made by human hands (or technology) and not "natural." The difference between "artificial" and "synthetic" relates to the end product. A humanoid, metallic robot, is *artificial*; an organism created through gene manipulation or combination is *synthetic*. Both are entities that give the impression of being a stand-in for something "natural" or "real."

    Artificiality as a concept, however, suffers from being too general in theory. After all, *everything* made or altered by humans is "artificial" if not found in exactly that form in Nature. One needs to distinguish between first order and second order artificiality. First order is anything not found in Nature. This is *not* the meaning of virtuality here. Rather, it refers to second order artificiality—creating something from materials or in a way significantly different from the way it has been created in the past. Thus, using carved ebony to make a piano key is first order artificiality; making the keys out of plastic so that they look, feel, and function like ebony keys, would be second order artificiality.

    A sub-category of "artificial" is "synthetic"—something created "unnaturally" but through the combination of different elements from separate sources that forge a *new* whole that can "behave" as the original or something very close to that. A recent example: synthetic biology that manipulates genes or combines organic materials to create "unnatural" forms of life. One could argue that the *product* of the "artificial" incorporates a greater degree of virtuality than a "synthetic" end result; nevertheless, the *process* is very similar.

    Another comparison between the two: "artificial" can be the creation of a non-natural artifact with natural materials (e.g., wooden chair), whereas "synthetic" is made from humanly created materials that do not exist in Nature (e.g., polyester clothing).

14. *"multi-dimensional"*: something existing in more than three dimensions. Whether Einstein's adding "time" as a fourth dimension, or current physics regarding 11-dimensional string theory, there might well be an entire "world" of additional dimensions that we cannot directly sense but yet affects our very being. This definition is close to #19 (not sensed/invisible).

15. *"indeterminate"/"ambiguous"*: these two meanings are not identical but are similar enough to constitute one definition. "Indeterminate" is what we know exists but find difficult to decide what it is qualitatively or its quantitative measure. This something need not be material; while there are indeterminate numbers (the number of microbes in the world), there are also indeterminate answers to philosophical questions. "Ambiguous" means unclear—something that could be one or the other "thing" (answer, interpretation, meaning). The difference between the two concepts is quantitative: "ambiguous" involves a few possibilities; "indeterminate" connotes a large, and possibly (close to) infinite, number of possibilities.

16.  *"incommensurable"*: that which cannot be counted or measured *in principle*. Something indeterminate or ambiguous does have a precise quality and/or quantity; we can never precisely know incommensurable "entities" e.g., Pi; infinity.

17.  *"deterritorialized"*/*"cyberspace"*: "deterritorialized" = not taking up space or not occupying a specific point in space; "cyberspace" = a specific type of "non-space" that actually does "reside" in a space (servers, computers).

According to quantum theory, an electron (revolving around the nucleus) is deterritorialized i.e., cannot be restricted to a specific geometric point or have an exact location in its orbit. Cyberspace, on the other hand, is a collection of physical (computer) media with specific locations, but as they are so spread out everywhere the user cannot determine their location.

Another seemingly related, but essentially different, definition: "rendered by a computer" (Castronova, 2005, p. 294). As opposed to the *spatial specificity* definition, Castronova's more *general* concept of being computer-produced cannot be placed in a taxonomical list such as the present one, because just about *anything* can be rendered within a computer—from the most realistic 3D replication (#1 above) to the most fantastical creation (#35 below). Defining "virtual" purely in terms of tools producing it (the computer) is tantamount to saying that "real" is anything perceived by the human brain; it includes too much and therefore tells us too little.

18.  *"atemporal"*: an indeterminate amount of *time*. This can have several meanings. First, something "outside of our world" where "time" might not exist and certainly *cannot* be measured (e.g., within other "universes"). Second, light: anyone external to a light photon can measure its speed (300,000 km/s), but when "riding" on the photon, time stands still i.e., that person cannot measure time for others on the outside. Third, the unconscious passage of time, as when we are daydreaming, sleeping, or even working with high concentration "in the zone," unaware of time passing. Overall, then, virtual atemporality can relate to objective, external phenomena as well as subjective, internal ones.

19.  *"not sensed"*/*"invisible"*: real and physical but too small, too large, too fast to be perceived by our (un)aided senses, or altogether invisible—whether temporarily or in principle (e.g., a star at the edge of the expanding universe so far away that its light will never reach us).

Benjamin called this the "optical unconscious" (1985, p. 243)—what the eye must have seen but the brain cannot discern or comprehend, due to the size or motion of the thing "seen." Muybridge's stop-motion photography (1877–78) finally answered the ancient question whether all the feet of a running horse leave the ground: they do. A photo of such a horse in mid-stride is virtual because it portrays something that the individual cannot perceive in any "actual" (conscious) sense. Similarly, infra-red and ultra-violet rays; odors that dogs can smell, and humans cannot. Objectively, they are real, but for humans, perceptually virtual.

There are also phenomena that in principle we cannot perceive—even with powerful technology. For instance, quantum mechanical laws do not allow for any direct observation of quarks and other sub-atomic particles of almost infinitesimal size. We can infer their existence, but they will forever remain "unseen" i.e., virtual shadows on the cave walls of reality.

Finally, "unsensed" virtuality is also connected to string theory (explicated in Chap. 5) that postulates the universe comprising more than the three (or four) dimensions—much like the residents of two-dimensional *Flatland* (Abbott, 1884), incapable of perceiving humans having normal, three-dimensional shapes.

20.  *"mis-perceived"/"illusion"*: perceived as real and physical but no longer in existence—or existing in different form from the way it is sensed. It is the exact opposite of the previous category "unseen" because we *do* see the object, or otherwise perceive it through other senses, but such direct perception is misleading because the object seen may no longer be there! The best example: light photons from stellar bodies reach us many light-years after starting their long journey; we might perceive this sometime after the heavenly body no longer exists.

Optical illusions are also part of this category—seeing something differently than what is in fact there. It is not an objective distortion (#5 above) but rather our perceptual apparatus being fooled, temporarily or permanently. For example, continuing to "see" bright light even after we avert our gaze from the light source. Indeed, "[t]he most striking virtual objects in the natural world are optical—perfectly definite visible 'things' that prove to be intangible, such as rainbows and mirages" (Langer, 1953, p. 48).

On the other hand (pun intended), illusions can involve persistent, personal "virtual mis-presentations," as in the "phantom limb" whereby amputees continue to "sense" the missing limb in exactly the same place that it used to be.

21.  *"over-sensed"*: having perceptions beyond the norm. People with synesthesia perceive things with more senses than can the average person, such as sensing a taste, an odor, or a certain color, whenever hearing a specific musical tone (more on this in Chap. 3).

This raises the interesting question of who precisely is having a virtual experience: the "over-sensed" synesthete who employs more senses than is considered "normal"; or the average person who "suffers" by comparison from not employing more than one sense at a time (the "not sensed" virtuality category listed above)? There is no clear answer, reinforcing the point already alluded to: virtuality is relative and subjective.

"Over-sensing" also occurs everywhere and through most of recorded human history through hallucinatory substances, as the drugged individual "perceives" the world in "heightened" fashion, sensing things that in fact are not there.

Finally, over-sensing can also be technologically induced through "hyper-presence": "a medium in which one feels *greater 'access to the intelligence, intentions, and sensory impressions of another'* than is possible in the most intimate, face-to-face communication" (Biocca, 1997; emphases in the original).

For instance, through augmented reality technology we could pick up an inter-locutor's subtle body language that we would not have seen (or even sensed) otherwise e.g., heightened blood pressure. In other words, virtuality—whether self-created, induced or mediated—need not always involve a diminution of experience; it can also offer a *heightened* sense of reality.

22. *"quasi-representational"*: similar to something real but in changed form or essence. Art does not necessarily attempt to realistically mimic the world, although in the pre-modern age it almost always did try to do this by repre-senting reality but in ways that added something to (or removed from) the original object—hence the first two letters in *re*presentation. The line demar-cating this type of virtuality from the next one is admittedly fluid; one could collapse the two into one categorical definition.

23. *"abstract"*: non-representational but vaguely suggesting something familiar and real. Although perhaps only a matter of degree, the difference between quasi-representational and abstract is still palpable. The latter goes a great distance beyond the real world; if a connection still exists it is more in the mind than on the canvas (or music sheet; or printed page) e.g., expressionist art, certain types of mimetic music, onomatopoeia words or Lewis Carroll-style semantic gibberish.

24. *"symbolic"*/*"signs"*: having no intrinsic significance other than what humans have arbitrarily as*sign*ed to it. We are capable of creating and/or imbuing symbolic shapes and forms with meaning as a result of agreed upon interpreta-tion. The symbolic notation is virtual, even if the denotation (and occasionally even its connotation) has real meaning. Examples abound: modern alphabet letters, numerals, shapes of public signs, computer code, and so on. Should the social consensus break down regarding denotation, the artificiality (virtu-ality) of the notation (the symbol's shape) becomes clear for all to see. Body language mix-ups are a good example of this e.g., Westerners traveling to Southern France should not "OK" with a thumb & finger circle, as it will be understood there as signifying "you're a zero!"

Augustine differentiated between "natural signs" (*signa naturalia*) and "given signs" (*signa data*). The former are not originally intended to offer knowl-edge regarding something else, and indeed usually have an identity in and of themselves e.g., smoke is a natural sign of fire, but we also use smoke signals. Conversely, signs that are culturally symbolic can use a natural object but that's usually an arbitrarily chosen sign to symbolize the object e.g., the Subaru logo consists of six stars chosen from the Pleiades (constellation of Taurus) that in fact exists.

25. *"incorporeal"*/*"spectral"*: referring to different immaterial phenomena. "Spectral" relates to anything without mass, as in gravitational waves, x-rays, and "light" (photons). "Incorporeal" denotes having non-physical "substance" within a physical framework and so able to influence the actual world e.g., the distinction between bytes and atoms: non-material electronic pulses (8 bits comprise a byte) are stored on physical magnetic or optical disks. The electronic pulses and the software that carries the instructions for computing

are "virtual" in that they have no atomic structure (besides being "symbolic").
Thus, we have computer terms using "virtual": virtual memory (storage virtualization), virtual machine (software executing programs), virtual reality (human
interaction with and "within" a computer-simulated environment), etc.

26. *"potential"/"latent"*: something not yet effectuated but can/will be. For Aristotle, entities are either actual or potential; every type of existence is a discrete
actualization of a wider potentiality (a baby can grow up to be tall or short etc.),
albeit the nature of the potentiality is not necessarily known until actualization.
Beyond biology, this is relevant to social life as well: "In an informational
sense, the virtual appears as the site not only of the improbable, but of the
openness of biophysical (but also socio-cultural) processes to the irruption of
the unlikely and the inventive" (Terranova, 2004, p. 27). In this sense, our social
world is deeply virtual as there are close to an infinite number of possibilities
inherent in the present from which myriad different futures can emerge through
a combination of an object's inherent potentiality and other elements added
through human intervention.[7]

27. *"non-actualized"*: something that could have happened but hasn't (yet). For
Bergson (1991), memories of the past constitute a virtual dimension of the
present that could have been—and can still be—actualized any time. That is,
elements of the recalled past can influence present action but if these memory
elements are not expressed through action, they remain virtual.

"Non-actualized" virtuality can also be due to concrete, social, external exigen-
cies. For example, da Vinci's helicopter sketch was technologically correct
but primitive materials science and lack of other technologies effectively
prevented actually building such an artifact. His helicopter was virtual, whereas
Sikorsky's modern helicopter was/is real.

Obviously, the past-virtual cannot be actualized when its window of opportunity is closed. A Napoleonic victory at Waterloo was "virtual" at the
time, but *actual effectuation* of the virtual potential ceased once the battle
ended, and certainly after everyone involved passed away. Thus, "non-
actualization" is weaker virtually when such actualization is still possible;
once no longer achievable, it becomes strongly (and permanently) virtual (#35:
fantastical/impossible).

28. *"imaginative"*: significantly different from what exists but still theoretically
possible. Fictional literature offers a vision of worlds (or at least elements)
different from what exists but not inherently impossible. Modern science
has also taught us that yesterday's "magic" (alchemy) is today's "science"
(nanotechnology). In other words, any employment of human imagination is
inherently virtual, until it bears actual fruit.

---

[7] Potential latency is not merely "forward" (in evolutionary or functional terms) but can be "backwards" as well i.e., reverse engineering. Paleo-biologists mull "de-extincting" dinosaurs by reverse
engineering chicken embryos (Horner & Gorman, 2009), or extracting dinosaur DNA from a
mosquito caught in amber that fed on dinosaur blood eons ago (Getlen, 2017).

"Imaginative" can take the form of daydreaming (usually reality-based, but not necessarily identical with any real situation), sleep dreaming (based on actual events but not in normal chronological order or even logical sequence), and even occasionally fantasizing (dating a "crush"). It does not involve hallucinating that usually leads to one of the definitions below of even greater virtuality strength.

*Mythology* is a final form of "virtual = imaginative" that could be included within several other definitions. Where myth belongs is receiver subjective. If people believe that gods are "real" in a corporeal sense, this would be "illusory" (#31); if they believe that gods are "real" in a theological/spiritual sense, then mythology would be "incorporeal" (#25), "metaphysical" (#33), and/or "otherworldly" (#34); modern audiences view mythological figures as "fantastical" (#35).

29. *"alien*: the noun, not the adjective—a foreign being that might exist but is so different from any known lifeform that we might not even recognize its existence if it were "staring us in the face." This is one level beyond "imaginative" beings from the previous category; as a product of our imagination, the former will probably remain within the realm of the "familiar" to us (e.g., dragons), no matter how strange.

30. *"ideal"*: a Platonic conception of the best form of any object, as in "ideal-type." Plato offers three types of "table": the pure form of a table, the carpenter's table, and the picture of a table made by an artist (Ulmer, 2002). For us, the second is real while the two others are different types of virtual manifestations: "ideal" for the first and "two-dimensional representation" (#8) for the third. This differs from "potential" because the "ideal" is never actualized.

Confusing matters somewhat is the fact that Plato views the "ideal" as more "real" than the actual. In other words, the previous paragraph presents the modern (or perhaps Aristotelian) approach to ideality and reality, whereas in Platonic philosophy (*Republic*, Book 10) the Ideal Form constitutes (real) reality, and our world is but the shadow on the cave wall, a pale reflection of the "real Ideal."

Two further aspects of Platonic Idealism = Reality need to be noted. First, for Plato, the Ideal is *permanently* real (being perfect, it cannot and should not change), whereas the material world we inhabit is *mutably* virtual (under the constant flux of earthly and sensory forces). Second, Platonism continued to hold sway during the Middle Ages as well, in religious garb. Here the Ideal = God, who is the paragon of perfect "reality," despite having (or because of) no perceptible elements. On the other hand, the human experience (individual life and social living) was virtual in that it could only attempt to approach a modicum of divine holiness. Moreover, in an inversion of virtuality (from the modern perspective) reminiscent of ancient paganism, shamanism etc., miracles were considered phenomena that uncover the *real* "reality" hiding behind our mortal comprehension. Another example of virtuality being perceived *subjectively*.

Very few people living in the contemporary era are Platonists in this regard. However, as we will see in Chap. 5, Idealism has interesting parallels in contemporary physics.

31. *"mistaken"/"illusory"*: something that people understand to be true (in the case of "illusory"—attainable) but in fact is not. This is different from an "illusion" (#20) that is more a perceptual, and not a conceptual, problem. The Ptolemian, geocentric theory (Earth as center of the universe) was mistaken; constant attempts of alchemists to turn lead into gold was illusory. Their virtuality lay in everyone believing incorrectly that something existed or was possible. This is qualitatively different from the next meaning.

32. *"false"/"fictitious"/"fake"*: giving a truthful impression but known or should be known (by the communicator) to be untrue. Eco provides an interesting example: the Museum of the City of New York displays the Manhattan purchase contract in pseudo-antique, English lettering—when the original was in Dutch! Eco opines (he apologizes for the pun): "it isn't a facsimile, but... a fac-different" (1986, p. 11).
"Fictitious" is not the same as "fiction," for the latter is understood by both writer and reader to be artifice, not real. Therefore, "fiction" belongs to "imaginative," as are such concepts as "make-believe," "fantasy," and the like.

33. *"metaphysical"*: beyond physical or corporeal matter. Just as several virtuality definitions can be said to be "real" in some sense, so too metaphysical phenomena might—or might not—be real. However, the seeming impossibility of finding direct physical evidence of the actual existence of anything metaphysical—e.g., God, soul, afterlife—renders anything metaphysical highly virtual.

34. *"other-worldly"*: the connotation (if not the denotation) of "metaphysical" is something in the realm of the "spirit." However, there is the possibility of something beyond our world that has a corporeal essence e.g., other universes simultaneously existing beyond our own. They would be virtual for us without any way of proving their existence or interacting with them.
Another "other-worldly" possibility are "previous" universes that preceded our Big Bang birth in an infinite time regress. Here too there is no way to know what they were like, with their Big Bust wiping the universal slate clean.

35. *"fantastical"*: impossible in principle/theory, at least insofar as present-day science and knowledge understands the world. The latter clause suggests that fantastical phenomena are not intrinsically impossible but are rather "virtually impossible"; what we consider impossible today might not be so tomorrow. Examples: time travel to the past or future, warp speed (faster than light), immortality—are all beyond our practical comprehension today but might not be inherently unattainable. Note the difference between *imaginative* virtuality and *fantastical* virtuality. We have no reason to believe the former cannot occur in the future; the latter is something that we think almost surely—but not absolutely/definitely—cannot occur. As Haldane put it: "...the Universe

is not only queerer than we suppose, but queerer than we *can* suppose" (1928, p. 286).

To be sure, it is not always clear whether to categorize things as "imaginative" or "fantastical." As Lem wrote, there are "three distinct kinds of dragon: the mythical, the chimeral, and the purely hypothetical" (1974, p. 85). This aptly sums up our entire list: it is not always completely clear how to precisely define, and into which "box" to place, each virtual phenomenon—but that difficulty does not in any way negate the essence of its virtuality.

## 2.4  Discussion

This is a long and complex list. Anyone could legitimately reject several of these definitions as falling outside what they consider the proper purview of "virtuality"—but others can just as persuasively argue for accepting precisely those "invalid" definitions/categories while rejecting others. Perhaps I even missed a critical definition or two.

Moreover, as I already suggested, there are two additional aspects to consider. First, examples offered in one definition could easily be ascribed to another, depending on which aspect one wishes to emphasize. For instance, virtual reality can be considered a case of "cloning/duplication" (#1), a "3D representation" (#7), or a "simulacrum (#12)—not to mention that it also enables "action-at-a-distance" (#2) and even simulation (#10). These are not hermetically differentiated categories but rather various possibilities and aspects of a multi-faceted phenomenon.

Second, any specific virtual phenomenon could be categorized differently depending on its relationship with other virtual phenomena. Just as something in the real world can have more than one identity depending on its context (I am a father of my child and a son to my parents), so too a virtual phenomenon does not necessarily have only one possible identity. For instance, the American wax museum reconstruction of the cinematic *Ben-Hur*'s chariot race, done in curved space to suggest panoramic VistaVision. Viewed as a presentation of what such a race might have looked like in ancient Rome, it is an "imaginative" [#28] "3D representation" [#7]. However, as an attempt to duplicate what the movie showed, it would be considered a "duplicate" [#1] of what appeared in *Ben-Hur*—far weaker virtuality.

A third complication involves cultural relativity: what will be perceived as something "virtual" in one culture, might not be considered as such in another. One example: Plato considered the "Ideal" [#30] as the real "real," whereas most of us see the "Ideal" as a virtual concept—at best an aspiration that someday we might be able to effectuate its potential [#26]. A different problem arises when the original itself changes over time. For instance, we are used to seeing classic Greek sculptures in pristine white marble, but in fact the ancient Greeks produced them multi-colored; the wear and tear of time "erased" the colors. For modern art lovers, which is the "original" and which the "substitute"? Objectively, white sculptures are the virtual substitutes; subjectively (for us today), the colored ones are the substitutes. Or perhaps both can be considered "original," given that natural erosion is the cause for the change in appearance?

I shall return to this cultural relativism point in Chap. 7 regarding fakes and forgeries [#6], but for now suffice it to note that given the somewhat amorphous nature of virtual phenomena, it should not be surprising that diverse cultures would view specific experiences or objects in different fashion.

Do these complications constitute an insuperable problem in discussing and analyzing virtuality? I believe not. As Gellner noted in defining "nationalism," another rather slippery concept: "Each of these provisional definitions... has some merit. Each of them singles out an element which is of real importance in the understanding of nationalism. But neither is adequate" (1983, p. 7). In short, we should not try to place the term in an arbitrary Procrustean Bed.

This does not mean that "virtuality" can be any and all things to all people, but rather that simultaneously it can be several things within certain parameters. One can accept the definition spectrum presented above in full or only in part; if the latter, this book's evidence will merely be a bit thinner, as some of the phenomena discussed in the following chapters will not be truly "virtual" for that specific reader. This won't negate the underlying arguments, unless the reader accepts only a very few (or none) of the virtuality types/definitions outlined here. To reduce confusion as much as possible, *regarding each of the virtual phenomena under discussion, throughout the book I will place its relevant definition/s in square parentheses, with a shorthand notation such as* [V2], [V4], [V21].

A central point (mentioned earlier in this chapter and expanded in Chap. 12) needs to be re-emphasized: although the "virtual" and the "real" tend to be viewed as two polar phenomena, they are not necessarily contradictory or diametrically opposed concepts. Several philosophers and critics, among them Proust, Bergson and Deleuze (Shields, 2003, pp. 19–21) also decry this dichotomization and instead call for differentiating the "virtual" from the "actual." Their approach is that "virtuality" is in some deep sense no less "real" than something corporeally concrete.

This also suggests another important point. The list of virtuality definitions presented above runs from weaker to stronger virtuality, from an *objective* perspective of corpo/reality and representation (physicality and realistic possibility). It does not—and cannot—offer a "measured" and orderly way to array virtuality from the *subjective* perspective of human perception and sensory and/or cognitive experience. In other words, objective virtuality and subjective virtuality do not work in parallel nor do they move in opposite directions. For example, Virtual Reality is "strongly virtual" in its objective manifestation, but precisely for that reason the user's subjective experience can feel very "real" (non-virtual)! On the other hand, a relatively mild form of objective virtuality such as a duplicate/clone ("mild" because it does not depart much from the original) will also provide the viewer with a sense of being in the presence of something very close to "real."

Another general complication of note is that many phenomena are doubly or even triply "virtual" (Plato's table). Modern technology offers relatively straightforward examples. Here is an example of triple (or more) virtuality in a sort of worlds-within-worlds egress: a virtual world (Second Life) screening room of "three-dimensions" (within our two-dimensional computer screen) in which a "real" TV channel offers avatar visitors full-length feature films showing animated or other phantasmagorical characters (Itzkoff, 2007).

Overall, then, defining the types of virtuality and assessing their degree of virtuality is a complex affair, because the issue can be viewed from several perspectives: objective versus subjective; mild versus strong; virtual versus actual versus real. These perspectives will be elucidated in the next chapter from the standpoint of "virtual" consciousness (the mind) within an organic "platform" (the brain), based on contemporary neuroscience and its understanding of the brain/mind's strange ways. Other chapters (especially towards the end of the book) will address these issues from an external perspective, especially how the virtual, the real and the actual may increasingly merge in our world generally, and personal life particularly.

# Chapter 3
# Virtuality and/in the Brain

## 3.1 Purpose of the Chapter

This chapter will argue that in so many different ways, the "reality" that we perceive is merely a virtual function (one is tempted to say "figment") of our brain's perceptual and cognitive apparatus. I shall show how the brain and attendant[1] sense organs repeatedly "fool" us into experiencing the world (literally and metaphorically) as it *isn't*. Of course, this is not to denigrate the human brain's incredible power; indeed, one might even say that if our own brain can "fool" each and every one of us, then it is even more powerful than we can imagine!

But first, why devote an entire chapter to illustrating the diversity of such "mind games"? There are two related answers. First, as already noted and/or to be explored in the coming chapters, two types of virtuality exist: (1) an external one determined by factors "out there" in our surrounding world, whether metaphysical (e.g., a Platonic Ideal form); spiritual (God); symbolic (paper currency); or material (objects); (2) an internal one that is a function of the way our brain works. Focusing only on the external types and areas of virtuality is akin to discussing Picasso's work in

---

The *Appendix: A Taxonomy of Virtuality*—listing by numbers 1–35, e.g. [V13] or [V34], some of the various types of Virtuality mentioned throughout this chapter—is freely available to the public online at https://link.springer.com/book/10.1007/978-981-16-6526-4.

[1] Why "attendant"? First, the sense organs are *conduits* of information to the brain, but the brain is the organ that does all the cogitation. Second, the sense organs (and other parts of the body such as limbs and intestines), are *integral parts* of the thinking process. This also means that external, artificial objects—whether limb prostheses (e.g., an artificial arm) or external machines (e.g., computer)—can be considered as part and parcel of cogitation, what McLuhan called "the extensions of man" (1964). For a serious argument in favor of the second approach, see Clark (2008). Later in this book I will return to this issue when discussing the inextricable nexus of reality and virtuality, body and mind (Chap. 12), and then again in my survey of the latest brain-to-machine communication technologies (Chap. 13).

© The Author(s), under exclusive license to Springer Nature Singapore Pte Ltd. 2021
S. N. Lehman-Wilzig, *Virtuality and Humanity*,
https://doi.org/10.1007/978-981-16-6526-4_3

terms of paint, canvas, models, etc., but not the artist's thoughts and internal creative process—not to mention the perceptual and interpretative processes of the museum audience.

Second, as will be shown throughout this book, virtuality is not merely an artificial and arbitrary construct of human ingenuity but rather a *natural* product of our very being. Humans are magnetically drawn to, and actively engaged in, all forms of virtuality in their life environment, not to mention actively creating and developing more sophisticated forms of virtuality and tools to help in this inventive process. Why should this be so? A probable reason: we are duplicating the same sort of phenomena that naturally occur in our own head. In other words, if our brains naturally "virtualize" then it should come as no surprise that we would project that sort of mental activity out of our cranium and project/activate it into our surrounding world. Therefore, *creation of external virtuality is an activity that complements our internal virtuality.* By laying out (many of) the internal workings of our virtualizing brain, one can better understand not just the *how* of external virtualizing but also the *why*.

This chapter makes no attempt to survey all of the brain's strange workings and it is certainly not meant to be a comprehensive primer on neuroscience. Its modest goal is to note the many ways by which we perceive the world differently from what in fact it "is"—and thereby go through our life in quasi-virtual fashion, sometimes very close to reality (but not quite there) and other times rather far from reality (but not completely divorced from it either).

## 3.2   Virtuality and the Brain: A Brief Philosophical Excursus[2]

Many neurobiologists, psychologists, philosophers and even physicists claim that in different ways, *everything* is "virtual."

We can start with Plato, Heraclitus, and Kant. Plato views what we usually call "real" as a pale imitation of the Ideal that constitutes the "real" real. Heraclitus (2001) offered a different take on the "reality" of the real: "No man ever steps in the same river twice, for it's not the same river and he's not the same man" (Fragment 41) i.e., the only consistent thing in the world is change, so that while at any specific second everything is "real," by the time we perceive that reality it is no longer the same reality. Here there is a "real," and we can perceive it, but at the moment of perception it is but a virtual "cousin" of its present (newer) reality.

Kant's position differed from both Plato and Heraclitus, bringing us close to the modern approach regarding perception and cognition, to be discussed below. "What objects may be in themselves, and apart from all this receptivity of our sensibility, remains completely unknown to us. We know nothing but our mode of perceiving them" (Kant, 1965, p. A 42/B59). Given the limitations of our perceptual system (Kant could account for only a small part of the limitations we know about today)

―――――――――――――――――

[2] Some of the points in this Sect. 3.2 will be expanded in Chaps. 5 and 13.

we can never fully know how far from true reality it is, although he/we assume/s that it exists—rendering all perception of our environment *ipso facto* somewhat virtual.

As opposed to these philosophers who focused on our perception and/or the essence of reality, contemporary biologists grapple with the question from the "inside": is "consciousness" something with a clear biological basis, materially/organically "real"? If so, then there is no brain/mind duality—the latter is an artifact of the former. Or is consciousness something "metaphysical": even if somehow related to the brain, perhaps there is a "mind" that still possesses a life of its own, colloquially called "free will"?

In either case, consciousness is "virtual," although in clearly different senses. If biologically based, the meaning of "virtual" is paradoxical: something that we believe exists as an independent "mental entity" actually is not independent [V31]. We feel that consciousness is "virtual" (non-corporeal), whereas it is intimately tied to our biological substrate. In this case, consciousness is "real" (in the corporeal sense) but it is "virtual" subjectively/psychologically. If this school of thought is correct, Descartes had it backwards—it should be: "I am, therefore I think." Thought that's independent of body is no more than a self-delusion, a mental attempt to virtualize that which is purely organic.[3] On the other hand, the second approach defines consciousness as being "virtual" by its very nature—a metaphysical entity [V33] (calling this a "*meta*physical some*thing*" would be an oxymoron) without corporeal substance, and yet able to affect not only the organic brain and one's body but also the world beyond the person [V25].

A third "approach" exists: even if consciousness is based on the brain's organic matter, this doesn't mean that we understand how it works or even where it comes from, because quantum physics (Chap. 5) cannot say whether our world is actually "material"! As Frank (2017) notes: "…after more than a century of profound explorations into the subatomic world, our best theory for *how matter behaves* still tells us very little about *what matter is*. Materialists appeal to physics to explain the mind, but in modern physics the particles that make up a brain remain, in many ways, as mysterious as consciousness itself."

Finally, a fourth approach: *Monistic Idealism* (to be expanded on in Chap. 5): consciousness is the basis of everything in the universe, so that human brain consciousness is "merely" part of a far larger phenomenon, from which it emerges. This is also called "panpsychism" (Hunt, 2021).

---

[3] The latest research suggests that this first approach is probably correct. According to the "global workspace model of consciousness" several parts of the brain individually analyze stimuli on an "automatic" sub-conscious level. It is only when a brain section "decides" that the stimulus is "important" and "relevant" to other parts of the brain and especially to the brain (person) as a whole, that it begins to communicate with other brain sections—and it is this "global" communication that sets off consciousness i.e., our awareness of thinking about the issue at hand (Ananthaswamy, 2010a). This is the "easy problem." The "hard problem" is how any purely biological process i.e., patterns of electrical activity, could lead to the subjective manifestations of our internal life that humans experience as "consciousness." Or to put it in terms of our topic here: how can "real" forces lead to "virtual" experience?

Contemporary physics has also added another fundamental element to the nexus between virtuality and the brain, but here the focus is on how the brain affects the external world. In brief, there is no objective reality other than what something with intelligence observes (and by extension, measures). This is certainly true on the atomic level: an electron "decides" whether to be a wave or a photon only when coming under "intelligent" i.e., human, observation [V2]. The age-old question of whether a tree falling in the forest makes a sound if no one is around to hear it, has been answered: "no"—if it were to do so at the atomic level (indeed, with no sentient being to perceive the tree, it would have fallen and not fallen at the same time). In short, *all of "reality" itself is no more than a product of human observation; until we look, it simply does not exist* (or, before we looked had not yet decided to exist in its present form) [V27]!

This even extends to what is considered to be an immutable part of existence: time. As Lanza (2007) avers, "even time itself is not exempted from the subjectivity of biocentrism. Our sense of the forward motion of time is really the result of an infinite number of decisions that only seem to be a smooth continuous path.… [t]he forward motion of time is not a feature of the external world but a projection of something within us. Time is not an absolute reality but an aspect of our consciousness."

Finally, one can note the famous question/surmise of which The Matrix movie is only the most recent manifestation: whether "reality = dream" i.e., how do we know that everything that we consider to be reality is not a figment of some higher Being's imagination? For that matter, as Descartes asked, can we distinguish between reality and our own dreams? As he notes in *Meditations on First Philosophy:* "…there are no certain indications by which we may clearly distinguish wakefulness from sleep," concluding that "it is possible that I am dreaming right now and that all of my perceptions are false" [V32] (Descartes, 1911/1931, pp. 145–46).

In short, from several standpoints we might be living in a universe of "virtual reality" in a sense far more profound than the recent nascent technology called by that name. However, we shall leave to the philosophers, neuroscientists and physicists these fundamental ponderings and conundrums regarding the "real = virtual." Such understandings of virtuality are important and certainly interesting in indicating the depth to which the phenomenon permeates the (perception of our) world, but they are both too wide and too narrow for our purposes. On the one hand, they virtualize everything—there is not much more to be said if we simply accept that our minds are virtual and/or that they transform the universe's "virtuality" into "reality" for us. On the other hand, they are completely divorced from the common understanding of virtuality—something emanating, but yet different, from our social world.

In this book, my goal is to explicate the different ways the common individual faces, and reacts to, different forms of virtuality in day-to-day life. Of course, all of us are a product of the sub-atomic world of quantum phenomena as well as being part and parcel of the universe's macro-physical cosmology, but the average person normally does not perceive, take cognizance of, or understand these concepts—and they certainly do not affect us directly in our daily doings.

## 3.3 The Brain: External and Internal Worlds

There are several ways of categorizing the virtuality definitions listed in Chap. 2. One way is to categorize them by *objective* versus *subjective* types of virtuality i.e., whether the virtuality is an artifact of something external to the person (e.g., *simulacrum, two-dimensional, simulation* etc.) or whether it is an internal product of our brain (e.g., *hallucinatory, imaginative, fantastical*). Most cases of virtualizing, however, involve *combinations* of both subjective, mental processes, and objective, physical phenomena. As Bergson argued: "So we place ourselves at once in the midst of extended images, and in this material universe we perceive centers of indetermination, characteristics of life. In order that actions may radiate from these centers, the movements or influences of the other images must be, on the one hand, received and, on the other hand, utilized.... Perception, in its pure state, is, then... a part of things" (Bergson, 1991, pp. 63–64). In order to fully understand virtuality, it is clear that we must take into account the workings of the human brain by itself and also in conjunction with external stimuli.

In psychological terms, this means that the human brain's activity can be roughly divided into two broad categories. The first is *perception*: absorbing and comprehending the external world. The second is *cognition*: internal, self-generated and/or reactive/responsive, mental activity. Of course, perception is also influenced by each individual's specific brain structure and chemistry. One person can perceive colors while another cannot—obviously, not because the first is surrounded by a colorful world and the second isn't, but rather because of a deficiency in the latter's brain and/or perceptual apparatus (retinal cones etc.).

On the other hand, "self-generated" mental activity also does not imply completely *sui generis* activity i.e., devoid of external stimuli. Rather, cognition is usually triggered by an immediate input from the environment that then leads to Proustian mental activity drawing on memories of prior experience, learned rules in the past, and other external phenomena that previously interacted with our brain. Moreover, on occasion cognition is aided or even generated by foods, drugs, and other ingestible products that change brain chemistry and thereby induce internal mental activity. Thus, perception and cognition are not "external" and "internal" in purely dichotomous fashion but rather are interrelated. As Borradori (n.d.) notes, "virtuality emerges as a constitutive component of experience, and as such, is neither mind-independent nor reducible to external sets of physical states."

Nagel (1974) illustrated the problem in his famous essay: "What is it Like to be a Bat?"[4] Here he points out that we can't experience "reality" as other creatures do because of their different perceptual apparatus and cognitive abilities. Moreover, our very concept of what constitutes a "mental state" must be circumscribed by the generic brain structure of all humans. For all we know, animals—not to mention much lower life forms and even plants—have radically different mental states i.e., the way they experience the world is completely alien to the way we experience it. Thus, human "mentalizing" is not only severely restricted but in principle we are

---

[4] For a more contemporary discussion, see "What it's like to be an ant" (Flyn, 2021).

also shut out from ever understanding how other creatures experience the world. For every living species, including *homo sapiens*, despite (and because of) our vastly superior cognitive abilities as well as our different sensory system, the external world is narrowly real and vastly virtual [V4; V16; V19] in that it cannot include other ways of experiencing it. This is because "our cognition and sense organs have been tuned by natural selection to those features essential to our survival, leaving us oblivious to the rest" (Spinney, 2020).

Nevertheless, despite the obvious nexus between perception and cognition, I shall discuss each separately for the sake of simplicity, pointing out some inter-connections as the discussion moves along.

One final but important note. There is one other central sphere of mental activity not mentioned here: the affective. Emotions do play a crucial role in the way the brain is structured and processes external stimuli, as well as generating internal responses (LeDoux, 1996). Recent research has concluded that in most (perhaps all) of our cognition an affective element is involved as well. This can be "primary," natural and inborn—found in most of the animal kingdom, and certainly all the later ones on the evolutionary tree (Tooby & Cosmides, 1990). Emotions such as fear, aversions and attractions (e.g., good and bad smells) are a function of the thalamus, that absorbs sensory input, directly affecting the limbic system which sends messages to various relevant body organs (e.g., fear = muscles contract, ready for immediate use). Emotions can also be "secondary" e.g., pride and satisfaction, that require more complex cognitive processing through the cortex—sometimes conscious, sometimes not (Damasio, 1994).

I will not devote space in this chapter to emotions because they are not directly related to "brain virtualizing"—although some do see a connection (Martin, 2014). Primary emotions are an automatic reflex reaction to perceived external stimuli, so that this will be covered by the sub-chapter on "Perception" below. Secondary emotions result from cognitive processes that interpret such stimuli, so that they are basically covered by the subsequent sub-chapter on cognition.

Nevertheless, I will not note the importance of emotion at several junctures throughout the book. For example, as will be discussed in Chap. 6, reading fiction is a matter of getting carried away into another world [V8; V28; V34]; this is accomplished by emotions, one of whose main functions is to focus our attention on the matter at hand, in this case being "absorbed in(to) a book." One more example for the time being: Chap. 13 notes how we can anthropomorphize and become attached to inanimate (or at least, non-living) entities—surely something that depends far more on reflexive emotion than rational cognition, as the latter "knows" that the entity is not alive.

Thus, the emotional element is crucial to a full understanding of the workings of the brain, but from the perspective of virtuality and/in the brain, most of it is covered by perception and cognition that initiate and/or mediate the emotional side of our mental world.

## 3.4 Virtual Perception

> If the doors of perception were cleansed every thing would appear to man as it is, infinite. For man has closed himself up, till he sees all things through narrow chinks of his cavern. (Blake, 1906, Plate 14)

Perception of our external world, making sense of the inputs, and then predicting what might immediately happen, is what the brain is primarily designed to do. As opposed to the Lockean concept of the newborn brain as a *tabula rasa*, we now know today that the brain from the start is constructed to perform certain tasks in highly efficient fashion. For example, language acquisition: we are all innate grammarians and vocabulary learners. Three-year-olds speak full sentences, mostly grammatically correct, before they have ever seen a dictionary or heard their first grammar lesson. Yet a child who has no verbal contact with the world (e.g., born deaf) will not automatically learn to speak; without external verbal stimuli, the brain remains in a state of frustrated linguistic potentiality [V27].

The same can be said for almost all areas of human skill and endeavor. Indeed, that is why humans are the only living creatures who spend so many years learning the rudiments of socially useful knowledge. Relative to the amount of information absorbed from external sources, the extent of our native, instinctual knowledge is minute.

However, the fact that the brain is "built" for external stimuli acquisition and comprehension, does not mean that the process is straightforward. In fact, it is anything but that. For learning to proceed in efficient fashion, the brain must take shortcuts—and it is this that starts us on the road to "virtualized perception": "Neuroscientists now know that what we really 'see' are actually copies, rough schemata, of the images our minds create by organizing patterns of light and color as they enter the eye and are transmitted to other parts of the brain" (Newton, 2006, p. 11). To paraphrase the common dictum: "what you see [or hear or feel or smell...] is *not* what there is!" [V4; V5].

There are a few reasons for this. The first, most important one is that the external world is simply too stimuli-rich for any brain—even the human brain, by far the most complex and absorbent of all—to handle. Every second of our waking life we are "bombarded" by surrounding stimuli, the vast majority of which we have to filter *out* in order to focus on what (for us at the moment) is the important stimulus. For example, if we are sitting in our living room watching a TV show, a short list of surrounding stimuli would include: the image and sounds on the screen; cars driving by; our daughter talking on the phone in the next room; the cat on our lap; logs crackling in the fireplace; distant music faintly emerging from our son's upstairs bedroom; and so on. And that's a relatively "quiet" scene; go out in public and the bombardment increases exponentially.

Obviously, in order to concentrate on our daughter's conversation, we block out most of the other stimuli; for the brain, "less is more." How does this occur? We have built-in, unconscious mechanisms for dealing with stimuli overload. The brain enables large amounts of data to traverse optical and cortical passageways without hindrance, along the way choosing specific data (approximately 30–40 "objects" a second) for a careful, more focused view. Metaphorically, it quickly moves a flashlight from one datum to another in order to decide what is relevant for the purpose(s) at hand (Cepelewicz, 2019). What, then, penetrates our consciousness i.e., what do we subjectively experience? One could argue that our brain offers us only a small fraction of reality for conscious cognition, as it has "removed" a large part of the true surrounding scene through "subjective filtration" (although at some rudimentary, subconscious level these stimuli are absorbed, but then quickly discarded from short-term memory).

However, this is just a small part of the perceptual picture. As powerful as is the human brain accompanied by our sensory organs, in fact we are incapable of picking up the vast majority of objective stimuli that surround us on a constant basis. One banal example: with the unaided eye, we cannot *see* things at great distances. Far more significant: we do not *perceive* radiation: gamma rays, infra-red, ultra-violet, x-rays, etc.[5] Indeed, light visible to the human eye constitutes less than one ten-billionth (0.00000001%) of the electromagnetic spectrum, stretching from gamma rays to radio waves. One can literally and figuratively say that humans are mostly blind to reality i.e., our immediate environment is far more virtual to us than the "real" reality [V4].[6]

Our *hearing* capabilities run from around 20 to 20,000 Hz—we cannot hear pitch beneath that level (infrasound, that elephants can pick up) or above that (ultrasound, that bats and whales can hear), nor can we hear sounds beneath the zero decibel (0 dbl) level. We cannot *feel* the bacteria crawling over our skin. We *see* only three basic colors—soon four with tetrachromatic vision (Lant, 2017), whereas some insects and birds have up to six different color receptors, enabling them to perceive colors that we can't even imagine. Conversely, "There's something strange about the way we [humans] see colours. We have prioritised distinguishing a few types of colours really well, at the expense of being able to see as many colours as we possibly might" (Higham, 2018).[7] Finally, regarding scent, we have only 6 million olfactory cells compared to a dog's 300 million, so that we cannot *smell* a good part of the world's odors (Williams, 2011).

In short, most of the world is beyond our sensory capabilities and perceptual apparatus—all connected to the brain. What we perceive (in the broad sense of the term, not just the visual), then, is reality in extremely truncated form i.e., most of the world is virtually beyond us, perceptually speaking, as we can sense very little of what is really going on [V19]. Indeed, some neuroscientists think that "our senses and brain evolved to *hide* the true nature of reality, not to reveal it... [as] it is too complicated and would take us too much time and energy" (Folger, 2018, p. 36). That

---

[5] Goldfish can see infrared radiation that we cannot; the same is true of bees, birds, and lizards regarding ultraviolet light.

[6] Modern technology is about to come to our rescue, with Artificially Intelligent Hyperimaging that will enable us to "'see' across separate portions of the electromagnetic spectrum in one platform" (Valdes Garcia, 2017).

[7] In fact, human color perception even seems to be "seasonal" (Engelking, 2015).

all this un-sensed and partly sensed reality is the "normal" way of our perceiving the world does not make the unsensed any less virtual; it is merely a "virtual reality" that we are not aware of as a matter of course.[8]

The virtuality of such phenomena, of course, is a relative matter. From the perspective of "objective reality" (a concept not accepted by everyone, as discussed in Chap. 5) or a hypothetically omniscient—at least pluripotent sensory—being, nothing is virtual. One need not go so far afield, though, for humans are also seriously deficient in some respective sensory capacities compared to other living creatures. A few additional examples: cats, dogs and bats hear much better than us; mosquitoes have a far stronger sense of smell; eagles see a lot farther ("Birds' eye view...," 2011); etc. Of course, they too perceive a virtual type of reality, for their stronger senses are counter-balanced by others that are weaker than ours. In short, all living beings are seriously constrained in how they sense the external world, so that parts of it are virtually "not there" for each.

However, the "problem" is not merely one of *degree*. Various living creatures tend to have sensory systems that perceive the world *differently in essence*. Bats use sonar to navigate, something quite different from a human's hearing; fish, insects, and birds transmit signals to their own species via ultra-violet light reflections; eels orient themselves through low-intensity electric impulses that are used for sensory perception; migratory birds use the earth's electro-magnetic field for orientation; ants (who are blind) communicate through chemical reception; etc.

Some humans too perceive the world in substantially different ways than (most) other people—for sensory or cultural reasons. Sensorily: perhaps the best example is something mentioned in Chap. 2: "synesthesia"—a crossing of the senses i.e., when one sense is stimulated it causes stimulation in another sense as well. Examples: always seeing a specific color when reading a specific letter (Simner & Ward, 2008), called "grapheme-color synesthesia" (e.g. red = "A"); hearing a musical note when looking at a specific color, "music-color synesthesia" (green = B flat); etc. This syndrome is found in up to 1 of every 23 people (Simner et al., 2006), with approximately 60 "cross-modal" forms of expression. In other words, there may be up to 300 million people in the world who perceive reality in an "abnormal" fashion—a number close to all United States residents! Obviously, if a "normal" person became synesthetic for a day, s/he would feel that s/he is perceiving an "unreal" world. However, what is important for our purposes is that the reverse is true too. If there is more than one way to perceive sensory stimuli (even for the human race alone), then there is no reason to call only synesthetic experience virtually "unreal." *All* perception is "virtual" in that it does not perceive the world in the full sensory panoply possible, even for that specific species!

---

[8] To be sure, the only way we know what we are missing is that we have developed technologies that "increase" our sensory abilities e.g., spectrometers. Indeed, it is quite possible that there exist other forms of stimuli that we are not aware of, and which will move from non-existent to virtual (i.e., potentially sensed) once we create the technologies capable of perceiving them. Thus, one could posit two levels of virtuality: strong, virtual phenomena that we are not yet aware of; mild, virtual stimuli that we have become aware of through our technologies, even if we cannot perceive them unaided.

Perception also has a cultural filter: "Categorical perception [based on any specific language] is a phenomenon that has been reported not only for color, but for other perceptual continua, such as phonemes, musical tones and facial expressions" (Roberson & Hanley, 2007, p. R605). In other words, the cultural connotation of language can "color" our perception of the specific stimulus.

Even if we limit this discussion of virtual perception to what our sensory apparatus is built to perceive, there is still another problem—the aging process that reduces the ability of our eyes, ears, nose, skin, and tongue to do what they are designed to do. Starting at the age of 40–45, most people need more light for reading and other focused tasks, as they suffer from reduced visual acuity (e.g., colors appear dim), and in general their eye lens becomes less elastic, losing its ability to focus on small print. Hearing loss (presbycusis) is also a common, indeed almost universal, age-related problem. The olfactory (anosmia) and taste senses, usually decline at a somewhat later age, over 65, although problems with taste are generally a function of olfactory diminution and not a weakening of the taste buds themselves (Doty, 2007). Finally, as many people age, they have reduced or changed sensations such as pain, cold, heat, pressure, and touch. All this means, of course, that as people grow older, they sense the environment in an even more virtual fashion than "normal" [V4; V5; V19].

This is one area of virtuality that has increased tremendously in the past century for positive and negative reasons: we are living longer than ever, and this entails living a greater portion of our life with weakened senses that undermine our perception. It also means that the phenomenon I discuss in later chapters—media and general technologies tending to virtualize reality—is not always true. Eyeglasses and hearing aids *de*virtualize our environment by correcting or supplementing our weakened senses so that we can perceive our surroundings at a level close to what we had in our early years. Nevertheless, the general trend is unfortunately (un)clear: the years take their toll on our senses, and thus we perceive the world in ever more distorted fashion [V5].

Returning to more "normal" modes of human, perceptual virtualizing, as noted concentrated focus is one way of "distorting" our perception of the world by removing "extraneous" stimuli. But even when we carefully focus on something, the human brain is sometimes at a loss. A good example is "change blindness": our frequent inability to detect a significant alteration of something right in front of us (Angier, 2008), such as pedestrians giving directions to a "lost tourist" researcher were distracted for a moment, and when they turned back to continue their instructions, they failed to notice that another person had taken the place of the "lost tourist." A second study found that people asked to count basketball passes in a video, tended to miss the fact that a "gorilla" slowly walked by in the video's background (Simons & Chabris, 1999; Chabris & Simons, 2010)![9] In short, as Dr. Jeremy Wolfe of the Harvard Medical School put it regarding human visual perception as a whole, most of it is "a grand illusion" [V20] (Angier, 2008).

We are all familiar with this next phenomenon but tend to consider it a niche-type, mental curiosity: "illusions." These lead us to see things that are not there, or to miss

---

[9] The video can be viewed at: http://viscog.beckman.illinois.edu/flashmovie/15.php.

seeing things that are there, or even to see something differently from what in fact is in front of us (Martinez-Conde & Macknik, 2008a). One type is *sensory* illusion. For example, we may be able to see two different figures within the same picture (the same sketch shows a rabbit looking one way and a duck looking the other[10]). Even more confounding from a virtuality standpoint are illusions as a function of reality. For example, the Kanisza illusion whereby we see a "triangle" only as a ghostly outline of the internal border of surrounding shapes,[11] illustrates how virtuality and reality are often inextricably bound (Chap. 12).

Illusions can also be auditory, as in the Shepard Scale whereby we think we hear a constantly falling (or rising) scale of tones, when in fact the notes are not going anywhere.[12] This is the auditory cousin of the famous, visual Penrose (continually rising) staircase.[13] Another form of illusion is self-induced: "illusory conjunction" (Deutsch, 2017). This occurs when we hear a combination of disparate aural stimuli, each emanating from a different area, and we recombine them incorrectly i.e., hear a melody that was not produced. It can occur visually as well e.g., presenting color A for object X and color B for object Y—and the receiver reports seeing AY and BX.

Another type of illusion is magic, where we are deceived into seeing something that we know could not have happened (the lady sawed in half) or exist (the girl levitating without external aid). In both cases the illusion is carried out by manipulating brain psychology in different ways (Martinez-Conde & Macknik, 2008b). Here the audience "asks" to be fooled—and then spends time trying (usually unsuccessfully) to figure out how the magician/artist "did it" i.e., why we could not "see" the manipulation. Thus, we are aware that our brain is capable of playing tricks on us (and that others can do it *to* us as well)—and find this more amusing than threatening.

What we perceive can also be a function of our expectations within a specific context—and these can override the evidence gathered by our own eyes to the extent that we will even imagine something that is not there (Zhaoping & Jingling, 2008). Magicians know this well: if they throw a ball in the air, and then a second and then a third, the audience will "see" a fourth ball thrown in the air when the magician uses the same motion, even when there is no fourth ball. This need not even be immediate nor initiated by an external party—we can do this to ourselves. For instance, a study found that "mental imagery… leads to the formation of a short-term sensory trace that can bias future perception" (Pearson et al., 2008, p. 982)—merely imagining something (even without external stimuli) can cause us to see things in a way that fits the internal imagery and not the external reality. A third and quite notorious form of mental distortion is "false memory" (Loftus, 1997): remembering something that did not occur (or did not occur the way it is remembered) [V5; V11; V32]. These can be "recalled" in a mere 42 ms—half the time it takes to blink an eye (Intraub & Dickinson, 2008).

---

[10] https://littlethings.com/lifestyle/rabbit-duck-optical-illusion.

[11] https://www.illusionsindex.org/i/kanizsa-triangle.

[12] https://web.archive.org/web/20180405072516/, http://asa.aip.org:80/demo27.html.

[13] https://www.illusionsindex.org/i/penrose-stairs.

A fourth phenomenon is somewhat bizarre: the brain can *create* for itself the illusion that parts of the person's own body exist when in fact they no longer do: the *phantom limb*. Amputees have the distinct impression that their lost limb still exists—they may not "see" it, but they certainly "feel" it, including great pain (especially if that's what they were feeling at the time of amputation). This is virtualizing of a high order: "replacing" a limb with a manufactured, and then keenly perceived, avatar [V6]. Paradoxically, the loss of a limb also alters (and distorts) the person's visuospatial perception around the area of the (former) limb so that not only is the former limb "phantom felt" but anything in the near area ("action space") of that "limb" cannot be easily touched by the person with any remaining limb (Makin, et al., 2007).

Something similar occurs with "chronic pain" and "referred pain." In the former, when the neurons in our brain's "pain map" are damaged they fire incessant false alarms, making us believe that a part of our body is hurt, when in fact the pain is merely being "felt" in our brain. Indeed, long after the body has healed, the pain system continues firing. Referred pain, on the other hand, is a condition in which pain signals in one area of the brain map spills over into nearby areas (similar to what happens in synesthesia); one part of our body is injured but we "feel" the pain in another part, some distance away (Doidge, 2015).

A fifth type is the most prevalent of all: the illusion of holistic perception. Once the human brain is imprinted with a "picture" of the world (it need not be visual), it will make an effort to re-present that "ideal-type" entity even when a subsequent appearance will have something missing or actually appear in a form that is not precisely real. For example, when we look at George Seurat's famous *pointillist* painting "Sunday in the Park with George," made of myriad tiny splotches of colors and not overlapping brush strokes, the viewer still perceives a "fully coherent" picture. Similarly, all digital music and pictures have "pauses" or "spaces" between the digitally produced sound bites or pixels, but our eyes, ears and brain either cannot perceive these "gaps," or refuse to accept them, and thus we "see/hear" continuous music/pictures. Is this a virtual or a real experience? It is both: partly virtual perception of a real artifact. The same phenomenon occurs when we watch movies (24/48 still frames a second that our brain interprets as a "moving" picture[14]), or view TV (older analog had 500–600 "lines" on the screen, HDTV has 1080 lines, and 4 K Ultra HD; all these still contain invisible "spaces" between the lines).

However, we don't have to be viewing an artifact for our eye/brain to be pulling a fast one on us; it happens every minute of our waking life in slightly different form. When we look at any visual scene, our eyes do not stay still but rather "saccade"—jump from one side to another in order to capture the full picture. We can only fully focus on a relatively narrow part of our vision, about 10°—the rest is "peripheral" where color vision disappears, and visual acuity quickly drops e.g., you can't read a book from your peripheral vision [V4]. Why then don't we continually "see" a

---

[14] The maximum number of images a second that the human eye/brain is capable of seeing is no more than twelve. Thus, 24 or 48 images per second is perceived as a very smooth "moving" transition.

jumpy world constantly sliding back and forth? Because while the brain "captures" the herky-jerky pictures, it ignores them until the eye settles on a specific "point" (location)—and then retroactively fills in the "time gap" [V 21] (the milliseconds taken up by our saccade) as if it were a static picture (Linden, 2007, pp. 94–96)!

Color perception follows the same pattern as our "fill-in-the-blanks" movie-viewing: the eye and brain take the full spectrum of colors and turn them into one: "white." There is no such color as "white"—it is an artifact of the brain combining the rainbow's color spectra into a "holistic" color that we call "white." Thus, white is a "virtual color," a product of our "collapsing and merging" all colors into one. Indeed, all colors other than the three primaries undergo the same phenomenon/process (for example, red + green = brown), so one could argue that most of our color perception is also a form of subjective virtualizing: "...colors are not properties of objects (like the U.N. flag) or atmospheres (like the sky) but of perceptual processes—interactions which involve psychological subjects and physical objects" (Chirimuuta, 2019).

All these perception examples show how the brain distorts, artificially creates and/or eliminates what the person should be perceiving. However, in several instances the reverse is also true: the brain can "perceive" things that our sensory apparatus should not be able to, or does not, absorb. An example of this is "blindsight," in which a major vision center of the brain is destroyed (operational stroke; etc.), rendering individuals functionally blind. Yet when presented with stimuli and asked to "guess" their color or shape (etc.) they will do far *better* in answering the visual acuity test correctly than a 50–50 guess [V21] (Trevethan et al., 2007). Indeed, when tested for the ability to see contrast, it was found that the "blind" eye of the study's experimental subject was superior to his normal eye! In other words, blindsighted people actually "register" what they cannot "see" (Martinez-Conde, 2007). This means that what their conscious mind can perceive is far less than what their unconscious brain manages to pick up—a sort of awareness without perception.

Another example of "indirect" perception is found among normally sighted people, colloquially called "subliminal messages." When shown stimuli that they did not commit to memory (because they were thoroughly distracted by another task) ordinary people could still accurately discriminate repetitive stimuli from new stimuli without being aware of the fact that they had "seen" the repetitive stimulus before (Voss & Paller, 2009).

In certain situations, our brain alters the way we perceive something, utilizing different sensory mechanisms of our body [V6]. One example among many: "the brain shifts its reference point toward interior perceptions whenever we lie on our sides... the brain is deliberately dimming the vestibular system," leading to the research conclusion that "our bodies themselves actually shape the way we perceive and understand the world" (Shore, 2021).

There are also two additional, related phenomena. First, only recently has an old question (from Locke and Hume to the present) been definitively answered: does an earlier image influence how we see a later one? The answer is yes. The consequence is that in perceiving the external world "objects have a remarkably persistent dual character: their objective shape 'out there,' and their perspectival shape 'from here'" (Morales et al., 2020). For us, neither is completely "real" [V11], as the former image (no longer existing in our field of vision) remains in our mind to "reshape" the new image that is presently being capturing by our eyes.

Second, déjà vu, whereby one senses that personal history is repeating itself (i.e., "I have seen, or been in, this scene before")—and we are disturbed by this sense, perhaps because we are conscious of the feeling but cannot get our memory to bring forth the appropriate former stimulus. It is as if our brain knows something that "we" do not, and refuses to tell us (Phillips, 2009). Even more telling is the discovery that blind people can also have a déjà vu experience (O'Connor & Moulin, 2006)—it is not merely a matter of vision.

Finally, we perceive (evaluate, judge) external virtuality differently than external reality. For example: "When an image of a person appears within a photo, that individual is perceived as being less real and having 'less mind'," than when seen the same way in real life (University of British Columbia, 2021). Indeed, if a person's photo [V8] appears within a poster, [V8 "doubled"] we perceive that person (as seen in the photo within the poster) as even less real than when the photo appears by itself. We distort [V5][15] the "perceptual meaning" (or substance) of virtual stimuli more than our perception of identical real-life stimuli—and when doubly virtualized, the interpretive distortion increases. It remains to be seen whether the same phenomenon holds for completely life-like 3D images within a virtuality reality framework.

To sum up this segment on perception (for the time being), it should be obvious by now that while the external world can have great stability and consistency, this is not what our brain perceives. In many different ways, humans (as all living creatures— but *we* are the only ones who can *talk* about our sensory experiences) exist in a kind of netherworld between the virtual and the real. When we manage to perceive external reality "accurately," it is a mere sliver of that reality for reasons of both attention/concentration and sense organ limitations. However, more often than not, what we perceive is not accurate at all, but rather a distortion of reality as result of flawed/weak sensory organs, others' manipulation of our perceptual system, and probably most prevalent of all, internal mental biases: "In truth, we are realising that our [visual] experience is closer to a form of augmented reality, in which our brain redraws what it sees to best fit our expectations and memories…. The same goes for our other senses" (50 Ideas to Change Science…, 2010). Or as Linden puts it more bluntly: "…this feeling that we have about our senses, that they are trustworthy and independent reporters, while overwhelming and pervasive, is simply not true. Our senses are not built to give us an 'accurate' picture of the external world at all…. In fact, our sensory systems are messing with nearly every aspect of our sensations from their quality to their timing" (2007, p. 83).

However, if "the information we receive by seeing and listening with our natural organs is just a small portion of what is potentially available… the question is: Can we claim, with a valid ontological justification, that the 'normal' is 'real,' whereas the transformed [perceived by our senses] is 'unreal'?" (Zhai, 1998, p. 8). Many

---

[15] Objectively, this is not distorted *external* reality but rather *internal* distortion of external reality. However, that reinforces a central point: the near impossibility of separating objective and subjective perception.

scientists answer in the negative, even going so far as to argue: "Every last scrap of our external experience is of virtual reality… including our knowledge of the non-physical worlds of logic, mathematics and philosophy, and of imagination, fiction, art and fantasy [that] is encoded in the form of programs for rendering of those worlds on our brain's own virtual-reality generator" (Deutsch, 1997, p. 121). One need not agree with such an extreme view; but it is hard not to accept that our world involves no small measure of distance between objective reality "out there" and the way we perceive it "in here." It is this "distance" that renders our experience more or less virtual.

## 3.5   Virtual Cognition

The mind is its own place, and in itself/Can make a heav'n of hell, a hell of heav'n. (Milton, 1667, Book 1, lines 254–55)

As noted, perception is but one-half of mental activity. Cognition, the ideas and thoughts that we generate, constitutes the other half—indeed, what differentiate us from the animal kingdom. However, these are not two different mental worlds—they are related in several ways and shed further light on our "non-real" mental life.

Physiologically, the human brain has three modes of operation depending on whether or not external stimuli are un/consciously generating "reactive" mental activity. First, when external stimuli impinge on us, we are usually aware that the brain is working because we make a conscious effort to think about those stimuli. Second, external stimuli need not be perceived by us consciously in order for cogitation to commence—they can enter our brain without us being aware of it. Third, cognition can even be *sui generis*; external stimuli are not needed at all. Our brain can be quite "self-starting" in ways that will be discussed below. In short, the brain is *constantly* working (not necessarily "consciously"), whether we have initiated the cognitive process or not.

What is the brain doing? Much more than simply reacting to (thinking about) the external stimuli. Through several types of mental processes, it is quite "pro-active" in deciding which stimuli to process, and how to do so.

One way we can see this connection between perception (absorption of stimuli) and cognition (evaluating stimuli) is through a widespread phenomenon called "framing." How I evaluate something is a function of how it is presented to me. There is no perceptual distortion here: we all read the same words in the newspaper and understand their meaning; we will have little disagreement about what the picture in front of us is showing in objective terms. However, just as our mind will focus on a narrow "spotlight" in order to make sense of a scene, so too our fellow compatriots will transmit messages to us in the way that they want us to understand them. Up to this point we are dealing with perception—but unlike the magician whose sleight of hand "reality-distortion" has little cognitive effect on the audience (the magic show will not change their opinion about the world nor their personal behavior), in other cases—professor, journalist, religious leader, politician, author—such "selective stimuli" affect not only perception but also the audience's ensuing cognition. To take but one example in a vast field, the newspaper reader will form quite a different opinion about an event if the journalist uses the term "freedom fighter" as opposed to "terrorist" (and all other possibilities in between: soldier, militant, guerilla, etc.) when describing the news item's protagonist. Assuming the veracity of the news report, all the readers understand the objective *action* of that actor the same way (shot a missile); not so the *subjective meaning* of the action.

Media framing is a matter of an *external* body influencing a person's perception and subsequent cognition [V9]. In the same way, each of us performs such "framing" on ourselves. The nomenclature we use to describe things not only affects the way others perceive the matter but also (usually) reinforces the way we perceive the matter too—unless we are consciously prevaricating. In principle, we could use a different connotative term for the same phenomenon each time in order to describe the full panoply of possibilities (yesterday I will use the term "terrorist," today I'll call that person a guerilla fighter, and tomorrow a freedom fighter), but we also value consistency in our world perspective, so that this almost never happens. Once we have settled on a steady "frame," it colors the way we perceive all future information regarding that issue.

In order to ensure a steady stream of information that is consistent with our individual frame we resort to a second mental trick: "appropriate" selection of the stimuli reaching us. Here there are two central (f)actors: people and informational sources that each person individually chooses to be exposed to, and our own further conscious selection of individual bits of information that we wish to access and store in our memory. Returning to the example of the newspaper, I might choose to read *The New York Times* instead of *USA Today* for all sorts of reasons. In that sense, I am master of my own informational destiny—and from there also to my cognitive makeup (facts, and evaluation of the facts, as "framed" by the newspaper I read). However, *The New York Times,* as almost every other informational source in the democratic world, is not monolithic. So, let's say that I choose to read that paper because of its "liberal slant"—and find myself one day faced with a *Times* news headline that clearly suggests the existence of evidence buttressing a "conservative" point of view. Or even more likely in these days of media surfeit, I come across similar evidence in another venue: internet, social media, TV, radio, cinema, etc. What, then, do I do?

There are several additional mental gymnastics routines that can be employed. The first is a continuation of selective exposure: I can decide not to read the article or watch the TV news interview. In other words, our filtration system does not rest after selecting for media (newspapers and not TV news) or venue within that medium (*Times* instead of *USA Today*), but rather continues to work within that selected medium.

Second, I can employ selective retention i.e., read the article and then forget it. This is easy to do because the brain in any case has to cull the vast majority of information it deals with on a daily basis (many neuroscientists believe that the main purpose of sleep is precisely this culling process). The main criterion for moving short-term memory to long-term memory is relevance—potentially how important/useful is this information for me in the future? If the news item does not fit my cognitive framework, it will usually be "scored" as unimportant (or even "dangerous") and will be discarded along with lots of other informational flotsam.

But let's assume for a moment that for whatever reason (e.g., the researcher providing the evidence is very famous), my mind was not successful in banishing it from memory. Here a third trick comes into play: selective memory [V4]. If and when the news item pops up again in my memory due to a new stimulus (e.g., the following week I find myself in a discussion with friends about the issue), I will attempt to recall from my memory other articles and data that provide evidence for my liberal point of view—and not mention the *Times* report at all.

However, in the end, if and when I have been continually bombarded by counterevidence I will (very reluctantly) change my mind. Thus, I might read the article and change my opinion, but only because of repeated exposure to evidence supporting the opposing point of view to what I [actually everyone] normally hold[s]). What this means, however, is that in numerous areas of life when such a change of heart does not occur (a much more common situation) we will perceive the world "virtually" (incomplete and distorted) in light of the factual evidence that we've tried to avoid.

Of course, it should be noted that this overall framing process is not exclusively a "media"-related phenomenon. Rather, it occurs constantly from cradle to grave: at home—parents trying to nurture a specific value system; in school at most levels—providing information that reinforces dominant social values; among our friends—who we choose in some measure because they share our *weltanschauung*; at work—where we gradually take on the dominant work culture; and so on. In short, the distortions resulting from our need for cognitive consistency will color the perception-cognition interaction over the course of our lifetime—enabling us to "understand," and act on, the world in very specific fashion. In theory and practice, two people can live very similar lives objectively but still have completely different cognitive experiences and ideas, depending on the way the messages reach them perceptually and their cognitive interpretation of such stimuli.

The bottom line: we all exist within a social environment whose reality is heavily distorted by the specific stimuli streaming into our minds and our idiosyncratic interpretation of them [V5]. To a significant extent, therefore, we can say that the inter-relationship of human perception and cognition virtualizes external reality.

If that were not problem enough, another one is even more prevalent—and has little to do with any specific "belief framework." Rather, it is a function of our cognitive need for logical development that leads us to try and make sense of the world by creating in our head some sort of "story"—an orderly progression of past events. Unfortunately, we don't always choose the most important elements but rather those that seem to us to have the best "fit" for a logical narrative. As Taleb notes:

> ... memory and the arrow of time can get mixed up. Narrativity can viciously affect the remembrance of past events as follows: we will tend to more easily remember those facts from a past that fit a narrative, while we tend to neglect others that do not *appear* to play a causal role in that narrative.... This simple inability to remember not the true sequence

of events but a reconstructed one will make history appear in hindsight to be far more explainable than it actually was – or is. (2007, p. 70)

We thus find ourselves in a paradoxical situation: precisely because we want to render our memory as precisely logical as possible, we will in fact virtualize it by deleting/ignoring aspects that seem not to be relevant to the "narrative" we have built in our heads about what happened [V4, V5]. Nor are we the only initiators; propagandists understand that "nostalgia doesn't need real memories—an imagined past works too…[because] we redeploy much of the same neural mechanisms that we would have employed had we actually engaged in the simulated action" (De Brigard, 2020).

This is not even the main legerdemain that the brain performs on itself. Neuroscience has discovered something quite startling: "there is a huge amount of activity in the [resting] brain," that researchers call a "default network" (Fox, 2008). In other words, even when we "think" (believe) that we are *not* thinking about anything, in fact our brain is hard at work. This suggests a seeming non-sequitur: the brain has a mind of its own! Indeed, many brain researchers (Carruthers, 2017; Rosenberg, 2018) have shown in different ways that the brain is able to make decisions and come to conclusions that are not based on any possible, *conscious* sensory perception (e.g., temporarily blinded) nor on any personal, conscious decision-making process. Thus, as Pinker notes (2007): "the intuitive feeling we have that there's an executive 'I' that sits in a control room of our brain, scanning the screens of the senses and pushing the buttons of the muscles, is an illusion." Or as Zimmer (2008) put it colloquially, we all seem to have a "zombie" sharing our brain with "us." Without pushing the point too far, at least metaphorically (perhaps literally as well?) we are of two minds: the conscious real and the unconscious virtual—the latter insofar as we do not control it; it's even beyond our personal (but not scientific) ability to know that it is there.

A specific and well-known example of this is dreaming. When asleep, we have no control over our dreams nor even when we will dream, the latter occurring at regular intervals during the brain's sleep cycle (the REM stage). Each of us is like a couch potato paralyzed in front of the TV set, waiting for the programmers to display their wares. However, unlike a television audience, after we awaken from our sleep it is very hard to remember even part of the previous night's dreams. Moreover, infrequently in our sleeping state we can also exhibit the zombie aspect of cognition: sleepwalking. Here the unconscious brain (and not the personal "I") not only makes abstract decisions but also manages to move our body.

What we have here, then, is an entire world of virtuality taking place within our brain, without conscious input [V28]. The dreams themselves—whether uplifting, nightmarish, or mood neutral—tend to be loosely based on real-life experiences, but also tend to have unnatural qualities, at times bordering on the fantastical [V35]. In any case, what is most germane for our purposes is that human beings spend a third of their entire life in a less than conscious state, and a good portion of that (altogether around one to two hours per night) is spent in a virtual world of our uncontrolled and unconscious making.

Between being awake in a conscious cognitive state (albeit with subterranean unconscious cognition) and unconsciously asleep—rigid or walking—we also indulge in a broad intermediate stage called reverie and/or daydreaming (Grotstein, 2000). This may be more or less focused by us: we have a problem to deal with and thus let our mind jump all over the place in search of a solution; or we wish somehow to bring sense and order to the swirl of images and ideas that are internally generated and externally stimulated; or we merely want to take a break from focused thought and let our mind wander without specific purpose. With daydreaming, we are somewhat aware of our part in consciously initiating the process for the purpose of temporarily escaping the constrictions of our social or physical environment, but the "product" is not an outcome of our "will." As opposed to what we might think, daydreaming constitutes a very significant part of our life; some studies have found that people spend *half* their time while awake doing some sort of daydreaming! (Klinger, 2009; Killingsworth & Gilbert, 2010).

Cognition is normally considered to be an analytical process in which the brain absorbs external stimuli/data (perception) and then evaluates its meaning, deciding what to do about it. However, if the brain were to start every analysis from scratch, it would have to spend inordinate amounts of time repeating its thought processes. We don't cogitate like that but rather rely on a critical component of the brain: memory. This includes previous experiences, decisions, outcomes, etc., that we can draw on internally in order to efficiently think about the problem at hand. Unfortunately, as seen earlier, while humans have a prodigious memory for myriad facts, data, etc., our memories tend to be malleable, so that in (great?) part they too are "virtual."

The virtuality of memory can be understood on two different levels. First, by definition a memory does not exist in the here and now but rather is a "simulacrum" of something that did exist in the past [V12]. In Bergsonian terms, a memory is "real" but not "actual" (Bergson, 1991). While this might seem to be a trite (or "soft") way of equating memory with virtuality, in fact it is not trite or soft at all because our memory can never resurrect the full perceptual panoply of the original experience. In our mind's eye, while recollecting our first puppy love kiss, we can usually reproduce the friend's face and certainly the approximate emotional feeling that we had at the time. However, we cannot really feel the lips, nor smell the other's scent, not to mention the difficulty in replicating the surroundings (weather, light, sounds etc.). So even if one does not accept any "exact memory" as being virtual in the trite sense, one has to admit that almost all memories do have a measure of virtuality in those aspects that cannot be remembered.

Notwithstanding this aspect of memory = virtuality, there is a weightier expression of the equation. Memory not only is incomplete by leaving things out, but it is "creative" in remembering things that never existed [V21]! Current research shows that all our memories wither over time and are prone to distortions [V5] (Krackow et al., 2005–2006). There exists broad consensus among psychologists that memory isn't *reproductive*—it doesn't duplicate precisely what we've experienced—but rather *reconstructive*. What we recall is often a blurry mixture of accurate and inaccurate recollections, along with what meshes with our beliefs and hunches.

Memory itself is not like a video-recording, with a moment-by-moment sensory image. In fact, it's more like a puzzle: we piece together our memories, based on both what we actually remember and what seems most likely given our knowledge of the world. Just as we make educated guesses in perception, our minds' best educated guesses help 'fill in the gaps' of memory, reconstructing the most plausible picture of what happened in our past. (Brady & Schachner, 2008)

And this "filling-in" occurs repeatedly regarding the same specific memory when recalled at different times. Thus, we can aptly describe memory as an ever-changing artifact, something that highlights our ability to *create* (i.e., distort: V5) fluid narratives of our experiences. We do this actively, albeit unconsciously, through a process of memory "reconsolidation," retelling the story but in somewhat inaccurate fashion through misattribution (e.g., the detail is correct but source, time and/or place are wrong), suggestibility (external misinformation that colors our memory), and bias (our own values/interests conflict with the memory) (Schachter, 2001).

Making matters worse is our tendency to "prettify" past experiences so that the memory becomes more pleasurable. As Salvador Dali (1942, p. 38) once said: "The difference between false memories and true ones is the same as for jewels: it is always the false ones that look the most real, the most brilliant." Of course, on occasion such falsification can work in the opposite direction: to intensify the ugliness of a traumatic experience (e.g., the rapist may actually have been handsome but the victim's memory will remember a beastly face). But this does not contradict Dali's quote nor undercut the point here about the intensification of memory's virtuality: "brilliant" in this case would simply mean "in sharper relief" or "exaggerated."[16]

The outcome of all this, then, is doubly virtual: not only is memory based in part on reality and part on fiction, but if cognition itself is heavily dependent on retrieving memory, then the cognitive outcomes must also be partly virtual i.e., not completely consonant with reality. And that is precisely what psychologists have found: we are constantly making errors in our thinking process, in part due to the brain's reliance on faulty memory of facts, experiences, arguments.

This occurs on two levels. First, it can take the form of a macro-problem i.e., remembering only a fraction of events from a large class of phenomena [V4]. For example, Tversky and Kahneman (1973) noted the "bias due to the retrievability of instances"—we will make decisions based on things we tend to remember better than others of similar nature. Here the *overall* picture of the phenomenon at hand is partly virtual because whereas some things are recollected correctly, others of identical qualities/nature are not recalled at all so that the overall picture is not truly realistic

---

[16] The phenomenon of selective memory, prettified memory, and distorted memory is so common that many pundits have come up with highly pithy aphorisms describing it. Among the best: "Nothing is more responsible for the good old days than a bad memory" (Franklin Pierce Adams); "Our memories are card indexes—consulted and then put back in disorder, by authorities whom we do not control" (Cyril Connolly); "The heart's memory eliminates the bad and magnifies the good; and thanks to this artifice we manage to endure the burdens of the past" (Gabriel Garcia Marquez); "A man's memory may almost become the art of continually varying and misrepresenting his past, according to his interest in the present" (George Santayana).

[V4] (somewhat akin to a jigsaw puzzle with only 70% of the pieces placed in their proper position, the others left blank).

Second, it is also—and perhaps primarily—expressed on a micro-level: incorrect recollection of memories [V4; V31]. This is not necessarily fabrication of an event that did not take place (as in the relatively rare "false memory syndrome") but rather unintentional distortion of details [V5], sometimes critically important e.g., wrongly identifying characteristics of an attacker (misidentification) or as we have just mentioned: mixing up place, time, or source (misattribution). There are several subjective reasons that memories may become distorted: bias leading to wrong association of the right facts with the wrong people; mixing up and/or combining memories of disparate events; receiving information about the remembered event from other (somewhat unreliable) sources. However, the main reason is objective: brain cells have a life span of around fifty years. Although the latest research Weintraub, 2019) clearly shows that neuron regeneration *does* occur in the hippocampus (the brain's memory center), this does not fully replenish what is lost or weakened. Thus, without constant reinforcement (who has the time and inclination to try and recall all our memories every day?), our memories weaken and become distorted as time goes on with the brain's cells gradual diminution and wearying.

Put another way, we live in the real world every moment, but each of those moments becomes more and more virtual [V4; V5] as we recede from that specific time. This means that our cognition, based in large part on memory (again, previous experience, information, education, etc.), also becomes virtualized to the extent that it relies on older memories. Over the course of a lifetime, we are all caught in a paradox. On the one hand, we accumulate more experience as time goes on, more memories, so that the *amount* of stored memory increases, in turn aiding in cognition. On the other hand, as we get older our memories get weaker—not just as a result of the extended time between the present and our earlier years (entropy), but also because our brain's memory-creation ability loses its former potency (aging). In short, the *quality* of each memory (both distant and recent) gets worse, thereby undermining cognition. Quantity and quality of memories, then, are held in some rough equilibrium over most of our lifetime, balancing verisimilitude and virtuality in relatively equal measure.

Finally, whereas most of the above discussion related to unconscious virtualizing, the fact is that all humans do this consciously as well: "lying." Whether white lies to smooth social interaction ("you look great!"), grey lies to puff up our self-esteem in the eyes of others ("I graduated with honors"), or black lies to cheat or harm others ("I can guarantee you 50% returns on your investment"), we are all purposeful reality-falsifiers: "deception [i]s the intent to distort someone's perception of reality" (Bhattacharjee & Winters, 2017). Worse yet, people are hardwired to be intrinsically trusting and therefore gullible (Raden, 2021), easily believing such distortions—especially if they please us or affirm our worldview ("alternative facts"). As communicators, then, we are inveterate virtualizers, whether as senders or receivers.

## 3.6   Extreme Mental Virtuality

… the Places inhabited by the insane and the exceptionally gifted are so different from the places where ordinary men and women live, that there is little or no common ground of memory to serve as a basis for understanding or fellow feeling…. (Huxley, 1954, p. 13)

Everything discussed in this chapter up to now relates to people who ostensibly have "a hold on reality." Unfortunately, many people live their life (voluntarily and/or non-volitionally) in a subjective state of almost complete virtuality; indeed, for such people the virtual and the real are close to being one and the same. These individuals can be divided into two main categories. The first is comprised of two sub-groups: people who are mentally ill from birth or commencing in their adolescent early years (e.g., on the autism spectrum; non-medicated schizophrenics); normal people who late in life decline into senescence (senile dementia, Alzheimer's). Of course, none of us would make the argument that this "characterizes" human perception and cognition; it certainly is a state that we wish to avoid.

Nevertheless, these "unfortunate"[17] souls are part and parcel of the human condition, so that at the least we have to note that a chronic and deep, perceptually/cognitively virtual state (disconnected from reality), far beyond the "normal" quasi-virtuality of the healthy human mind, is also part of the overall picture. While there are occasional differences in the *definition* of mental illness between nations, each of whom may even have some different, "local" *types* of mental illness (Watters, 2010), there is little reason to think that the overall proportion of mentally ill differs significantly around the world, although *treatment* of mental illness can vary radically from culture to culture. I shall not try here to survey the full panoply of mental illness but rather note several conditions that place the victim in various types of a strong "mental virtual reality."

It is estimated that approximately 21 million people in the world suffer from schizophrenia (Schizophrenia: Fact Sheet, 2021), although modern medicine is now able to bring many around to reality-based perception and at least partial, cognitive functioning. As opposed to conventional wisdom, schizophrenia is *not* "split personality" (itself a different but far less prevalent malady[18]). Schizophrenics suffer from a range of severe mental disorders, including: hearing voices that other people don't

---

[17] They are "unfortunate" from the conventional perspective of mentally healthy people. However, an argument can be made that their "reality" is no less real to them than ours is to us: "The visions and sounds schizophrenic patients see and hear are just as real to them as this page or the chair you're sitting on" (Lanza, 2007). In some cases, the patients themselves do not accept their condition as a handicap. For example, those with Asperger's Syndrome claim to have an "Autism Culture" and that autism is simply a natural difference from the "neurologically typical" (http://www.neurodiversity.com/positive.html). One could say that this approach "normalizes" the neuro-virtual.

[18] "Multiple personality disorder," or MPD, is a mental disturbance classified as one of the dissociative disorders in the fourth edition of the *Diagnostic and Statistical Manual of Mental Disorders (DSM-IV)*—renamed dissociative identity disorder (DID)—in which "two or more distinct identities or personality states alternate in controlling the patient's consciousness and behavior." "Split personality" is not an accurate term for DID and should never be used as a synonym for schizophrenia (Multiple Personality Disorder, 2010).

hear [V21]; the belief that other people are broadcasting the sufferer's thoughts to the world [V31]; or extreme paranoia i.e., convinced others are plotting to cause them harm [V20] (Schizophrenia, 2020).

In the U.S. 2021, approximately 1.9% of the entire population (6.2 million people) suffer from Alzheimer's disease (Alzheimer's Disease..., 2021), and there is little reason to think that the numbers are different elsewhere where the average lifespan is above seven decades. Alzheimer's causes a gradual loss of mental ability, almost always among the elderly, severe enough to undermine normal activities of daily life until there is a total disconnect with the environment. Senile dementia (progressive neurodegeneration typical of older adults) is not the same disease as Alzheimers but the symptoms follow much the same pattern.

Beyond schizophrenia and Alzheimers/dementia, the list of mental illnesses that cause a disconnect with reality is unfortunately rather very long.[19] Some are widespread: *autism, catatonia, delirium, delusion, depersonalization* (intense feelings of detachment from one's body or thoughts), *psychosis*; others relatively rare: *Alice in Wonderland syndrome* (*AIWS*: perceiving living things and inanimate objects as substantially smaller or bigger than in reality); *Capgras delusion/syndrome* (belief that someone has been replaced by an impostor); *Pareidolia* (belief that random stimuli are very significant); *Fregoli delusion/syndrome* (different people are a single person changing appearance or disguise); *Morgellons Disease* (mistakenly "feeling" bugs crawling on their skin); *Charles Bonnet Syndrome* (visually impaired people seeing things they know aren't real).

A methodologically rigorous, scientific analysis of mental illness estimates that 6% of all Americans suffer from *severe* mental illness (Kessler et al., 2005) i.e., one in sixteen has lived (temporarily or permanently) in a state of "psychological virtuality." Moreover, research (Reuben & Schaefer, 2017) suggests that almost everyone will suffer a psychological disorder sometime in their life (albeit temporary for most people).

Beyond mental illness, there exists a very large category of people who occasionally or permanently divorce themselves from reality: drug users of mind-altering/enhancing substances [V5; V20]. The huge number of drug users around the world is not proof of the advisability or "normality" of such a mental state. However, the incidence of drug use—since human records have been kept, across almost all cultures[20]—suggests that using mind-altering drugs cannot be considered "unusual": "Almost every society in the world has sought to use plants or fungi

---

[19] This list includes only those mental illnesses that alter the patient's perception of reality in significant fashion. There are even more problematic conditions affecting the (state of) mind but that leave the individual able to perceive the surrounding world similar to normal people. These are usually called "disorders" and not diseases: depression, bi-polar disorder, ADHD, acute stress disorder (if long-lasting, then usually post-traumatic stress disorder), anxiety disorders, bulimia/anorexia nervosa (eating disorders), neuroses (e.g., obsessive/compulsive), seasonal affective disorder (SAD), phobia (social or environmental), Tourette's syndrome, kleptomania & pyromania, etc.

[20] "Drug use and abuse is as old as mankind itself. Human beings have always had a desire to eat or drink substances that make them feel relaxed, stimulated, or euphoric. Humans have used drugs of one sort or another for thousands of years. Wine was used at least from the time of the early Egyptians; narcotics from 4000 B.C.; and medicinal use of marijuana has been dated to 2737

to alter or transcend everyday human consciousness" (McBain, 2021). Indeed, that many seek and/or manufacture new forms further suggests that it is a regular feature of humanity. Some even argue that the drive to alter one's state of mind is natural and even primary (Siegel, 2005; Weil, 2004, p. 15)! Certainly, our brain chemistry is geared for such mental alteration, given the huge number of substances that can produce perceptual (visual, taste, sense of time), cognitive (hallucination), or affective (euphoria, mellowness, empathy) virtual effects.

Another sub-category of "mind-altering" substance use is far larger, divisible into three sub-groups. First, those who drink alcoholic beverages, coffee, tea, and other energy boosting drinks, as well as mind-sharpening stimulants [V21] or mind-deadening sedatives—the occasional armament of most of the world's adult (and increasingly, adolescent) population. Second, the fast-growing populace of pill-poppers, trying to restore some mental balance or to enhance/extend cognitive facilities [V21]: amphetamines, mood improvers, concentration aids (e.g., Ritalin, especially by non-ADHD adults), and so on. Third, a large and continually growing group employing several types of mental techniques for perceptual/cognitive strengthening and/or enhancement: bio-feedback, yoga, transcendental meditation, and the like.[21]

Overall, the general phenomenon of working *on* the brain (and not merely *with* it), is a clear sign of our awareness that the human brain is an organ not able to absorb reality to the fullest extent biologically possible. By ingesting substances that enhance, cure, avoid, modify and/or distort cognitive functioning, humanity acknowledges that left to its own devices the human brain is incapable or deficient in grasping a fuller panoply of external reality. In short, some want to go beyond their limited reality to absorb an "enhanced subjective reality"; others already have too much reality and seek solace by experiencing a modicum of unreality; the rest of us are quite satisfied to continue living in a perceptually distorted reality, in most instances quite oblivious to how much "virtuality" exists within that reality.

One final form of mental virtualizing should be mentioned: the placebo effect. This is the opposite side of the coin: instead of ingesting drugs to *induce* an altered state of mind, the placebo effect entails ingesting a non-drug that *reduces* an abnormal state of mind or body—virtual medicine [V32] to alleviate a real condition!

---

BC in China" (The history of drugs, 2002). Other evidence puts early drug use even further back in time, with archeological evidence of the use of psychoactive substances dating back at least 10,000 years (Merlin, 2003). Regarding cultural incidence of drug use, "a cross-cultural survey of 'relevant ethno-graphic literature' involving 488 societies in the 1970s indicated that 90% (437) of these groups 'institutionalized, culturally patterned forms of an altered state of consciousness'" (Merlin, 2003, p. 295).

[21] One can legitimately argue that this third category does not have virtuality elements as these practices merely bring out of us what is already potentially there [V26]. However, given that "normally" we don't (aren't capable of) reaching such perceptual and cognitive heights, they do lie outside our regular experience [V27] and thus "over-sensed" [V21].

## 3.7 Conclusion

The brain and its connection to reality/virtuality has extreme and normal manifestations. At the extremes lie two fundamental, philosophical and biological questions: whether consciousness is "real" or a virtual artifact of organic brain processes; whether physical "reality" is objectively real, existing of its own accord, or can "be" only through the observation of an intelligent entity (in this case, humans). We have also surveyed the myriad ways that the malfunctioning brain can distort reality—whether sui generis or externally induced.

Normal brain functioning can be divided into two phenomena: (1) perception—the interaction between the brain (and bodily senses) and the objective, "real" external world; (2) cognition—the internal "mentalizing" of the brain, divorced (to some extent) from the external world. Such mentalizing is heavily imbued by processes, thoughts and other mental activity that are divorced—and separate us—from objective reality. Our environment might be real, but humanity's entire perceptual and cognitive apparatus can only partly display and evaluate such reality. The remainder is psychologically virtual: "realities are internally real, no more and no less" (Zhai, 1998, p. 34), not only due to such distorted or partial perception but also (mainly?) because "reality and imagination are completely intermixed in our brain... the separation between our inner world and the outside world is not as clear as we might like to think" (Dijkstra, 2021).

Precisely because we "naturally" live in a virtualized reality, we don't think about it nor are we generally aware of how much our perception and cognition continually "fool" us. We might not care about the serious disjunction between our perception of reality and what "really" is. However, ignoring this state of affairs has consequences: it leads us to fear forms of external virtuality or prevents us from assessing such virtuality in terms of the true human condition. By not comprehending the essential similarity between "artificial" virtuality (e.g., video games) and "natural" psychological virtuality (distorted perception and cognition) we continue to falsely perceive ourselves as living a mental existence more "naturally real" than it is, and creating an external reality more "unnaturally virtual" than it is.

The present chapter has by no means exhausted the relationship between the mind and virtuality; the latter is not solely a matter of neuroscience and psychology. From the core base of our virtualizing brain, humanity has created an external "social" world within which virtuality is deeply embedded. In the following chapters I will occasionally return to the human brain in the context of how it perceives and relates to "strongly virtual" phenomena of our own making—central areas of human life that have expressed the universality of human virtuality from time immemorial. If the present chapter mostly dealt with virtuality and psychology in the "natural" state of the brain's relationship with its environment, the chapters in Part II of the book will look at how our brain also artificially imagines and/or creates virtual elements in our surroundings and how we deal with such phenomena. As Ralph Ellison aptly noted: "A lot of living is done in the imagination" (Chang, 2006, p. 386). To which one can add: imagination enables a lot of real-world living as well.

# Part II
# The History and Evolution of Virtuality in Human Society

# Chapter 4
# Religion and the Supernatural

## 4.1 Introduction

Belief in God and other religious beings is a virtual experience—so patently true that it borders on tautology. For what distinguishes religion from other forms of belief (e.g., ideology) is that phenomena beyond the physical world lie at its core [V33], explaining things that can't be understood by the normal powers of human cogitation. As Berger (1969, p. 2) pointed out, the main underlying assumption of religion is "that there is an other reality, and one of ultimate significance for man, which transcends the reality within which our every day experience unfolds."[1] This is something uniquely human, incorporating two different abilities: "We have the power not only to try to connect to a deeper reality, but to envision what this reality might be" (Bloom, 2010, p. 221).

Not all such beliefs or visions are "religious." The difference between religion and the supernatural involves "quality" and scope. As Boyer (2001, p. 137) notes: "Religious concepts are those supernatural concepts that *matter*." Part of *not* significantly mattering is supernatural belief's lack of "packaging" as an all-encompassing system. Religion offers a holistic system where the supernatural plays an important, but not necessarily dominant, part. Poltergeists, dybbuks, ghosts, vampires, demons, the evil eye et al. [V35], are not usually—and certainly not *necessarily*—part of a

---

The *Appendix: A Taxonomy of Virtuality*—listing by numbers 1–35, e.g. [V13] or [V34], some of the various types of Virtuality mentioned throughout this chapter—is freely available to the public online at https://link.springer.com/book/10.1007/978-981-16-6526-4.

[1] This quote could easily have come from a quantum physicist instead of Berger, the sociology of religion scholar. However, Berger believes that there's a larger "meaning" in this "other reality," whereas a physicist would claim that there is no "meaning" at the sub-atomic level; indeed, as we'll see in Chap. 5, what happens there is even "illogical." The paradox: hard scientists impute sub-atomic "irrationality" to the unseen world; theologians see a rational Hand of God behind the world's seeming chaotic flux!

© The Author(s), under exclusive license to Springer Nature Singapore Pte Ltd. 2021
S. N. Lehman-Wilzig, *Virtuality and Humanity*,
https://doi.org/10.1007/978-981-16-6526-4_4

comprehensive world view that incorporates supernatural elements.[2] They also do not have a clear purpose to regulate human behavior in moral fashion; most such "stand-alone" supernatural beings tend to sway humans *from* the proper path. Thus, while religion almost invariably includes the supernatural, it offers far more than that, whereas supernatural belief is generally focused on a discrete phenomenon without larger social ramifications.

This chapter will focus more on religion than on the supernatural as a "stand-alone" phenomenon. However, the latter is abundant within the former, occasionally coming close to fulfilling the "comprehensiveness" criterion for religion.[3]

## 4.2  Levels of Virtuality in Different Types of Religious Systems

Supernatural belief is found in all societies (Humans 'Predisposed'…, 2011) and almost all religious systems. As Darwin noted: "a belief in all-pervading spiritual agencies seems to be universal" (1883, p. 613). Although Buddhism does not accept any split between body and soul, and certainly does not include non-corporeal entities (it's more a "philosophy" than a religion), many practicing Buddhists *do* believe in spirits; Buddhist temples clearly exhibit "religious" behavior vis-a-vis the Buddha (supplications, candles, prostration, and prayer). In other words, anthropomorphic human psychology almost always trumps theological philosophy. I say "almost always" because there are *individuals* within religious societies who are atheists, and this has been the case even in antiquity (Drachmann, 1922).

Religion's almost universality past[4] and present is expressed in extremely diverse fashion. Put another way, every culture seems to believe in some metaphysical force "guiding" the world, but the precise nature of that force has been understood in many ways. Here are the five major approaches (each having its own sub-varieties):

*ClassicalPolytheism*: Many gods; each person chooses which to pray to (and how many). Polytheists don't worship all the gods equally, but rather tend to focus on one or a few particular deities, sometimes each at different times. Such religious practice

---

[2] Fabulous creatures such as Bigfoot (Lavers, 2009; Blu Buhs, 2009) differ from supernatural beings such as the devil and dybbuk. Fabulous creatures are believed to be an integral part of the "real" world—just "hiding out of sight" [V19]. The latter type are invisible in principle [V25]. This is a distinction without a difference. Both highlight humanity's belief in the "reality" of what is in effect virtual i.e., "not there" [V31].

[3] The difference between "cult" and "religion" is numerical: relatively few adherents and lacking "seniority." A cult existing for centuries invariably also increases its membership, becoming a "religion."

[4] One cannot precisely "date" the start of religion. Nevertheless, for reasons of evolutionary psychology, buttressed by some (albeit inconclusive) archeological evidence, it probably began to appear between 100,000 and 50,000 years ago—around the same time that *homo sapiens* began to develop deep culture, almost certainly a result of radical change in hominid mentalizing that simultaneously gave rise to religious belief (Mithen, 1999; Boyer, 2001, pp. 322–325).

has a relatively high level of virtuality, even if traditionally represented by corporeal idols [V6].

*Pantheism*: Everything in the universe is connected, so that in a sense we are all part of "God." Within pantheism one can discern three main streams. The first ("Monist-Naturalistic" Pantheism) holds that the universe is made up exclusively of physical substance (matter and energy), of which we are all a part. Its spiritual foundation is of reverent feeling regarding the awesomeness of the universe, and not immaterial "spirits" wandering around. Of all religious approaches, this probably has the least amount of virtuality, if any at all.

The second is called "Monist-Idealist" Pantheism; it too believes in only one type of "substance"—but here it is immaterial-mental i.e., spiritual. The universe, ultimate reality, is formed by a single consciousness [V25]. This is perhaps the most intense form of virtuality within the religious world—everything is "thought." The third approach is a hybrid form of intermediate virtuality: "Dualist-Pantheism" with two forms of substance: physical and spiritual-mental (e.g., reincarnation, whereby a soul leave its body, returning in another body [V6]).

*Mono-Polytheism*: This approach believes in many gods but that "ours is dominant" e.g., the biblical *Elohim*, which is a *plural* of God ("*el*" in the singular}, who demands sole allegiance from the Hebrews as their exclusive God, more powerful than other local gods [V16]. The existence of many gods makes this approach relatively highly virtual, especially when the main God becomes incorporeal.

*Monotheism*: only one God (e.g., Allah) [V30]; all others are false or fake or non-existent [V32]. In most cases, God does not have physical form [V25; V33], although Christianity blurs this with the Trinity as well as images of the Almighty [V7; V8] (forbidden for Jews and Moslems). Here the level of virtuality is very high in most (not all) cases.

*Dualism*: This posits two main deities, with two main variations. First, Manichaeism: two forces—a God of the spiritual world of light and good [V30], and a Satanic god in the material world of darkness and evil [anti-V30]—are at constant loggerheads for dominance; second, Gnosticism whereby a minor force/penumbra of the Deity brings evil to the material world, as opposed to the dominant Deity (Godhead) who created a perfect, non-material world. The level of virtuality is approximately equal in both Dualism versions, the main difference being the way in which the split came about.

Overall, it is clear that not all religions incorporate the same degree of virtuality. Within many, perhaps most, religions one can find elements that are not virtual at all—for example, rituals that are palpably concrete (e.g., circumcision, baptism), even if for the believer they may have symbolic meaning as a stand-in for something divine [V6; V22]. Certainly Monist-Naturalist Pantheism holds a non-virtual position, with everything being part of the physical world. Conversely, even the most material elements of life can be "meta-physicalized" by religion. For instance, the *process* by which a clearly flesh-and-blood person arrives in this world may be other-worldly, such as the Hindu belief in reincarnation. Obviously, numerous religious relics that are substantively mere artifacts are thought to be imbued with supernatural powers. Religion, therefore, is not necessarily a belief system that relies exclusively on otherworldly imagination. As Vasquez (2010) argues, religion also involves social practices and the use of dynamic materials by flesh and blood people who are embedded in historically relative societies and natural environments. White-house (2000) goes farther, arguing that in essence there exist two quite different types of religion: "imagistic" (icons, rituals, relics) and "doctrinal" (belief, dogma, creed). Nevertheless, there is no denying that even material, "imagistic" expression is perceived by the religious adherents as possessing or representing an essence beyond the concrete here and now [V33].

Moreover, even at the core of specific religions i.e., Deity/Deities, there are marked differences in the "quality"/intensity of their respective approaches to God/s. The Old Testament's Yahweh had very clear human characteristics [V6] (Jewish commentaries even assigned female and male traits to the Almighty); Jesus was a flesh-and-blood person, but also is considered by Christians to be a sort of god as the Son of God. Finally, many Buddhist adherents believe that gods die and are reincarnated similar to people—not that far from Christianity's belief in Jesus Christ's Second Coming or the Jewish belief in the Messiah as the Son of David.

Third, within each religion, especially those that have lasted centuries and longer, one can find religious sects, offshoots etc., each of which has a greater or lesser modicum of belief in virtual phenomena. Even certain "mainstream" beliefs of a specific religion can raise or lower its general level of virtuality. One example: the "afterlife" is not mentioned in the Old Testament but became an integral component of Jewish religious belief in the post-biblical period, continuing until today among the most "traditional," Orthodox group.

However, to say that religion is virtual by its very nature is not to say that it is "false." For the believer it serves a purpose no less "real" than the pleasure an atheist receives upon undergoing a different type of virtual-"spiritual" activity (e.g., aesthetic awe). Here we return to the highly important differentiation made in the previous chapter between *objective* and *subjective* virtuality. The more "unreal" the theological element, the greater the internal-subjective *belief* in its virtual *reality*, leading to feelings of religious transcendence.

## 4.3  Origin and Functions of Religion[5]

Where does religious belief come from? What are its functions? The answer depends on whether one is "truly religious" i.e., believes in a Deity (or several) and other religious system accoutrements, or one does not. For the true believer, religious

---

[5] There is a legitimate reason for differentiating between "origins" and "functions." If the following explanation of evolutionary psychology is correct regarding the origins and mental basis of religion, then the functions that religious belief/practice serves are not the *cause* of religion but rather its *by-product*. For instance, overcoming existential angst is not necessarily the *reason* that religious belief came into being, but it is a need that religion seems to be well placed to fulfill.

belief derives from the reality of the supernatural world. If God/s exist/s, it is only natural that "He" (or They) would imbue us with the capability of comprehending "His"/Their existence. A religious adherent's "belief" in God derives from the "miracle" of the world around us, as well as daily miracles that are obvious to anyone with their eyes open (remission of a terminal disease; narrowly surviving a natural disaster). For the true believer there is not a whole lot more to say about the origins of religious belief.[6]

For those neutral, skeptical, or disbelieving religious "truth," the question is a very difficult one, as religious belief is found in great variety among all societies known to humanity. Thus, either one believes that humanity from time immemorial has been extremely stupid and all-too-easily deceived, or one must confront the following conundrum: despite the invisibility of God [V19] and other religious "agents," why have so many human beings believed in religion and continue to do so, even in the Age of Science, the great "demystifier"?[7]

Contemporary science offers two possibly related answers. Two preliminary points: (1) these are "emerging theories" i.e., analyses based on psychological, anthropological and neurobiological evidence, but still tentative (Boyer, 2001; Atran, 2002; Barrett, 2004; Keleman, 2004; Bering, 2010); (2) they are intrinsically tied to virtuality precisely because of religion's subjective-psychological nature, a *belief* system—not merely inventing virtual phenomena (God, spirits, etc.) but also reflecting the workings of the mind virtualizing.

One theory posits that religious belief, or belief in the supernatural in general, is an accidental adaptation of the mind. During the human brain's evolutionary development, a differentiation was created between the inanimate physical world and psychological reasons/"motivation" underlying the actions of "living matter." These two types of comprehension are by and large not learned but seem to be mostly innate (even infants understand the concept of "motive" in living things).

However, once the world of objects is perceived as being separate from the world of minds, we can envision a mind (read "soul") without a body. From here it is but a small step, if at all, to the concept of God and human afterlife—a mind without a body [V25]![8] Moreover, our brain's "motivation program" extrapolates further than it should, inferring purpose and desires where they do not exist, leading to different

---

[6] Indeed, there is not much to say either about the origin of God or the world. For many religions, God is timeless [V18] and thus predated the universe that He decided to bring into being at some point.

[7] Admittedly, the assumption that religion and science are antithetical in principle is not altogether correct. Not only are there contemporary religious scientists, but early modern science itself was started in large part by very religious individuals e.g., Isaac Newton being the most notable, as well as many of the (British) Royal Society founders who believed in alchemy or witchcraft. Indeed, paradoxically the religious stricture against trying to manipulate spiritual "agencies" opened up space for the empirical investigation of "non-religious" invisible forces such as gravity (Copenhaver, 2015).

[8] Most religions have no problem with cremation; even burial symbolizes the primacy of soul over body. The exception: embalming, the body being kept in its original state as long as possible—and contemporaneously, cryogenic (deep freeze) preservation, hoping to resuscitate the deceased when a cure or treatment is found for what killed them.

forms of animism (e.g., the earthquake killed my beloved because of harm done to the environment) and Intelligent Design (e.g., even if Darwin's evolutionary *process* is empirically true, there must be an invisible hand guiding the "ascent" of mankind).

Thus, one can view the belief in an unseen Deity not as a psychological "quirk" but rather as another example of a fundamental characteristic of human psychology. As Guthrie (1995) argued, human beings constantly indulge in "systematic anthropomorphism," attributing human characteristics to objects (ergo his book's title: *Faces in the Clouds*) and teleological purpose to events and natural phenomena ("I was destined to meet my soulmate"). Significantly, children believe this even more than adults; it is innate to human psychology. The underlying reason for this is the human need for "effectance motivation," our basic need and desire to master our environment. By imputing motivation to the inanimate (or invisible) forces around us, we believe that we have a better chance of directing those phenomena over which we have no direct control (Waytz et al., 2010).

Finally, many people combine the two "programs" in bizarre ways: supernatural "minds" that take over living beings, to form a hybrid where the latter are moved not by their own will but by the will of the former: dybbuks, demons, poltergeists etc. [V29].

Why would all these mental gymnastics occur? Mainly because humans find it almost impossible emotionally to accept that the world might have no purpose or teleological logic, not to mention the total extinction of our personal existence.

Freud suggested three uses of the Gods (i.e., religion): "…they must exorcize the terrors of nature, they must reconcile men to the cruelty of Fate, particularly as it is shown in death, and they must compensate them for the sufferings and privations which a civilized life in common has imposed on them" (1927, p. 19)—akin to Marx's religion as the opiate of the masses. We employ religion as a spiritual narcotic to lessen our psychological and spiritual pain in a universe seemingly bereft of underlying purpose.

Personal "extinction" is no less difficult for us to countenance. We understand that our corporeal body will not last forever—indeed, it didn't exist for most of human history until each of us was born (which is why many religions also believe in reincarnation i.e., we also existed in the *past*). However, by separating the underlying life force ("soul") [V25] from our body, the latter's loss is easier to handle, especially if there is an overall unseen mind ("God") with whom our souls can "join" after death.

A second theory focuses on the *social* origins of religious belief and its *utilitarian* benefits: religion enables strong attachment to a group of believers, thus increasing the members' survival chances—an adaptive phenomenon of macro-evolution within a group setting.

Humans constantly walk a tightrope between self-aggrandizement, survival, and procreation on the one hand, and group cohesion for safety (police, army) and nourishment (group hunting) on the other hand. Something is needed to bind the group together (e.g., common rituals, food ceremonies) while laying down rules (e.g., Thou Shalt Not Kill) to ensure each member's survival. Religion does a good job of providing social glue—increasing the chances of believers' survival compared to their less socially-connected, non-believing compatriots. Moreover, compared to other forms of social solidarity (guilds, clubs etc.), religion's *comprehensiveness* renders it almost uniquely perfect for maintaining group solidarity; there is no "out"—one can't escape the Divine, as one could leave a guild.

As a result, religion could well be a fundamental factor in longevity (Veenhoven, 2008), as it seems to offer survival (and by extension, reproductive) advantages to believers—the case of virtuality extending reality.[9]

There is a growing body of scientific evidence for discrete benefits deriving from religious belief. For instance, Newberg and Waldman (2010) have shown how prayer and/or meditation alter neural connections in the brain, producing long-lasting states of tranquility.

In sum, religion might have originated through a double process: psychological and social-physical. Psychologically, religion provides adherents with answers regarding mysterious forces beyond our ken, perhaps giving them a greater will to live, or at least an optimistic outlook on life that increases lifespan, especially through the years of fertility (Maruta et al., 2000). Practically, religious ritual could have reinforced social solidarity in several ways: a requirement to undergo a "difficult" experience (e.g., difficult adolescent membership rite), in order to display one's commitment to the group; consistently setting aside specific times for the group to pray together and/or participate in recurring religious rituals. Of course, the individual could pray or perform other religious rites privately, but in all cases the collective was preferred—occasionally a necessary condition for the proper undertaking of the religious event e.g., Judaism's minimum ten-person "*minyan.*"

This is paradoxical from the standpoint of virtuality. Obviously, collective prayer is the antithesis of virtuality; people meeting in one place of worship reinforces the feeling of *physical* proximity. Yet the prayer services are usually structured so that through group chanting, singing, and swaying, the congregants are more easily "transported" into the virtual realm of spiritual feeling and even on occasion undergo transcendence [V33]. There is a (nether)world of difference between praying out loud by oneself and doing so in choral unison with hundreds of others. One must add to this the *place* of worship, what Langer called a "virtual realm" of "religious space" (1953, p. 97) e.g., awe-inspiring cathedrals [V25].

If we combine the first two functions of religion, then, we find a dual, reverse process at work. Just as belief in virtual beings enables people to face the difficulties of real life (and especially death), so too the very *mass*ivity (pun intended) of collective prayer, usually in large, enclosed (or at least bounded) meeting places, strengthens the subjective virtuality of religious experience (the strong connection between virtuality and reality will be expanded in Chap. 12).

A final psychological note: studies have identified a broad network of frontal and parietal (side and top) brain regions that underlie religious beliefs (Harris, et al., 2009). Indeed, as Kapogiannis et al. (2009, p. 4876) concluded: "Our results are

---

[9] Religious adherents tend to have more children than non-believers, especially among fundamentalist groups (Kaufman, 2010). If this held true in the past as well, that would also explain the spread and continuity of religion. Moreover, the more devout the parents the less chance of their children leaving the fold. It might even be the case that children of religious families are more *genetically* susceptible to indoctrination (Blume, 2009).

unique in demonstrating that specific components of religious belief are mediated by well-known brain networks, and support contemporary psychological theories that ground religious belief within evolutionary adaptive cognitive functions."

We are left with an exquisite paradox (given religion's general antipathy to evolutionary theory): religion might well have played a key role in our Darwinian "ascent" from the apes! Although religion deals with metaphysics and evolutionary theory is involved only with biological processes, the former may have had an important hand to play in the evolution of our own species. Once again, virtuality (in this instance, of religion) is linked to "real" (i.e., material) phenomena of the human condition.

This also explains a conundrum: despite numerous and recurring predictions that religion will slowly wane under the onslaught of modern science and secularity, belief in the supernatural (whether "officially religious" or otherwise) maintains its hold on wide swaths of humanity. For instance, the current belief in Intelligent Design as an "explanation" for, or counter-narrative to, Darwinian evolution—despite overwhelming evidence as to the latter's veracity. Their "problem" with Darwin's system is its non-teleology: there's no "purpose" behind the "ascent" of species, an idea extremely hard for people to accept, when they perceive motive in every other sphere of life (Newberg et al., 2002).

## 4.4  Religion and Virtuality: Historical Progression?

From a virtuality perspective can one discern any progression from less to more over the millennia? Given space limitations, I will merely suggest some general quantitative and qualitative trends.

Early religions tended to be local as well as offering palpable representation of deities in the form of idols [V6]—a low level of virtuality. Over time, two related developments occurred increasing virtuality: the idol-gods became a single, invisible God [V25] who was territorially unlimited [V17]. Of course, this is a generalization with several notable counterexamples. As noted, Buddhism started as an abstract philosophy, but was ultimately transformed by its adherents into a quasi-idolatrous religion. More generally familiar (at least to Westerners) is Christianity—a virtuality regression compared to its forebear Judaism (Trinity versus Yahweh) but a progression vis-a-vis paganism.

On the one hand, Christianity moved away from classical polytheism—indeed, its greatest proselytizing success was among pagans. On the other hand, unlike Judaism its concept of the Trinity did not completely abolish all concrete manifestations of the divine (which might be part of the reason that Jews were less amenable to conversion to Christianity). Catholic churches are full of statues, paintings and tapestries of Jesus, the Disciples, and assorted saints (and sinners).

Moreover, within a few decades after Jesus' death the Jews no longer could serve the Lord through temple animal sacrifice (when the Second Temple was destroyed by the Romans). Animal sacrifice was henceforth replaced by the far more virtual "spiritual prayer." Christianity never practiced animal sacrifice, but it did not altogether let go of religious corporeality: the Sacrament of the Table (Eucharist) is a commemoration of the Last Supper and a transubstantiation of the flesh sacrifice of Jesus himself ("this is my body; this is my blood"). Indeed, it is meant to be a "real" phenomenon: "Christ is really (not just in sign or symbol), truly (not just subjectively or metaphorically) and substantially (not just in his power) present in the Eucharist" (Eucharistic theology, n.d.). Notwithstanding official theology, however, the Eucharist is still mildly virtual in its quasi-substitution of Christ's body (V6).

Protestantism widened Christianity's virtualization by removing graven images from the house of worship but did not significantly reduce the Eucharist's corporeality. Almost all Protestant denominations (excepting Unitarianism) also did not move closer to pure monotheism: the Trinity remains a central element of its Christian theology.

Monotheistic belief, however, expanded enormously with Islam—a wholly virtual/abstract conception of the Deity (Allah), very similar to the earlier Jewish deity Yahweh ("I am that I am..."). This influenced not only the peoples of the Middle East but also made significant inroads into southern and eastern Europe, northern Africa, and a bit later the Asiatic Far East (Indonesia today is the world's most populous Moslem country). Overall, therefore, one can view the historical trend as being one of greater virtualization at the core of several, central world religions: the Deity.

There is another way of viewing the picture. Religious belief in ancient times was almost invariably holistic, encompassing not only the believers' entire immediate, external world but also prescribing and proscribing all human activity. Lacking any "scientific" (or naturalistic) understanding of how the world works, ancient humans sought religion for dealing with the surrounding mysterium—and a road map for surviving all its vagaries and vicissitudes. Such an all-inclusive approach is monotheistic Judaism, with its 613 Commandments (and innumerable ancillary/concomitant required actions) that regulate every moment of the traditional Jew's daily life. In short, ancient religion was not at all virtual in the way it controlled and/or guided all or most of the adherents' real-life behavior.

Christianity and to a lesser extent Islam reduced the behavioral element in religion. Prayer and especially belief were sufficient for salvation, whereas morally acceptable activity was now a function of broad religious principles and not necessarily very specific commandments. In addition, whereas in the ancient past political and religious leaders were either located in the same body (e.g., the demi-god Pharaoh), or within the same general Establishment (priests), this increasingly became divided, a la Jesus's "Render unto Caesar the things which are Caesar's, and unto God the things that are God's" (*Matthew*, 22: 21)—today's "separation of religion and state." Thus, positive (realpolitik) law replaced much of natural (virtual-religious) law.

Somewhat closer to our era, from the late Middle Ages onwards in the Christian West, and even earlier in the broad Moslem Middle East—both based on earlier Hellenistic natural philosophy—the spread of science narrowed the religious purview even more. Increasingly, there was less need for religion to explain the world's mysteries, although it still served the important function of providing general moral guidance and a measure of existential comfort. Here, however, one has to distinguish between organized religion and general religious (supernatural) belief. The former (especially Christianity, and later Judaism) was seriously weakened as scientific facts began to highlight the problematics (some would say, falsity) of biblical sources and official religious dogma.

None of this dampened belief in supernatural forces that continues its hold on the human psyche, as noted earlier. It might even be that belief in non-religious, supernatural forces has *increased* in the modern era parallel to the decline in religious practice. The supernatural has become a mainstay of contemporary culture (Hill, 2011): extra-terrestrial life on Earth (UFOs, alien invasion),[10] quasi-humans (vampires, zombies[11]), parapsychology (e.g., extra-sensory perception and telepathy), astrology, sundry superstitions, and so on. The modern audience is obviously receptive to this type of content: in an international, non-random, self-selected sample (scientifically, not to be taken too seriously) surveying supernatural phenomena, 96% claimed to have had at least one paranormal experience ('Spooky survey'…, 2006). Even the afterlife has been thoroughly researched scientifically, using systematic methodologies (Roach, 2005) or historical-archival materials (Casey, 2009).

In short, supernatural belief—whether "religious" or "secular"—is akin to a balloon. One can try to push in from one side (decline of religious belief) only to have the bulge appear on the other side (acceptance of the paranormal): the Devil exists, angels look after us, aliens lurk in every corner, houses are haunted. Humans insist on retaining the existence of a virtual "world out there," in different forms of belief and/or practice (performing religious rituals; watching movies about the paranormal; superstitiously avoiding "obstacles"). "Spirituality" can even be found among the most anti-religious (or non-religious) group: atheists.

## 4.5  Atheism: Anti-religion; Anti-virtuality?

Religious unbelief is not dichotomous but rather entails a spectrum of certainty: atheists are sure that God does not exist; non-practicing agnostics either aren't sure about the Deity or are too busy to care or to give the issue much thought, choosing to focus their life on the here and now. There are two other, main intermediate categories: people who do believe in some Divine force, but don't accept the supernatural

---

[10] The idea of the existence of alien life [V29] elsewhere in the universe is a mainstream subject of scientific study (e.g., the SETI program searching the heavens for "intelligible noise"). A quasi-supernatural variation involves the belief that the Earth was "seeded" with life from outer space (panspermia and exogenesis). "Mainstream" supernatural belief holds that aliens landed here in ancient times, providing humanity with advanced technologies (von Daniken, 1968), or have appeared more recently in UFOs and the like, at times kidnapping humans.

[11] Some analysts make a direct connection between the cultural popularity of zombies and our world of increasing virtuality: "zombie killing is philosophically similar to reading and deleting 400 work e-mails on a Monday morning" (Klosterman, 2010, p. AR1).

particulars of any specific religion; those who do not believe in a Deity but do participate in religious ceremonies or practice some religious rituals, more for social or ethnocultural reasons.

Atheism is the most interesting (and difficult "virtually") category, going against the normative tide of human activity through the ages, until at least the early modern period. In denying God, then, have atheists "freed" themselves from virtual spirituality or transcendent meaning? "Not necessarily," as can be seen from a recent study: more than a quarter of 275 bona fide U.S. scientists hold a spirituality that (they feel) is consistent with science (Ecklund & Long, 2011).

Bloom (2010) provided a general explanation, based on ideas and quotes from some of the most prominent and vocal contemporary atheists:

> Even those who explicitly reject religious belief show signs of the transcendent impulse. They are not blind to the attraction of a deeper reality; they just resonate to this attraction outside the bounds of organized religion. ... [C]onsider the view of some prominent modern-day atheists. ...Richard Dawkins wrote a book about the transcendent appeal of scientific inquiry. Sam Harris is well known for his attack on the monotheistic faiths, but he is strongly enthusiastic about Buddhism... . And Christopher Hitchens, author of *God is not Great*, has spoken about the importance of the "numinous" – which usually refers to the experience of contact with the divine – and has argued that one can experience it without religious or supernatural belief. (p. 215)

Geraci (2010) made a similar point, from a different direction: "...popular science authors in robotics and artificial intelligence have become the most influential spokespeople for apocalyptic theology in the Western world. Apocalyptic AI resolves a fundamentally dualist worldview through faith in a transcendent new realm occupied by radically transformed human beings" (p. 8). Among other things, they promise a real afterlife through uploading our minds into cyberspace or an artificially intelligent brain that will absorb our memories and accumulated knowledge. His conclusion: "Ultimately, the promises of Apocalyptic AI are almost identical to those of Jewish and Christian apocalyptic traditions" (p. 9). This is seconded by Lanier (2010) who noted that the (religious) Rapture and the (cybernetic) Singularity have a common denominator: both involve the Death of Humanity and our transmigration into another post-human realm. In short, some of the most prominent contemporary cybernetic researchers and leading-edge practitioners have traded in the old religion that promised an Afterlife for a new religion that promises Immortality within (or conjoined with) the Computer [V13].

While the writers that they talk about (not merely "scientific popularizers" but central, research cyberneticists: Minsky, Moravec, Warwick, Kurzweil et al.) do not represent all types of atheists and/or hard-headed scientists (among others, a few hard-core spiritually [*not* morally!] nihilist atheists exist), the point of these three commentators' is clear. Even when people deny the virtual existence of divinity and other theologically based deities, beings and worlds (e.g., Heaven and Hell), or at least are mainly concerned with hard scientific facts sans religion, there is something within the human mind that drives us to seek out other forms of virtual/"spiritual" existence or expression. There are several speculative reasons for this.

First and foremost, the Mind/Body conundrum mentioned in Chap. 3: how does "consciousness," a purely mental state of being (or so we feel), emerge out of a palpably physical body? Or as Schrödinger (1944) phrased the problem: if our body goes through constant change through our lifetime—almost every cell undergoes constant decay and renewal—how does the "I" remain more or less intact? If what makes me feel "me" (in my brain) is purely organic matter, then how does my identity stay constant? Religion's answer is to split body from "soul," but this does not answer the question, for the term "soul" is but a semantic cover for what we do not understand.

Certainly, for those (atheists and others) who don't accept the existence of a soul, the conundrum remains. Indeed, they are left with one of two alternatives. Either each of us is literally being fooled into believing that I "am" (i.e., consciousness is essentially a masquerade or an illusion, whatever that means), or there is something "out there" above and beyond the pure materiality of the world that will ultimately explain consciousness and our feelings of spiritual (not necessarily religious) uplift.

A second possible answer also lies firmly in the field of neurobiology. If indeed the human brain is predisposed to some sort of "spiritual" feeling (not necessarily in the accepted "religious" sense of the word), then all humans—atheists among them— have a need for some sort of "transcendent" experience, or at the least are particularly susceptible (compared to other animals) to such an experience if and when properly stimulated. There are numerous possible sources for such stimulation: awe-inspiring sights (pictures of incredibly distant, vast nebulae); aesthetically magnificent cultural creations (Michelangelo's Sistine Chapel); an exquisitely "beautiful" (parsimonious) mathematical proof; natural wonders of our world (Grand Canyon); running the marathon that produces an endorphin-stimulated "high"; ingesting hallucinatory substances; and so on.

Moreover, atheists are no less prone to existential angst regarding their personal future fate (after death), although perhaps they are a bit more willing to face the possibility that the answer is not cheery. However, as most are steeped in modern science, they can turn to either cryogenics for possible succor, putting death on hold, or human–machine (cyborg) hybridization, enabling the mind to continue living after the body dies. This is not spirituality per se, but it is a transubstantiation (to mix metaphors) of virtuality in real life into virtuality in the "after"-life.

Thus, one should differentiate between two aspects of the question: during one's lifetime; the afterlife. Regarding the second, both believer and nonbeliever have the same *need* for "hope"—but the *means* to fulfill that need ("immortality") is significantly different. Regarding the first, here too the difference between the religious person and the atheist is not the *need* or *desire* for a virtual experience of transcendent emotional/spiritual uplift, but rather in what each believes to be the *source* of such an experience. The former seeks a source external to our natural world; the latter believes that it can easily be found within our natural world. But the common denominator of need/desire is stronger than differences regarding the source. As Geraci noted: "religious ideas infuse the thinking of all persons, including those who have rejected institutional religions and even the explicit promises and beliefs of those religions.

What the conscious mind rejects in one format, it reformulates subconsciously and adapts to new conscious thoughts" (2010, p. 44).

One final, general point remains regarding "belief": it obviously involves something virtual, for if it were concrete one wouldn't have to "believe" in it but rather "know" it. Belief is also universal, but not necessarily in the spiritual or transcendental sense. Everyone believes in some "thing": there is a future, even if it obviously has not occurred [V26]; or the metaphysical superiority—and certainly existence—of an ideal value, moral system, ideology, philosophy [V30]; scientists believe that quarks exist, even if no one has ever seen one [V33]; and so on. As Peterson, put it: "Ignorance is a condition of human existence and belief is a necessary means of coping with ignorance.... The assumptions we make about the world directly regulate our emotions and they provide hope and inhibit anxiety" (Lewis, 2007). Thus, even if one does not agree with Geraci regarding everyone's "religiosity" (in the secular sense of the term), the even wider concept of belief—accepting or hoping for the truth of something that one cannot see or prove (yet) [V16, V19, V25]—is something that we all share.

## 4.6 Religion Online and Online Religion

I conclude this chapter with a very recent phenomenon. It is no surprise to discover that religion—so permeated with virtual elements—has enthusiastically embraced another major virtual phenomenon in the contemporary world: cyberspace. Indeed, this is fighting fire with fire: with modern science undercutting religious belief, religion has exploited modern technology to regain adherents—not the first time that religion has indulged in such technological "exploitation." As O'Leary noted (1996), early Protestantism's religious renewal and growth relied heavily on the invention of that era's revolutionary new medium: the printing press.

This manifests itself in two different, albeit reinforcing, ways (Helland, 2000; Dawson & Cowan, 2004). First, "religion online": the use of the internet to gather information about religion, to proselytize, to communicate with co-religionists, to organize learning groups (Hackett, 2006)—what Howard (2010) called a "virtual *ekklesia*." Here the emphasis is on *cyber*space—steering information to continue traditional religious practice. The second is more radical: "online religion," using cyber*space* for actual religious practice e.g., group prayer on Zoom.

As a generalization, religious adherents tend to be more conservative than non-religious people. Thus, one would expect them to avoid anything that could weaken the group's social glue. Prayer specifically, and religious practice in general, have always functioned to strengthen social bonds by dictating (or at least preferring) collective, physical proximity. Nevertheless, online religion is growing (Young,

2004),[12] suggesting that growing numbers of tradition-minded people no longer view "collective, religious activity at a distance" [V2] as an oxymoron.

Indeed, one could speculate that religious people might be *more* open and attuned to virtual collective activity than others because the very nature of religious belief contains the core idea "entanglement"—the instantaneous influence of (or communication between) one atom on another, despite being separated by great distances that would seem to preempt such immediate influence. Similarly, from many religionists' perspective God cannot be seen, He is immaterial, and (at least older religions held Him to be) up in the Heavens—and yet despite this God obviously can "make things happen" down here. If space and distance are no bar to the Deity, why should it prevent believers separated from each other from serving God together?

Or serving God individually, when and where collective action is not necessary but the activity is logistically/economically impossible? Some examples (Hackett, 2006, p. 71): pilgrimage to far-off holy sites is not possible for all adherents; cyber-pilgrimage is. Webcams can be used to view potential religious events e.g., Christians awaiting Jesus's return to the Wailing Wall in Jerusalem at the eve of 2000; Jews faxing virtual supplication notes to be placed in the same holy venue. Even physical rituals can be carried out virtually, as the Hindu *puja*, in which Durga dolls are created out of clay.

Finally, the incorporeal [V25] and metaphysical [V33] nature of virtuality can lead those working in high-tech (internet etc.) to greater spirituality and non-denominational (even cross-denominational) practice (Kinstler, 2021). Artificial intelligence, especially, lends itself to "theologizing": a contemporary, quasi-omniscient, deity, despite its being a wholly human-built artifact.

In sum, the internet and artificial intelligence specifically, and virtual technologies in general, serve as a tool of religion, a place in which religion can be practiced in extended fashion, *and* as a metaphorical catalyst for, and/or buttress underlying, the essence of religion. As cyberspace is mostly virtual, connected by physical nodes of computers, smartphones, and the like, this is not much different than the world of religious virtuality connected by the physical (and quite secondary in importance[13]) bodies of believers. If the Virtual is divine for cybernaut surfers, it can be equally argued that the Divine is virtually real for religious adherents.[14]

---

[12] There is even a scholarly journal called *online: Heidelberg Journal of Religions on the Internet*: https://heiup.uni-heidelberg.de/journals/index.php/religions/issue/view/2388.

[13] Not every religion would necessarily agree with this. As a Catholic theologian recently put it: "Bodies are not optional. There's a reason God created us as bodies. Why can't we go to Confession by Skype? Because the body must be present" (Burns, 2020).

[14] The Catholic Church now officially recognizes the connection between these two spheres of influence, having designated a patron saint of the internet: St. Isidore of Seville (http://www.scborr omeo.org/saints/isidores.htm).

# Chapter 5
# Mathematics, Philosophy, Physics and Cosmology

## 5.1 Preface

On the face of it, Mathematics (including Geometry and Statistics) and Philosophy are abstract disciplines, whereas Physics and Astronomy/Cosmology are scientific fields with greater empiricism and "materiality." Nevertheless, they cannot be easily sundered, if at all. Prior to the modern age, "scientists" were called "natural philosophers." Ancient and medieval science was based heavily on philosophizing and observation—thinking and theorizing underlying their (distorted) perception of reality. By comparison, modern science has been driven primarily by empirical experimentation. Moreover, mathematics cannot easily be divorced from physics and/or the study of the heavens; the great astronomical discoveries of Kepler and Galileo, as well as the later Moderns (e.g., Einstein) relied heavily on mathematical computation to predict, arrive at, or explain, their discoveries.

No single chapter could hope to cover the entire corpus of each of these disciplines—and certainly not all of them together! Therefore, this chapter will focus on those aspects directly related to virtuality, with an important caveat: I make no claim to the "truth" or falsity of any theory, pronouncement, approach, or other form of expression by the philosophers, physicists, mathematicians, or cosmologists to be mentioned. The main question is not "to what extent is reality actually virtual?" but rather: "how do these theories reflect our perception or understanding that reality might in some way be virtual?."

In its day, a theory might seem crazy and impossible, or eminently rational and probable—but neither is any proof of its "truth." What should interest us more—if only because none of us can decisively prove the "correctness" of each theory—is the extent to which a theory is taken seriously by other deep thinkers and/or the general public. It is here that one finds a reflection of how much (or little) society is cognitively

---

The *Appendix: A Taxonomy of Virtuality*—listing by numbers 1–35, e.g. [V13] or [V34], some of the various types of Virtuality mentioned throughout this chapter—is freely available to the public online at https://link.springer.com/book/10.1007/978-981-16-6526-4.

imbued with virtuality. We may never know how "real" is our world (whatever that means); what counts is the extent to which we think it is based on something real, virtual, some combination thereof, or something in between. As Descartes argued: "...there are no certain indications by which we may clearly distinguish wakefulness from sleep.... It is possible that I am dreaming right now and that all of my perceptions are false" [V32] (1911, pp. 145–146).

## 5.2  Mathematics

At first thought, it might seem that arithmetic and mathematics[1] are purely virtual enterprises, as numbers are symbolic signs [V24] whose meaning is variable [V15], depending on their placement next to other numbers (1 = 1 when standing alone; 1 = 10 when standing to the left of a 0). However, math is also intricately tied to the "real" world—and conversely harbors aspects of virtuality well beyond its form of symbolic notation. The world of math can be roughly divided between "applied math" and "pure math." Although these are not completely different types of intellectual work—pure math has been found to have real-world applications[2]; applied math can lead to perplexing real-world observations that engender new types of pure math— they are distinctly different in their immediate purpose and usually in their product as well. Applied math solves specific, *real*-world problems in engineering, technology, physics, and architecture; pure math is pursued for its own sake, the *virtual* intellectual-aesthetic beauty inherent in mathematical equations and systems.[3]

The origins of math were distinctly reality-based. Ancient civilizations would count with objects: three goats over here, five camels over there. At some point, they moved to "pebble counting"—substituting a commonly recognized object in nature for *different* things to be counted: seven stones would reflect seven goats, camels, pots. (Indeed, the Latin word "calculus" means a stone used for counting.) This was the start of decontextualized math in its earliest arithmetical form, divorcing the counting medium from the object to be counted [V6]. The next obvious step was to use an *artificial* counter [V13] for the same purpose: manufactured "tokens" (e.g., abacus).

---

[1] Arithmetic is the study of quantities, generally as the result of combining numbers. Mathematics is the more complex study of quantity, and of space (geometry), dynamic change (differential calculus), and structure (topography).

[2] The famous nineteenth century mathematician Lobachevsky is reputed to have claimed that "there is no branch of mathematics, however abstract, which may not someday be applied to phenomena of the real world" (Rose, 1988).

[3] To be sure, mathematicians are not the only ones who feel this way. For instance, the famous researcher and popularizer of evolutionary theory, Richard Dawkins, offered the following: "The feeling of awed wonder that science can give us is one of the highest experiences of which the human psyche is capable. It is a deep aesthetic passion to rank with the finest that music and poetry can deliver" (1998, p. x)—reinforcing the "transcendental" issue discussed in Chap. 4.

The succeeding stage, however, demanded a big conceptual leap—from a concrete number system to an abstract-symbolic one. Not surprisingly, this revolutionary advance occurred at around the time of a related revolution: writing. Numbers and letters are based on the same idea, each in its specific realm: the "abstractification" of counting and language [V23] through the use of symbolic notation [V24]. Just as written words represent the sound of speech symbolically, so too "...number sequence...became abstract by gradual detachment from the objects it counted" (Menninger, 1969, p. 86). This was the first step on the road to greater abstraction: "The history of mathematics not only shows dramatic shifts in the conceptualization of numbers but also a loosening of the bonds between numerical concepts and the concrete world" (Berdayes, 2010, p. 89).

Many of the first arithmetical notations were based on the body, especially fingers and toes, a form of "token counting." Later, when raised fingers were represented on a writing medium (clay tablet, papyrus), the number borrowed directly from the body: one straight finger became a "1"; three straight fingers looked something like a "3" (imagine a reverse Ǝ) [V22]—seen most clearly in the first three Roman numerals: I, II, III. This is the reason that most numerical systems are decimal, based on 10 (fingers). In principle, there are many other number bases that can be used with mathematical "efficiency"; the Babylonians' number base was 60 whereas today our computer notation is binary: 0 and 1 (representing a closed and an open switch on the microchip).

Once numbers were disconnected from their concrete origins, the path was open to several different approaches. First, widening their use to represent non-real quantities (e.g., imaginary numbers). Second, searching for truth through pure ratiocination instead of empirical observation. Third, viewing mathematics as the fundamental foundation of reality: matter as mathematical equations; the universe as computer (not necessarily only as a metaphor). I will start by focusing on the first ramification. The second one cannot be disassociated from the historically simultaneous emergence of writing, whose symbolic form also freed people (especially the Greeks) from an overly concretized type of thinking (expanded on in Chap. 6). Finally, this chapter's sub-chapters on Physics and Cosmology will deal with the third approach.

As math became more abstract in its symbolic notation, it slowly distanced itself from the real world. This was not inevitable. Think of arithmetic (as taught to children through the early grades) as inextricably connected to reality, no matter how complex. For example, although it's somewhat complicated, dividing 1,056,892 by 434,452 (=2.4327) still provides a straightforward "real world" answer—even if it is not simple to visualize 1,056,892 units of any object or to think about the size of 0.4327 of a specific object. The same regarding mathematical equations and other math forms such as factorials: the numbers on each side of the equation (or the results of a factorial) can represent real-world quantities. How, then, did math metamorphose into something increasingly abstract [V23]?

The history of mathematical development can be divided into four conceptual stages or eras (Katz & Barton, 2007):

1.  *Geometric stage*: For the Babylonians and later the Greeks math concepts were largely geometric, using space, shape, size, and the relative position of figures. This is a highly "concretized" type of math, relying on real-world perception. True, Plato viewed geometry, "knowledge of that which always is" (1950, p. 527b), as emanating from the eternal perfection of pure mathematics, but he was not a geometer. The Pythagorean School did play around with some abstract aspects of geometry but discovered (to their horror) certain "philosophical travesties" such as incommensurable lengths [V16], thereupon abandoning abstract numbers [V23] in favor of "real-world" geometric quantities e.g., distance, length, area of figures.

    Two thousand years later—when Descartes introduced the concept of coordinates, coupled with algebra—geometry began its analytical stage with equations and functions (leading to infinitesimal calculus later in the seventeenth century). In addition, developments in the world of art regarding perspective led to another modern branch of abstract geometry: projective geometry [V7; V8].

2.  *Static equation-solving*: Here the objective is to find numbers satisfying certain relationships. The move away from geometric algebra dates back to the Alexandrian Greek Diophantus (third century CE) who was the first to recognize fractions as numbers. However, algebra didn't decisively move to the static-equation-solving stage until the eighth century Persian, Al-Khwarizmi, wrote the foundational book "Al-Jabr" (whence our word "algebra"),[4] eventually introducing the decimal number system to the West, as well as linear and quadratic equations. Decimals are a more abstract form than fractions to represent unit parts, if only because the fraction retains a real number; it is easier to envision an eighth of a pie (1 of 8 small pieces) than 0.125 of a cake.

3.  *Dynamic function stage*: where curvilinear motion is the underlying idea. Along with major advances in astronomy—the science of *moving* bodies—a major new category of mathematical problems emerged: finding a solution that involved an entire curve. The appropriate mathematical notation was invented by Fermat and Descartes, and from there Leibniz and Newton developed the revolutionary mathematical tool of differential calculus. With calculus coming into being to resolve real world problems, math moved well beyond what the mind could "picture" by itself [V6; V22]. Calculus was based on "algebra-izing" geometry, but the result pushed both geometry and math into the realm of the abstract [V23], away from a definite set point to paths of motion. Eventually, this led to chaos theory: deterministic but fundamentally unpredictable [V15; V16].

4.  *Abstract stage*: Especially in the nineteenth and twentieth centuries, a new major branch emerged: "abstract math," "speculative math," or most commonly: "pure math" [V28; V30]. It focuses in large part on "structure," attempting to find common characteristics of groups of numbers and/or the relationship between

---

[4] Much of his work built on earlier Indian and Greek sources (including Ptolemy), but his place as one of history's foundational mathematicians is secure, as can be seen by several terms originating in his work e.g., algorithm, a derivation of his name.

"sets": set theory, category theory, number theory, and so on. Even geometry underwent revolutionary change in the nineteenth century with Riemann's curved geometry that seemed to bear no relation to anything in the real world [V34], but subsequently proved to be the foundation for Einstein's relativity theory of curved space–time as well higher-dimensional physics.

Part of this historical evolution towards increasing mathematical abstraction involved the creation/discovery of new types of "numbers"—especially zero, negative numbers, imaginary numbers, and infinity [V19]. The concept of zero did not exist in the ancient world (e.g., Roman numerals do not have a sign for "zero") because it was believed that "Nature abhors a vacuum" and therefore "nothing" cannot be a concept representing anything in the real world [V35—back then]. Even worse, zero leads to mathematical absurdities [V31]: Zero multiplied by anything provides the same result—zero! And if we divide anything by "zero," we can mathematically prove anything we want!! The ancients stayed away from zero like the plague (Seife, 2000).

Zero was first created as a concept in fourth century CE India, leading Indian civilization to discover the Babylonian "zero as numerical place holder."[5] As Hinduism by this time had embraced the yin and yang of Creation and Destruction, its adherents had no philosophical aversion to "nothingness" or "void" [V25]. Quite the opposite: the cosmos emerged from Nothingness (the god Shiva was the ultimate void) and re-achieving nothingness was the goal of humanity [V30]. In addition, the Hindu "universe" was infinite, for beyond our own universe lay uncountable other universes—a concept not to be found in the West until very recently.

The consequence of holding these theological-philosophical ideas was twofold. First, the Indians expanded the Babylonian "placement" sign for zero into a digit that had meaning in and of itself. Second, by the twelfth century Indian mathematicians had accepted with equanimity that anything divided by zero would result in infinity. Thus, two of the most virtual "numbers" that we can conceive—zero [V19] and infinity [V15]—emerged from a civilization that was philosophically and psychologically ready to accept the reality of the incommensurable [V16], one of many examples illustrating the interrelationship of mathematics and philosophy.[6] As noted in Chap. 4 (fn. 8), a variation of this nexus—math and religion—appeared in Europe: "Medieval mathematics was everywhere 'enchanted', its numerology animated with allegorical, generally theological meanings" (Hobart, 2018).

Nor was this all. The Greeks viewed math as integrally tied to geometry, both being reality-based enterprises. Mathematicians in India, on the other hand, because of their philosophical openness to virtual numbers, began to look at mathematics as a

---

[5] Seife (2000) pointed out that around 300 BCE the Babylonians began to use a notation for "empty space" between numbers in order to distinguish (in their specific, 60-base number system) between 11 and 101. Thus, 11 looked like 11 whereas 101 looked like 1 × 1. However, this "×" (two slanted wedges) was more a "place card" than a number that meant "zero = nothing."

[6] See Grabiner (1988) for a masterful survey and analysis of how mathematics has influenced many other fields of endeavor, such as analytical philosophy, political philosophy, metaphysics, theology, social science, as well as such specific issues as the existence of free will.

system that can stand on its own two abstract feet—a short step to another major math advance: negative numbers. The idea is simple: if numbers are based on geometry, then subtracting five objects from three similar objects makes no sense, but for the Indians the arithmetical manipulation of 3 minus 5 made as much sense as 5 minus 3, for these were abstract digits without necessarily having any real-world counterpart [V22].

The West eventually jumped on the infinity bandwagon, but complicated the concept when Galileo raised the paradox of a set of infinite numbers "more numerous" than another set of infinite numbers. Then in the nineteenth century Georg Cantor (1874) further deepened the mysterious nature of infinity by illustrating through his "Diagonal Argument" how infinity cannot be counted at all [V16].[7]

The second "impossible" mathematical concept was the imaginary number. At least from the twelfth century, Indian mathematicians had argued that there could not be a square root of a negative number for the simple reason that when you square a negative number (e.g., minus 3 times minus 3) you end up with a positive number (plus 9). Descartes not only agreed but even came up with a derisive term: *imaginary* number (the notation $i$) [V31]. Some went even further. "Leibniz likened $i$ to the Holy Spirit: both have an ethereal and barely substantial existence" [V25; V33] (Seife, 2000, p. 135). Nevertheless, other mathematicians soon found uses for imaginary numbers such as solving polynomials. Ultimately, $i$ would reveal the deep connection between zero and infinity—a step towards a "unified field" of virtual math!

The history of nineteenth and especially twentieth century mathematics is one of increasing abstraction (Cantor being one of several leading mathematical "virtualizers"; Riemann's non-linear geometry [V28] constituted another breakthrough), paralleling the same trend within physics. I will not survey the late modern development of pure math as it is far too complex and technical; moreover, the survey of modern physics later in this chapter will illustrate the complexity and ever-deepening "other-worldliness" of physics and math, two sides of the same increasingly virtual coin.

Two additional, important, related, contemporary fields reflect math's virtuality. The first is statistics. Similar to geometry, statistics seems to be a distinctly non-virtual enterprise, largely based on, and describing, real world phenomena. However, statistics involves virtual elements in three ways. First, at least in its formative decades (seventeenth to nineteenth centuries, when the state—hence "*stat*istics"—was the main "customer") it was a *gross* way to describe society because it artificially chose to measure certain categories and ignore other demographics; the emerging picture was at best only partial and at worst it distorted social reality [V4; V5] (Davies, 2017). Second, in the twentieth century it was used for socio-economic-political purposes in more refined fashion, but now in many cases by sampling a population i.e., a "virtual" representation of the whole through a minute part of it [V6]. Third and most recently, statistics has increasingly been used to *simulate* [V10] future outcomes

---

[7] A closely related but still different meaning of infinity is "without end." Classic examples: the Möbius strip; Escher's drawings that offer the artistic illusion [V31] of a loop without end.

in fields ranging from nuclear physics to engineering to finance—and many social sciences in general—by running myriad inputs through a computer thousands of times, each run with a slightly modified combination of input variables. Collectively, the results present a realistic picture of probable outcomes [V26]—usually gathering around a narrow point that can be considered the "prediction."

The second area is computer programming. Even today when computer functions have expanded significantly beyond math outputs, the underlying foundation of software programming involves constructing formal systems that employ specific rules, combining them to produce "programs" [V24] that run the computer. The actual practice of programming uses very similar techniques to what mathematicians use—formal systems, axioms, symbols, rules, logic[8]—even if the specific tool (code) is not identical to the notation and other tools of the mathematical trade.

This explains why the word "virtual" in its contemporary incarnation is used within the field of computers far more than in any other: virtual memory, virtual machine, virtual storage etc. [V17]. If math began its career thousands of years ago as an *applied* discipline ("counting"), today it has returned to its applied roots with a vengeance as our world becomes "computerized" i.e., symbolic code as the means to manipulate reality.

## 5.3 Natural Philosophy: Philosophical Physics in Antiquity

For most of human history, one could not clearly differentiate "philosophy" from "science" (at least scientific "physics" and "chemistry"). Thus, the following sections are divided to a large extent chronologically and conceptually, not by modern disciplinary categorizations.

Empirical observation of natural phenomena commenced thousands of years before the Greek natural philosophers appeared on the scene, perhaps first in Sumer where "scholar-scientists" studied astronomy and cosmology, as well as also zoology, botany, and mineralogy (Kramer, 1959, p. 2). Subsequently, Babylonia and Assyria also added to the corpus of empirical knowledge. However, "the same practitioners of this rudimentary form of science were also engaged in astrology, the reading of animal entrails, the interpretation of omens... a mixture of logic, superstition, myth, tradition, confusion, error, and common sense. No distinction was made between religious and scientific thinking" (Logan, 2004, p. 85). In short, this was "science" based on unreal phenomena [V32].

The Greeks were the first to focus on the natural world without recourse to religious dogma and/or prior, received wisdom. Moreover, they seem to have created the first culture to think about the scientific method itself.

---

[8] Until the nineteenth century, logic was generally considered to be a matter of dealing with things immediately at hand i.e., one "did" logic about concrete phenomena and processes in one's direct experience. George Boole's *symbolic* logic system changed this forever. In Boolean logic, statements are described in—and only have value through—the relationship between abstract symbols [V23; V24].

Parmenides' philosophy—one of the earliest Greek philosophers to deal with the nature of reality—embodied two contradictory, virtuality aspects: 1- Our normal perception of the physical world's reality is mistaken. We constantly tend to see change and movement all around us when in fact these "processes" are mere appearances of a "true" static, eternally immutable reality. Thus, the Parmenidean distinction between perception (i.e., appearance) and reality leaves us living in a world of virtuality: what we "see" is not "what is" [V20]—as described in Chap. 3.

However, Parmenides was not a supporter of virtuality in two other respects. First, he denied the possibility of a "void," of nothingness. Aristotle would expand this idea into his principle that "nature abhors a vacuum"—arguing that a vacuum would enable anything within it to move at infinite speed. Second, as just noted, Parmenides was an extreme materialist: the universe is an indestructible, unchanging whole. Spectral, incommensurable, metaphysical phenomena do not exist.

Further development of this idea can be found in the atomic theories of Leucippus, Democritus, Epicurus (Greeks) and Lucretius (Roman). They saw nature as entirely composed of unchanging and inert material atoms. What we perceive as dynamic process, they argued, is but the atoms' change of position in space and time. The underlying properties of matter do not change, but rather merely the relative *relationship* between the various substances.[9] However, in order for any relationship to transpire, noted Lucretius, there must be open space, a void, so that the form of an object can change—for example, from liquid to gas or from liquid to solid. Thus, from the standpoint of virtuality, their position is an inversion of Parmenides: reality is material (non-virtual) but it also incorporates a lot of vacuum—space de*void* of substance [V25].

Nevertheless, in one important sense they agreed with Parmenides: the weakness of our senses. Democritus argued that our perception is completely subjective— different people have diverse impressions so that each person lives in a somewhat different, slightly distorted [V5; V20] world relative to what other humans experience.[10] For these atomists there was another difficulty to accurately perceive reality: true physical reality resides at the level of atoms, obviously far too small to be perceived by our body's sensory apparatus [V19].

This dualism led Plato and then Aristotle to further differentiations between the "real" world and the world of artifice, imperfection, quasi-reality, although each understood such a divide very differently. Plato argued that there is a disjunction between the world of Perception and the world of Ideas, the former being but a "flickering shadow on the wall of the cave" (1950, pp. VII 514a,2—517a,7), as he put it in his famous, metaphorical parable. In other words, what we perceive in our world is but a distorted reflection of a higher reality that exists as a perfect Form or Ideal. This comes close to arguing that the unseen Virtual is Real [V30], and what

---

[9] McLuhan (1965, p. 86) offers a fascinating, if speculative, argument: "They saw the analogy with what the alphabet had done to language and likened their atoms to letters." New words (objects) come into being merely through a rearrangement of letters (atoms).

[10] He argued that in principle we *can* reach the truth through our intellect, but not all humans are capable of pure cognition or have the same cognitive capabilities.

we perceive as "real" is only virtually so [V5; V31]. Aristotle's dualism avoided the metaphysical for the heavenly: our earthly, sub-lunar world in which we reside is imperfect, whereas the celestial (nevertheless, material), unchanging heavens, are where perfection resides. In other words, our world is not any less "real" than the heavens, but it conforms less to the standard of perfection, precisely because it is in constant flux.

The main opposing school to the immutable atomism of Parmenides et al. emanated from Heraclitus who viewed the world as a jumble of opposing forces locked in constant, disordered conflict. At base is Fire, the most ephemeral element; what we perceive as fundamental "stuff" is not material substance but rather an ongoing, natural *process*, with all "things" being products of the interplay between the four traditional Greek elements—water, air, earth, and fire—with fire the catalyst for the others' constant change. Heraclitus argued that such changeability is so pervasive that one couldn't step twice into the same river. Thus, reality is at bottom not a conglomeration of *things*, but rather of ongoing *processes*. To continue his aphorism, the river is not an *object*, but an ever-changing *flow*—just as the sun is not a *thing*, but a fiery flame. We exist in a world of process, of ceaseless change [V15].

In terms of virtuality, this approach does not claim that our world is somehow "unreal" but rather that its composition is indeterminate [V15] and incorporeal [V25]. This is not at all the same as virtuality in the Platonic Ideal. Plato didn't accept Heraclitus's process principle, but it forced him to posit reality's bedrock somewhere outside our material world. Thus, process philosophy offers a somewhat mild version of virtuality ("indeterminate flux," yet not "unreal"), but it has influenced others to develop stronger kinds of virtuality philosophies. These two schools of thought—"process philosophy" and "Idealism"—constitute the foundation upon which later virtuality-oriented philosophies are based, and ultimately contemporary physics as well.

## 5.4 Idealistic/Process Conceptions of Reality in the Modern "Atomistic" Era

With the decline of Greco-Roman civilization in the early Christian era, natural philosophy in the West went into hibernation for close to a millennium, superseded by a theological mindset. Even when questions of "reality" were broached by some thinkers, it occurred within a religious framework of thought. This Western medieval, *theological* mentality was substantively different from the earlier Greeks who were "natural [i.e. Nature] philosophers": instead of trying to find the basis of reality in the present, physical world, "theological philosophy" looked for ultimate causes and the basis of reality in the world of *metaphysics* [V33]—outside science.[11] As Chap. 4 was

---

[11] Although similar to Platonism, there were crucial differences: (1) Plato's system had no God; (2) Plato felt that the Ideal could be comprehended, albeit with great difficulty (by philosophers),

devoted to virtuality and religion, I will focus here on the modern era (pre-twentieth century) and return to the medieval period later regarding cosmology and physics.

The underlying disagreement between the *atomic-materialists* and their two opposing schools—*incorporeal-process* philosophers and *neo-Platonists*—continued throughout history. The atomistic approach was the dominant one during the Greek and Roman eras, as well as through most of the modern period (until the past century or so). The medieval period, on the other hand, was basically Idealist, although one can find several Christian and Arab sources upholding atomism in the Middle Ages. However, as Grellard and Robert point out, "medieval... atomist doctrines become more and more confused when one wants to embrace them as a whole" (2009, p. 2). Thus, there is more than an element of truth to the common expression "the exile of the atom" during the medieval period.[12]

Atomism's coherent revival as a full-fledged school (sometimes called "corpuscular"[13]) occurred in the seventeenth century with noted theorists such as Hobbes, Gassendi, Descartes, Galileo, Boyle, and Newton. Some of these were "pure" philosophers (Hobbes); others combined philosophical speculations with math (Descartes) or with empirically-oriented science (Galileo); still others mixed mystical theology—mostly Christian Kabbalism—with science: Bacon, Bruno, Leibniz, Newton (Glinert, 2017, pp. 142–143); a fourth group attempted to remain within the exclusive confines of empirical science (Boyle). Overall, though, this was a gradual but clear move from "natural philosophy" to "modern science," a process still not fully accepted today by everyone in the modern world, as seen in continuing controversies regarding Darwinian Evolution and other "problematic" (from a religious standpoint) scientific theories.

From the lengthy Hellenist period all the way to the modern era, non-theological philosophizing was generally and heavily imbued with an atomist mindset. This is an important point: the *philosophy* of the (strongly) virtual was *not* universal or even predominant. *Theological* philosophizing was overwhelmingly "virtual," so that philosophers of an anti-religious, or even a non-religious, bent would naturally gravitate to philosophical systems that focused on the palpable here-and-now: "One simple name for the plague that Lucretius brought... is atheism" (Greenblatt, 2011, p. 32). Not all atomists were atheists (far from it), but atomism/materialism did tend to undercut a religious, metaphysical *weltanschauung*.

Atomism was opposed by two "natural philosophy" schools. First, process philosophy. Reawakening in the modern era, it became an influential, well-defined school of thought, albeit still a minority until the late nineteenth and the twentieth centuries

---

whereas religion traditionally views God's work in the world as not completely within human understanding.

[12] This is largely true in the West. For the *continued* espousal of atomism in the Indian and Islamic worlds from ancient times through the medieval period, see "Indian Atomism" and "Islamic Atomism" in http://en.wikipedia.org/wiki/Atomism.

[13] Atomists held the atom to be indivisible; proponents of corpuscular theory believed the atom to be divisible into smaller entities. Contemporary physics shows the latter to be correct.

when it was able to free itself from the shadow of such "materialist"[14] giants as Descartes and Locke (dualists, believing that material and immaterial things both exist), and Hobbes (monist: only material things exist). The leading exponents of process philosophy were Leibniz, Hegel, Bergson, Peirce, William James, and last but certainly not least: Whitehead (1929).

Their process approach was not monolithic. Leibniz's philosophy was heavily based on a *physics* conception of the world (the "monad": clusters of minute processes acting as centers of force); Hegel's approach (thesis-antithesis-synthesis, repeat the process…) was steeped in history, whether of thought or Nature; Bergson and White-head viewed process through an *organic* prism; James focused on the *subjective-psychological* side of process. Yet their philosophies all had a few common denom-inator propositions: (1) time and change are among the main categories of ontology, the nature of Being; (2) process is more (or at least not less) fundamental than "things" for understanding the world; (3) process can be described in several ways that place flux, and not stability, at the core of reality: contingency, novelty, creativity, emergence, passage, alteration, change, re-composition.

How does process philosophy reflect virtuality? First, much of ongoing change leads to a "distortion" (in the neutral sense) of what came before [V5]. This occurs when the change is caused by *extrinsic* factors ("force" B impinging on force A). Second and inversely, process carries within it the idea of latency or potentiality [V26]—change caused by *intrinsic* forces from within the thing being looked at. Whether a biological organism, a cosmological entity, or a seemingly "inanimate," geological object, there are always inherent forces (respectively, nuclear fusion, cellular metabolism, atomic decay) that cause a transformation every moment.[15] Third and as a result, all components of our world are in a sense either continually indeterminate [V16]—for what they are at present is not what they will be in another moment—or even inherently ambiguous [V15], for such change can influence the identity of the subject under study e.g., water turning to ice. Fourth and finally, as already mentioned, process theory posits the non-materiality of the world [V25].

The second major, modern era, philosophical opponent of atomism-materialism was neo-Idealism with several schools of thought. Herewith a very schematic survey running from the "purest" forms to more hybrid philosophies:

(1)  *Monistic Idealism*—This pure type of Idealism states that consciousness, not matter, is the *exclusive* foundation of all existence—indeed, consciousness *creates* matter [V25; V33]. Monistic Idealism can be found in many religions and several schools of Eastern philosophies such as neo-Confucianism (Wang Yangmin) and most recently Hindu Idealism (Amit Goswami). In the modern

---

[14] The term denotes thinkers who believed in the underlying "materiality" of our world i.e., based on matter. It does not denote or connote value-laden, economic "materialism."

[15] A major philosophical issue regarding such "transformation" is whether change is teleological or not i.e., has a final purpose, either intrinsic (e.g., an acorn growing into a tree) or extrinsic (e.g., animals in the service of humanity). This issue relates to virtuality as potentiality [V26]—but discussing such a complex subject would lead us too far astray.

age, it was propounded in the West most famously by Hegel[16] who argued that "being" is a product of an ongoing, dynamic, historical process and interplay of consciousness(es), from which the world's great diversity emerges. Schopenhauer used substitute terms: "will" (on the micro level) and "energy" (on the macro-level), as the driving force behind reality.

(2)  *Subjective Idealism* - Otherwise known as *immaterialism*; what we perceive as objects are merely aggregations of sensory data within the perceiver [V31]. Berkeley argued that nothing can exist independent of a perceiving mind: "the objects of sense exist only when they are perceived" (1710, p. 21, par. 45).

(3)  *Transcendental Idealism* - Kant argued that it is our mind that forms the world we perceive as having both space and time [V7]: "In the true sense of the word... I can never perceive external things, but I can only infer their existence from my own internal perception..." (1781, parag. A367). As opposed to Berkeley's "Dogmatic Idealism" (Kant's characterization) that denies the reality of the external world, Kant's Idealism is more modest: "It must not be supposed, therefore, that an idealist is someone who denies the existence of external objects...; all he does is to deny that they are known by immediate and direct perception.... (Kant, 1781, parag. A367). In other words, external reality may well be "real" and even correspond precisely with our perception of it, but there is no way of ascertaining this in any absolute sense. This approach places reality and perception in an ambiguous position: the former might be all there is, somewhat correctly represented by the latter (no virtuality, or at most, reality well mediated by our sensory apparatus [V3]); *or* the latter might be all there is, falsely representing the former [V5; V31]; *or* the latter might only be able to offer us an approximation of the former [V4]. The Kantian approach, then, leaves the relative strength (or existence) of virtuality up in the air, depending on which of these three possibilities one accepts.

Similarly, but from a different vantage point, Wittgenstein focused on human language that distances us from understanding reality. More specifically, as Hacker explained: "in doing philosophy, we come to realize the character of the grammatical and linguistic scaffolding from which we describe the world, not the scaffolding of the world" (Garvey, 2010). Philosophizing, then, can tell us a lot about the way we think about the world [V31], but nothing about the world as is. As per Kant, this does not mean that the world is not "real" but rather that we can't get at this "real" essence, at least not through philosophical ratiocination.

(4)  *Objective Idealism* - Here we find a common-sense acceptance of Realism (i.e., material objects do exist), but Objective Idealism also posits that the mind and concomitant metaphysical things as morality, culture, ideas do not emerge from the material world [V33]. Rather, the mind "bootstraps" these ideational creations from within itself; whereas external reality is palpably material, the internal world of thought is divorced from such materiality. Thus, the external world is not virtual at all, but the internal world of cognition and creativity has

---

[16] Hegel also called his philosophy "Absolute Idealism."

elements of virtuality insofar as these mental processes have no foundation in the material world [V25].

Based on the way we experience the world and our judgment (an extension of our experience) most of us intuitively accept the atomist-materialist approach, as it is extremely hard to touch a table and not believe that it is a "thing." This is most probably why atomism has traditionally been the dominant school of thought; even Great Minds are drawn to common sense.

## 5.5  Idealism's Resurgence in the Contemporary Era: Quantum Mechanics

Why, then, did the situation change in the last century and a half? In a word: science. To recapitulate: philosophy and science were almost indistinguishable in the Hellenistic Era; they were rent asunder in the medieval period through theological philosophizing that shunted aside most scientific thinking; the early modern period (fifteenth through nineteenth centuries) gradually reversed this trend with empirical science disengaging from speculative thinking and religious philosophizing. However, the late modern era witnessed another reversal with philosophical speculation, purely hypothetical theorizing, and empirical observation/experimentation reuniting—through osmotic filtration (and intellectual infiltration) of scientific discoveries into the world of philosophy broadly defined. Now connection between philosophy and science became indirect as major scientific developments (theory of evolution; quantum mechanics; relativity) were impressed on the minds of philosophers and other social thinkers who then translated those scientific ideas into philosophical concepts.

This started with Darwinian Evolution and its clear connection to process philosophy. Darwin not only offered a logical and empirically based theory of nature's diversity, but also demolished the Aristotelian world view of natural *stasis* that held sway for millennia. Species did not permanently keep their biological identity; rather, they changed ever so gradually through eons of mutation and adaptation, at some (unidentifiable) point becoming something new. Each species at any point was virtually the same manifestation of its forebearers—but not quite identical to them [V5; V26].

This offered "process philosophy" a solid scientific foundation, but of course hardly dented the atomist point of view. After all, the animal kingdom may be constantly changing but it is still based on solid, organic matter. The main revolution—the return of Idealism, mixed with new "process" elements—emerged not from Biology but from twentieth century Physics, gradually manifesting itself in different ways.

First, Einstein's Theory of Relativity ("Special" and "General," the latter taking gravity into account) showed how there is no privileged frame of reference in the universe; even time and distance/length are dependent on the inertial speed of the observer. A person traveling close to the speed of light will "age" much slower than her identical twin on "stationary" Earth. This constituted the epitome of process theory, but Einstein also heralded the return of Idealism, for relativity theory offered the famous equation $E = mc^2$ positing the equivalence of mass and energy: corporeal matter and non-material energy were one and the same "thing"! True, this did not prove that matter does not exist, but it did suggest that the "materiality" of matter (the indivisible atom) is not as fundamentally material as we tend to think.

This insight was given further credence in the "replacement"[17] of classical physics with Quantum Mechanics (QM) that revolutionized not only physics but our general conception of the world, paralleling the relativistic paradigm change: "Twentieth century physics has... turned the tables on classical atomism. Instead of very small *things* (atoms) combining to produce standard processes (windstorms and such), modern physics envisions very small processes (quantum phenomena) combining in their *modus operandi* to produce standard things (ordinary macro-objects)" (Rescher, 2008).

QM has several "virtual" aspects. First, the very nature of reality, as the above quote hints at: reality (what we normally describe as a physical thing) actually consists of statistical fluctuations that usually coalesce into probability patterns—the quintessence of "process"! Thus (to somewhat oversimplify), we live in a relatively stable world only because the vast majority of atoms "do" what is "expected"—more than enough of them to lend "stability" to the macro-world despite statistically "outlier" atoms that don't behave precisely as "expected" [V15]. Second, the Heisenberg Uncertainty Principle that maintains the impossibility of ever knowing precisely the position *and* the velocity of an atom or sub-atomic particle. In other words, a completely accurate description of reality-in-the-here-and-now is unknowable *in principle* [V4; V9; V16] because any human observation/measurement of the atomic phenomenon affects it in a way that changes its behavior and even its essence (e.g., from "undecided" to discrete something). In short, the attempt to observe and understand atomic behavior invariably alters that same behavior! Even if the atom were not virtual, even if it were made of material substance, our involvement renders its "natural" state unknowable [V16].

The issue is even more complex. A third virtual aspect of QM, related to the foregoing position-velocity problem of the Uncertainty Principle, is that atomic particles have no unique, stable identity—they can be both a wave (process) and a particle (matter). Such wave-particle duality—the world as "real" and "virtual" at one and the same time [V15]—is also called complementarity. Returning for a moment to the previous point, the particle never has to decide "what it is" *unless and until some form of consciousness impinges on it* i.e., it is "observed" by a conscious entity (e.g., humans), at which time it is "forced" to become either a particle or a wave. As the noted French theoretical physicist d'Espagnat declared in his famous paper "The Quantum Theory and Reality": "The doctrine that the world is made up of

---

[17] QM replaced classical physics on the micro-atomic level but did not do so at the macro-level of human-scale life and beyond (stars, galaxies). Indeed, to this day physicists are to trying bridge the gap between the foundational laws at these two levels.

objects whose existence is independent of human consciousness turns out to conflict with quantum mechanics and with facts established by experiment" (1979, p. 158). Fourth and (not quite) finally is the highly unusual phenomenon called "quantum entanglement" (Buhrman et al., 2001). It "works" like this: two sub-atomic and even two macro-atomic molecules interact physically and then are separated. Even if they are a very large distance apart, they can somehow "communicate" one with the other—faster than the speed of light, something that should be impossible (*nothing* i.e., no thing, can travel faster than the speed of light). This theoretically contradicts Einstein's theory of relativity, and indeed he called entanglement "spooky action at a distance" [V2] (Einstein, 1971).

To sum up provisionally, QM has been hugely successful in predicting hundreds of empirical-experimental outcomes, but even physicists in the forefront of QM research find it perplexing at best and impossibly illogical at worst (Einstein's stance). No one can say with certainty that QM is the ultimate description of reality, but (un)clearly at this point in time our quasi-comprehension of reality incorporates a very heavy dose of virtuality in various guises.

## 5.6 Virtual Astrology and the Virtualization of Astronomy

For a very long time, humans have sought answers to several existential questions. Among the main ones are three in particular: where did the world come from? who or what controls its workings? how does the world work? To answer the latter question, most "looked downwards" trying to discover the invisible, basic elements or phenomena of reality—"natural philosophy" that ultimately led to physics, as described above. Conversely, the first two questions shifted their investigators' gaze "upwards" to the stars and firmament, but differentially.

One general approach interrogated these existential issues from a religious perspective, finding answers in Heaven.[18] The second approach looked not at "Heaven" but at the stars and whatever else could be observed in the firmament. This was astrology and astronomy, evolving at different points in time and place.[19] Eventually, in the contemporary era, this culminated in "cosmology" (broadly defined): the *scientific* investigation—theoretical and empirical—of the nature of the universe and even beyond.

In one important and rather banal sense this triumvirate was a more virtual enterprise than the classical study of "physics" because the subject matter was physically beyond the reach of the observers. This meant that even the most rigorous investigation of phenomena in the heavens had to be performed at a distance [V2], and in

---

[18] Technically this is not universally correct. Pantheism rejected a supernatural Being; Nature on Earth had as much "spiritual" content as the cosmos; "Animism" perceived spirituality in all forms of life (animals, trees) and inanimate objects as well (e.g., rocks).

[19] "Creation mythology" would normally be included in this list as a forerunner of cosmological investigation, but these were almost (quasi-)religious—discussed in Chap. 4.

the modern era with new observation technologies (e.g., telescope) it would also be mediated [V3]. However, some of these investigations (especially astrology) were virtual in a more fundamental sense, based on a mostly mistaken conception of (heavenly) cause and (earthly) effect [V31].

The "systematic" observation of the heavens is of hoary provenance. Cave wall and bone marks suggest that humans have been tracking lunar cycles as far back as 25,000 years ago, later recording how and when the moon's peregrinations influenced earthly bodies of water. We have solid evidence from Mesopotamia four millennia ago (1950 BCE), and indirect indications well before that, for what we today call "astrology"—the search for meaning in the heavens—as a foundation of belief and a guide to human action. It evolved to serve secular and theological purposes: understanding the seasons and gaining some control of nature to improve agricultural efficiency (and seafaring safety) as well as aligning religious practice with celestial handiwork to gain favor with the gods (Campion, 2008).

Despite the contemporary, negative connotation of astrology as a form of "voodoo prognostication," ancient astrology clearly constituted an "advanced" (for that time) form of scientific field of study, even if some astrological-theological conclusions were figments of human imagination. For the heavens *do* affect life on earth: lunar cycles influence sea levels, the sun's angle in the sky affects the seasons (not to mention its "disappearance" every night leading to colder temperatures). Our modern mind views this not as astrology but as astronomy, but the ancients had little comprehension of the underlying forces at play, not to mention the distances involved and the "environment" of outer space. In any case, this form or proto-astronomical astrology was only partly virtual, a form of "action-at-a-distance" [V2], as agriculture, seafaring and other concrete earthly activities were certainly influenced by the march of the near heavens.

Astrology became more virtualized because of two different developments. First, expanding the realms of earthly activity that was "influenced" by the stars. If the heavens had a direct impact on the tides, why shouldn't they also be influencing human affairs? Thus, astrology became a central means of governmental, and sometime later, personal divination: when to go to war, where to build the castle, whom to marry. Unfortunately, whereas one could empirically observe the causal connection between the lunar cycle and the tides, there was no clearly analogous relationship between the heavens and the social world. Entire systems had to be invented, the most famous and longstanding of which is the zodiac [V24; V25]. Thus, from relatively early on we find a bifurcation between scientific "astrology" (today's astronomy) and its non-scientific, astrological counterpart. From here astrology moved from secular goals (government "strategy") to religious purposes that turned the relationship of heaven and earth in the opposite direction: astrologers/priests trying to influence the *residents of heaven* (the gods, connected to the stars or constellations[20]) through human action on earth [V33].

---

[20] For instance, in Mesopotamia the names of planets were associated with central deities: Saturn with the destructive god Ninurta; Mars with war god Nergal; Jupiter with Marduk the creator; and so on.

Upon Alexander the Great's conquest of Egypt, horoscope astrology appeared for the first time—using the specific zodiac sign of the individual's birth month to predict that individual's ensuing fate. Ptolemy, the Egyptian astronomer and astrologer par excellence, established the astrological system that continued unchanged for well over a thousand years—a parallel virtuality to Christian theology throughout most of the medieval period.

Astrology eventually became the underpinning for all "scientific" knowledge, gradually expanding to include almost all the known sciences: anatomy, medicine, zoology, botany, as well as the inanimate realms of chemistry and mineralogy. Their elements, components, and properties such as colors, metals, and drugs (chemicals) were respectively associated with specific planets that "ruled" over them. The effect was to take even those fields of systematic observation that had a modicum of "scientific" basis and turn them into pseudo-science, especially regarding its connection to the exterior, "higher" world. This was one way in which astrology "virtualized" the early sciences—creating a mistaken picture [V31] of the interaction between each subject and the supposedly "superior" subject: astronomy/astrology.

Another way involved the issue of free will: if the movement of the heavenly bodies set human destiny, and the latter could even be predicted, then there was little that individuals could do to escape their fate. A person's life had no pluripotent latency; humans went through the motions in pre-determined fashion without the possibility of actualization [V27]. Indeed, this was the case not only regarding the person as a whole but also for the individual's constituent parts. The zodiac came to be regarded as a sort of human body simulacrum [V12], with each part corresponding to the zodiac signs e.g., the head was placed in the first sign of the zodiac: Aries, the Ram; the feet in the last sign: Pisces, the Fish; and so on. Thus, even the most corporeal of all sciences—medicine—was virtualized through substitution [V6], with illness being ascribed to the influence of planets, stars and constellations.

Interestingly, astrology denying self-actualization through free will was recognized as problematic by several leading Islamic theologians in the medieval period. Avicenna and Alpharabius argued that astrological methods were not empirical but rather speculative, and that such "predictive" powers were not consonant with orthodox Islamic views because they suggested that God's Will can be precisely known (Saliba, 1994). This didn't stop astrology from being further developed by other Moslem writers (Masha'allah, Abu Ma'shar, and Al Biruni), as well as Jewish theologians in the Moslem lands (ibn Ezra). Indeed, the Jewish Kabbalah (e.g., *Sefer Ha'Yetzirah*) was influenced by the zodiac—in Hebrew called *mazalot*, the same root as in the word "luck" (*mazal tov* = good luck).

During the Renaissance and the Enlightenment traditional astrology's grip weakened among the public, and it began to change focus. Astrology did not go completely out of vogue; indeed, the all-time most famous predictive astrologer, Nostradamus, appeared in the sixteenth century. However, the new astrologers refocused their efforts on mathematical aspects of astrology (e.g., its precise calculation), and less in interpreting horoscopes. This marked a return to its origins of not distinguishing between astronomy and astrology. In any case, traditional astrology declined for the same reason that religion weakened: the rise of empiricism and science—plus a related reason to be discussed shortly.

Astrology made a surprising comeback in the early twentieth century that has continued apace to this very day, notwithstanding (or perhaps because of) some new wrinkles. First, under the influence of the Freudian Revolution the more "serious" astrologers began to replace prediction with psychological interpretation [V11]. Along with one's *internal* subconscious affecting our character and behavior, they claimed, there are also *external* (astrological) influences determining our personality (Leo, 1919). However, the main resurgence of astrology is in the popular media: many newspapers carry a daily horoscope, in addition to constant mention of celebrities' zodiac sign and how that "explains" behavior, marital matching, and other sundry aspects of life [V32]. Viewed purely as entertainment by many readers, but taken seriously by numerous others, astrology has once again become part of the public arena—albeit not influencing the public agenda as in the past.

In the early modern period, astrology also declined because of scientific astronomy's rise. Modern scientific observation and investigation of the heavens showed how the astrological system was based on a deeply flawed understanding [V31] of the firmament's innumerable objects—huge, large, and small.

Astronomical observations and systematizations can be found in almost all major civilizations from the third millennium BCE, starting in Mesopotamia (Sumer, Assyria, and Babylonia) and Egypt where several pyramids and temples had precise solar orientations. Astronomy included diverse observations: the seasons of the solar year, the lunar cycle (monthly as well as the 18-year, repetitive "Saros" lunar eclipse cycle), the periodicity of the visible planets and fundamental distinction between planetary motion and the relative stability of stars. The Greeks then took astronomy to a higher level of exactitude—by the fourth century BCE developing three-dimensional geometric models to explain planetary motion and theorizing that the Earth spins on its axis in precession (i.e., gradual change in the orientation of Earth's rotation axis). In the third century BCE we even find Aristarchus of Samos positing a heliocentric system (planets revolving around the stationary sun) but this remained a distinctly minority opinion until Copernicus almost two thousand years later. The reason for the delay: Ptolemy of Alexandria wrote the *"Almagest"*—a monumental, almost universally accepted tome that mistakenly [V31] posited a geocentric view of the world, a universe far smaller than earlier Greeks had suggested.

Until the Renaissance, then, astronomy was mildly virtual in several ways. Why "mildly"? First, the planets and stars could not be touched [V25], but their existence, position and movement indirectly signified and guided real-world phenomena and human activity: farmers knew when to sow and reap based in part on the upcoming season that the heavens portended. Second, as already noted, the sun and moon directly impacted life on Earth [V2]. Third, the constant desire for greater astronomical accuracy drove technological invention—whether different forms of sundials or even the remarkable Antikythera Mechanism (a proto-computer with differential gears) dating back to the second century BCE, that calculated the movements of the moon and sun, and possibly the planets as well. Thus, while philosophers such as Plato (*Timaeus,* 33B-36D) and Aristotle (*Metaphysics,* 1072a18-1074a32) dabbled in astronomy in speculatively "theoretical" fashion [V31], others based their astronomy on rigorous mathematics, clever engineering and even innovative technology. In sum, although we cannot easily separate the virtual aspects of ancient astrology from astronomy in most ancient civilizations, among the Greeks astronomy began to appear as a far weaker form of virtual activity—an emerging proto-science.

This trend of separating scientific astronomy from theology and metaphysical/philosophical speculation (*a la* Aristotle) continued in the medieval Arab world where astronomy advanced several more steps: the first observatories were built, experimental devices were developed to prove hypotheses (e.g., the moon only reflected sunlight and did not emit its own), some Ptolemaic problems were resolved with new data, and so on. From a long-term historical standpoint, however, the most important "contribution" was the preservation and translation (into Arabic) of Greek knowledge, and from there into Latin from the twelfth century onward–thereby catalyzing the Copernican "revolution"[21] and subsequent astronomical advances that led to heightened virtuality in the ensuing centuries.

Until Galileo's telescope,[22] the naked eye performed all astronomical observation; after Galileo, astronomy was undertaken almost exclusively through mediated technology [V3]—optical telescopes, and later on sophisticated spectrometers and radio telescopes. Moreover, the new tools of astronomical observation also revealed entire "worlds" that heretofore were beyond our cognizance: stars and galaxies too far and dim to be seen by the naked eye, newer types of bodies in the firmament that we had no idea existed (quasars, neutron stars), and other phenomena that human eyes could not "see" directly without technological aid (gamma rays, neutrinos, black holes) [V19]. However, these same powerful technologies gave astronomy a much greater exactitude, thereby *reducing* and at times *eliminating* the amount of incompleteness [V4], distortion [V5], lack of three-dimensionality [V7], unsensed/invisibility [V19], and illusory [V31] elements of astronomical observation/theorizing that characterized earlier astronomy.

In sum, modern astronomy subtracted from, but also added to, virtuality in astronomy in very different ways. However, precisely because our recent tools have become so extraordinarily powerful, we are paradoxically turning astronomical cosmology into something almost completely virtual in several strong senses of the term.

Galileo's discoveries were manifold: Jupiter's larger moons, craters on our moon, the existence of sunspots. Most important of all: noticing a complete set of phases for Venus, thereby supporting Copernican helio-centrism. Kepler, relying on Brahe's very exact celestial observations, moved astronomy from mere empirical observation to theoretical prediction. Kepler hypothesized that the motion of heavenly bodies is

---

[21] Copernicus' heliocentric theory had several antecedents in the Greek and Moslem worlds. Thus, it had a "revolutionary" influence on modern astronomy but could not claim to be "original."

[22] Hans Lipperhey, a Flemish spectacle-maker, had already applied for a patent in October 1608 (which was refused!). Galileo unsuccessfully tried to get his hands on the invention, but by 1609 had created his own version (20 power magnification). Soon thereafter, he published "Starry Messenger," revolutionizing astronomy.

based on physical forces,[23] lending themselves to precise, mathematical prediction based on the three laws of planetary motion that he developed.

Newton's universal law of gravitation, based in part on earlier theorizing by Copernicus and empirical testing by Galileo, reinforced the connection between astronomy and physics. Based on mathematical computation, he showed gravity to be the fundamental force underlying the motion of all physical bodies [V2], thereby providing the first general explanatory theory for all astronomy. He did not successfully explain the mechanism causing gravitational pull between two bodies; Einstein's General Theory of Relativity did that. Nevertheless, Newton's universal law of gravitation was a monumental step forward in modern science.

It was also a major step placing virtuality within the scientific mainstream. Gravity is an *unseen* force [V19], not only in the distant heavens but right under our feet, influencing all material objects in our world. To a large extent, this was the scientific equivalent of religion's spirits (whether ghosts or God)—gravity's effects and influence can be felt in all places and at all times even if it cannot be perceived by human senses. This constituted an indirect blow to religion. By clearly proving that mysterious forces—heretofore understood in metaphysical-religious fashion—could be powerfully described in naturalistic fashion, he turned out to be a central actor in science's neutralization of previous theological explanations. By replacing previous mistaken [V31] theories (Aristotle's bodies moving in the direction of their "intrinsic nature") and metaphysical [V33] speculations (an unseen spiritual world underlying reality's movements) with a proven concept, the science of gravity in general significantly weakened our sense of virtuality inherent in the physical world, despite gravity itself being an unseen force working at a distance [V2; V19].

By the nineteenth century, astronomy and physics had become completely interconnected. For instance, it was discovered that sunlight was composed of the same spectral lines found in gases on Earth, each corresponding to specific chemical elements. This proved that terrestrial elements also formed the basis of the sun (helium and especially hydrogen)—an observation that cemented the non-uniqueness of earthly composition in the broad, astronomical scheme of things (other than the presence of life). Here, then, was another form of de-virtualization—astronomical objects/bodies were *not* "other-worldly" [V34], incorporeal [V25], or incommensurable [V16].

This was a double-edged sword regarding virtuality, for the ability to measure things precisely began to uncover all sorts of phenomena that we had no inkling of previously [V19]. More powerful telescopes discovered that the sun was a mere dot within the Milky Way galaxy that has over a hundred billion stars. Then Hubble discovered the existence of other galaxies, not only great distances away but rapidly moving away from us as well.

---

[23] Aristotle also explained celestial motion as based on physical forces ("On the Heavens"), but his theory was qualitative, not predictive. Moreover, he argued that heavenly material was ether—perfect, incorruptible and immutable [V30]. Where Aristotle distanced astronomy from physics, Kepler united them into a holistic theory (Applebaum, 1996).

Today we also know that the number of galaxies and stars in the known universe is indeed "astronomical": approximately 100 billion to 1 trillion *galaxies* and between 10 sextillion to 1 septillion stars (Cain, 2009)!! The heavens that used to awe us with hundreds of visible stars turns out to be a mere speck in a universe largely unseen [V19] and to a large extent even immeasurable [V16] because there are multitudes of galaxies and stars *that we will never be able to see;* they are so far away and receding so rapidly that in fact their starlight cannot reach us. In other words, whereas powerful telescope technology [V3] enables us to perceive a large part of the heretofore invisible universe, astronomical technologies (and accompanying mathematical computation) have also "virtualized" large swaths of the universe that will be out of sight (mediated or otherwise) forever [V19].

The unseen universe not only comprises distant stars. The new astronomical technologies (infrared and x-ray) reveal a panoply of unseen radiation and waves emanating from, and dispersing throughout, the universe: gamma rays, x-rays, microwaves, radio waves, infrared and ultraviolet radiation, and most recently discovered: gravitational waves. These are not only invisible to humans [V19] but also are spectrally incorporeal [V25].

Finally, there's the "black hole," caused by the inward collapse of a very dense star. The ensuing huge deformity of space–time creates an "event horizon" boundary where matter and light (indeed, any form of "information") can only move inward; once past the boundary nothing (except for quite weak, thermal "Hawking radiation") can escape so that we can never know what is happening therein—ergo ubiquitous black holes found in the center of most galaxies (Kormendy & Richstone, 1995). More than any other astronomical phenomenon, the black hole seems to have caught the public's imagination: a place, if not of "nothingness" then at least of "unknowableness" [V19; V34].

## 5.7 Cosmology

Cosmology is the science of the universe's origins, shape, movement, and development. Compared to astronomy, cosmology tends to be more speculative given the difficulties involved in direct "observation."

Whereas in the past, cosmology was based on theology and occasionally philosophy, the twentieth century discovery of the universe's ever-quickening expansion brought cosmology into the realm of science. If the universe is not static, as had been previously thought, then what caused (or is causing) this expansion? The Big Bang: a massive explosion 13.7 billion years ago of the infinitesimally small, extremely hot, and very densely packed "universe-point." It took several hundred thousand years for the universe to cool down enough for atoms to coalesce into elements, and from there to star formation after several hundred *million* years. In short, in the Beginning the universe looked nothing like it does today (it did resemble very closely the description in Genesis, I: 2: "the Earth was formless and void [*tohu ve'vohu*]") [V15].

The implications of this are profound—and profoundly troubling. If there is anything that has always been considered a constant in the universe (however understood), it was the "Ultimate Laws" of physics. Although natural philosophers in the ancient world and modern scientists in the modern world have differed on what the laws of the universe actually "said," everyone accepted that the basic laws of physics were and are universally applicable and timeless. Unfortunately, if the laws of physics are not immutable, then all research becomes an act in a play without possibility of an end [V18] or too many endings [V16]. Only the reductionist/universal-law approach offers us potential closure.

That might no longer be possible. A small but growing number of contemporary physicists are concluding that even the laws of physics are to some extent dependent on context and time (Frank, 2010) [V15]. There are several reasons for this startling theory. First, as philosophical background, the demise of religious belief, particularly among scientists regarding the non-existence of "God," undercuts the inevitably of the idea that the universe is necessarily ruled by one overarching Law.

Second, contemporary physics is raising more questions than providing answers (not necessarily something bad in principle). Indeed, the situation seems almost akin to the latter stages of the geocentric Ptolemy Theory—in order to keep the "system" in sync with the evidence, more and more complicated revisions and addenda had to be added, until the whole superstructure was swept away in the heliocentric Copernican Revolution. The situation today is reversed: the possibility that *a universal law does not exist* i.e., "the fundamental laws are not fundamental" is increasingly buttressed by experimental evidence:

1. "Clock ambiguity": quantum theory and its accompanying mathematics break down completely in the early stages of our universe i.e., laws inherent in the early universe could not "predict" its later configuration [V27; V34].

2. String theory: the central way of reconciling the micro-laws of quantum mechanics and the macro-laws of gravity. Unfortunately, the non-empirically testable nature of string theory leads to an unimaginable number of universe configurations [V15; V16], each possibly with different "laws."

3. "Backward causality": universal law is intricately connected to time's unidirectional arrow, from past to present to future. Astoundingly, we now have *hard, experimental proof* that the future seems to "reach back" and influence the present (Merali, 2010). This is not "time travel" (at least not in the science fiction sense of the term), but rather a form of chronological entanglement [V18]. The idea of feedback from the future is so counter-intuitive that it completely upends our common sense as well as scientific conception and perception of space–time reality. Our most fundamental *sense* (both meanings) of the way things work may be illusory [V31].

4. Spatial differentiality: recent astronomical evidence (not yet yielding irrefutable proof) suggests that the "immutable" *laws* of physics were somewhat different at the start of the universe—nor are they "today" identical in different sections of the universe (Brooks, 2010)! "As we peer back to the very beginning itself, when all matter in the observable universe existed together at virtually a single

point, the very nature of space itself, and possibly even time as well, may have been dramatically different" (Krauss, 2005, p. 118) [V34].

How is all this tied to virtuality? If there are no universal laws, then whatever reality we happen to be in at the moment is not completely part of a larger picture, but rather an "isolated" part that has a tenuous connection with everything else—chronologically and spatially. To use a simile, if we always believed that we're a seamless part of a gigantic canvas called the universe, painted on by one artist, it's possible instead that our time and place is at most a section of a gigantic mosaic, each piece and epoch—whether incredibly brief (e.g., immediately after the Big Bang) or extending over eons (our era)—created with different materials based on different "styles." Thus, in a most fundamental way reality might be virtual in the profound sense of being both "distorted" [V5] and "indeterminate" [V15]—as if we exist in a universe with an "identity crisis" incorporating multiple (behavioral) personalities.

However, the relativity of physics in our place and time within the universe is a minor aspect of cosmological virtuality; there's a far "worse" problem—most of the universe is missing [V4]. Based on the mass of the galaxies (and other celestial phenomena) and their speed of recession, the amount of matter in the universe should be several times greater than actually exists. As a result, two additional "things" were added to the universal pot: dark energy and dark matter—"dark" because as of now we have not been able to "see" them [V19]. Dark energy comprises 73% of the universe's total mass-energy, with completely invisible dark matter adding 23%. The result: *we are able to observe only 4% of all matter in the entire universe!*[24] In short, 96% of the known universe is virtually invisible [V19] i.e., without direct, measurable evidence [V16], and immaterial [V25].

What about the vastness of empty space? "Empty space is not empty at all, but involves a boiling, bubbling brew of... particle-antiparticle pairs popping in and out of nothingness... yet another hidden universe lying just beyond our perception" (Krauss, 2005, p. 109) [V17; V19]. Anti-particles are precise mirror images of particles, but with a reverse charge. However, despite having the same "materiality" as do particles, anti-matter is obliterated upon contact with matter; luckily for us, the universe started off with more matter than anti-matter, so that it is now composed largely of matter.

Nevertheless, anti-matter might exist abundantly in far-off "anti-matter galaxies" due to the huge distance between particles because of the universe's cosmic inflation over 13 billion years. Unfortunately, such anti-matter galaxies would look exactly the same as do galaxies of normal matter. The result is a doubly virtual situation: the possibility that whole sections of the universe are constructed in "obverse" duplicate fashion [V1; V22] but that from afar we cannot determine whether or not [V15]—and if we were somehow to check from up close, the "meeting" would cause the antimatter to be destroyed (along with the matter we sent there).

There is even more to the universe's complex and multifaceted, partly (largely?) virtual character. The following paragraphs are admittedly more speculative than most of the previous discussion—and in most cases, there are scientists who heartily disagree with this or that ensuing cosmological theory or model. Again, the point is

---

[24] A competing theory, MOND, posits no dark matter or dark energy; see: Carmeli (2017).

not whether these theories and scientific speculations are "correct," but rather that in a reversion to ancient "natural philosophizing" (albeit today based on some indirect evidence or attempting to explain problematic data) contemporary scientists do not hesitate to create weird, virtual models and theories to explain the real world. At the least, this reflects virtuality in our *zeitgeist*; at most, it suggests that we live in a world based on far more profoundly virtual phenomena than we ever imagined.[25]

Ever since Einstein and his revolutionary cohort, physicists have been trying to derive a unified field theory to combine all four basic forces—especially macro-gravity with the other three that work on the micro-quantum level. One of the most promising approaches to this is also one of the strangest ever created by the scientific mind: Super/String Theory (ST). This theory (actually five different ones—each mathematically "correct"!) posits that we live in a world of multiple dimensions: either eleven (ten of space + one of time) or twenty-six. The underlying idea is that reality is built on fundamental, incredibly tiny ("Planck length"), vibrating, one-dimensional "strings," each with a unique resonance frequency. Various combinations of vibrating string frequencies determine each fundamental force of physics. String Theory involves bosons that have no matter; Superstring Theory adds fermions that do have matter. Both lead to a multi-dimensional world [V14], but by dint of its immateriality the former has more elements of virtuality [V19; V25] than the latter [V19 alone].

It is not clear that ST can ever be experimentally verified (although some experimental outcomes could falsify it)—adding yet another aspect to its virtual nature, at least within contemporary economic-technological capabilities [V16].[26] As Nobel laureate Anderson argued: "string theory is the first science in hundreds of years to be pursued in pre-Baconian fashion, without any adequate experimental guidance" (2005). Moreover, ST suffers from over-prediction. In trying to explain how our three-dimensional world could emerge from an 11 (or 26) dimension universe, ST offers an almost infinite number (from $10^{100}$ to $10^{500}$) of possible configurations [V21; V26]!

Whereas ST is relatively new, the idea of extra dimensions has been bandied about by philosophers since at least the seventeenth century (e.g., Henry More), and became quite the vogue in the nineteenth century with books such as *Flatland* describing its whys and wherefores (Krauss, 2005). Additionally, the concept was used as the hook for several classic literary works for "children"—from Lewis Carroll's *Through the Looking Glass* in the late nineteenth century to C. S. Lewis's Narnia series in the

---

[25] I will not include here technical material such as mathematical equations and convoluted qualitative descriptions, nor every "sub-theory" or theoretical caveat that scientists have raised regarding these theories. Rather, I will schematically describe the central theories and their underlying explanations, focusing on their virtual attributes. These cosmological hypotheses are not a product of idle conjecture by bored scientists but rather are *serious attempts to create internally consistent systems to explain and/or eliminate problematic evidence and mathematical inconsistencies in current models of the macro-universe and micro-quantum world.*

[26] Given the incredibly small size of ST phenomena, a huge amount of energy would be necessary to test the theory. In principle, we could build such measurement machines, but that would be exorbitantly expensive.

mid-twentieth century.[27] In the 1970s, however, ST placed extra-dimensionality in the scientific mainstream with a mathematical foundation. Nevertheless, ST is not clear about whether these dimensions are huge or very tiny ("compactified") and how many there are [V15].

We are left with the question of whether the entire ST paradigm is not virtual in character: "if…these dimensions remain forever hidden, ephemeral theoretical entities, inaccessible to our experiments, if not our imagination, then in what sense are these extra dimensions more than merely mathematical constructs? What, in this case, does it mean to be *real*?" (Krauss, 2005, p. 208).

This question takes on added meaning given that ST could also possibly serve as a bridge to another cosmological phenomenon (mentioned by a few astronomers in the past[28]) itself not experimentally verified: multiverses (MV) i.e., the existence of entire *universes* outside of our own [V34].[29] Referring to the "philosophical implications of potentially large extra dimensions," Krauss suggested that "branes [membrane entities in which dimensions are located] could represent possibly infinitely large alternate universes that could exist, literally, less than a fingernail's width away from our own. Each of these universes could have laws of physics that might be dramatically different from our own as well as a dramatically different life history" (2005, p. 224). In other words, the almost infinite number of possibilities inherent in ST is paralleled by a similar "number" of universes in the overall MV that might be observable: "a collision of our expanding bubble [universe] with another bubble [universe] in the multiverse would produce an imprint in the cosmic background radiation—a round spot of higher or lower radiation intensity. A detection of such a spot with the predicted intensity profile would provide direct evidence of other bubble universes" (Vilenkin & Tegmark, 2011; see too Vilenkin, 2006).

To make matters even more confusing or exciting, a "bubble multiverse" is not the only type possible. Greene described (2011) other forms of "universes" that meet our contemporary mathematical and theoretical requirements e.g., a holographic universe in which we are mere projections on a very distant, surrounding surface [V7]—mirroring "real" processes occurring in a different reality. In other words, everything that we "are" and do might be virtual, whereas the impossible-to-see, unfathomably distant is the "real"! (Shades, or shadows, of Plato…).

It is precisely here that we return to cosmogony: how did it all start? If our universe is the only one around then that question makes eminent sense. However, if "reality" comprises an infinite number of universes, then the formation of our universe could be a by-product of the long-ago interaction of other universes. Indeed, our universe could well have been a "birth bubble" emerging out of another universe, the same way that new anti/particles could emerge *creatio ex nihilo* from a space

---

[27] Without steering this discussion too far afield, it is worth noting that extra-dimensionality is also suggested in Picasso's Cubist paintings.

[28] Giordano Bruno was burned at the stake in 1600 for, among other things, suggesting an infinite-space multiverse.

[29] With the coinage of "multiverse," the word "universe" undergoes depreciation—no longer the totality of all existence but rather everything within the entity that we are able observe.

"vacuum." While this still does not answer the question of how the *multiverse* came into being, it does push the genesis question much further back in time (assuming there ever was a beginning of time) and to a far distant/different place [V17; V18]. The concept of multiverses paradoxically brings us back full circle—albeit with a completely different sensibility—to the ancient world. Cosmological virtuality back then involved other worlds that could not be seen but could be vividly imagined: mythological (gods), theological (God and angels), and eschatological (resurrection of the dead). Contemporary physics also suggests the same: many universes, some similar to ours and others perhaps radically different; none that can be perceived, but that can be imagined and even mathematically grounded. Moreover, the ancient and modern virtuality mindsets deal(t) with the past (Creation; Big Bang), present (a hands-on God; multiverse impingement on our universe), and future (Paradise; universal collapse and another cycle of the Big Bang).

This is not to suggest that the way contemporary physicists think speculatively-scientifically is similar to the ancient magical-mystical mindset. However, it does suggest that thinking of the "world" in highly virtual terms [V28; V32] is something that all humans are prone to do—each generation in its own distinctive fashion. As Krauss observes: "Ultimately our continuing intellectual fascination with extra dimensions may well tell us more about our own human nature than it does about the universe itself. We all yearn to discover new realities hidden just out of sight. So much so, that we have continually reinvented them throughout human history, whenever the world of our experience has seemed lacking" (2005, p. 2). In short, if ST is correct and/or the MV exists, then we really do *live* in a physically broader, virtual environment [V14; V17; V19; V34]. If it doesn't exist, then even the most "rational" of people (physicists, astronomers, mathematicians) are *creating* for the rest of us virtual descriptions of reality [V28; V31; V33; V35].

In the end, however, we might not need multiverses to understand just how virtual our world is. We need merely turn to the latest findings regarding the fundamental foundation of matter. Thus, I conclude this chapter (almost) with a relatively lengthy quote from the frontiers of quantum mechanics, a rigorous science that has arrived at quite a startling conclusion (my emphases):

> *Matter is built on flaky foundations. Physicists have now confirmed that the apparently substantial stuff is actually no more than fluctuations in the quantum vacuum.* The researchers simulated the frantic activity that goes on inside protons and neutrons… [that] provide almost all the mass of ordinary matter. Each proton (or neutron) is made of three quarks—but the individual masses of these quarks only add up to about 1% of the proton's mass. So what accounts for the rest of it?

> Theory says it is created by the force that binds quarks together, called the strong nuclear force. In quantum terms, the strong force is carried by a field of *virtual particles called gluons, randomly popping into existence and disappearing again.* The energy of these vacuum fluctuations has to be included in the total mass of the proton and neutron….

The Higgs field is also thought to make a small contribution, giving mass to individual quarks as well as to electrons and some other particles. The Higgs field creates mass out of the quantum vacuum too... virtual Higgs bosons. So *if the LHC confirms that the Higgs exists, it will mean all reality is virtual.* (Battersby, 2008)

And as we now know, the Higgs boson has been confirmed!

## 5.8 Conclusion and Coda: Physics and Philosophy; Physics and Mind

To conclude this chapter, I will first close the loop between contemporary math, physics, cosmology, and philosophy—and then offer a highly speculative rumination.

Modern physics seems to have circumnavigated a full historical circle. Among the Greeks, physics was a purely intellectual enterprise: natural philosophy. True, many of their ideas germinated from keen observation of the natural world, but very few indulged in empirical testing, and certainly not in "controlled experimentation." This is very similar to where the most cutting-edge physics/cosmology finds itself today: "at its most abstract, theoretical physics leaves ordinary empirical science behind and enters the sphere of philosophy" (Haldane, 2011). Horgan (2010) borrowed from Geertz, calling modern physics a form of "faction": "imaginative riffing on all sorts of phenomena" [V28].

Moreover, many physicists today have reverted to Platonic Idealism, not in the normative sense of "perfection" but rather of immateriality [V25]. Tegmark, an MIT physics professor (I mention his serious institutional affiliation only because his claim might seem "crazy"), argued that "our universe is not just described by mathematics—it is mathematics" (2007). In short, ultimate reality is "grounded" in the ineffable (oxymoron intended)—just like the Pythagorean School that claimed Nature is Numbers! His colleague at MIT argued that the universe is essentially a giant quantum computer—"It from Bit": Matter from Information (Lloyd, 2006, p. ix). More recently, astrophysicist Scharf (2021) widened this: "information... isn't just a way to probe the fundamentals of nature; it may be part of the fundamentals"— harking back to the nineteenth century when "Maxwell's demon gave information physical reality" and to the twentieth century when Shannon "gave further physical weight to information by linking it to order, entropy, and thermodynamics" (Perkowitz, 2021). Finally, looking at the laws of nature, Nobel laureate Weinberg described himself as "pretty Platonist" (Overbye, 2007). Thus, although empirical science is not dead nor even passé, in physics/astronomy/cosmology empirical observation or experimental tests are finding it increasingly difficult (in some cases impossible) to provide answers to the deepest questions; "philosophical science" has taken over.

And now to my speculative coda regarding the relationship between contemporary math, quantum physics and human psychology.

First, a phenomenon that scientists have begun to study: fuzzy thinking. "Human thinking…often fails to respect the principles of classical logic…. Physicist Diedrik Aerts…has shown that these errors actually make sense within a wider logic based on quantum mechanics…[that] turns out to be better than classical mathematics in capturing the fuzzy and flexible ways that humans use ideas" (Buchanan, 2011). The human brain cannot deal with the huge number of factors that go into a fully logical decision—certainly not fast enough for timely decision-making in real-life situations—so we use the best data, or our sharpest, immediately accessible memories: "Situation A has a few similarities to situation B; when I last had to deal with A, I successfully reacted by doing C; thus, C will hopefully work OK in my present situation B."

Is the brain, then, *based* on quantum phenomena? Aerts' answer: "This is not to say that the human brain or consciousness have anything to do with quantum physics" (Buchanan, 2011). Well, yes and no. On the one hand, it is true that the similarity between quantum logic and brain logic does not prove the latter is based on the former; correlation is not causality. On the other hand, if the brain is "organic" (not altogether clear: Frank, 2017), and *all* matter stems from quantum particles, then one can surmise—as does Nobel Laureate (in physics) Penrose (2016)—that indeed our brain logic has somewhat of a quantum foundation.

The argument can be extended more widely. As this chapter has shown, our world, and the physical laws therein, can be described in several ways as "virtual." Given that this is the case on the vast macro level (dark matter) as well as the infinitesimally small micro level (almost completely non-corporeal quarks), should it then be so surprising that our brains are also *attuned* to "virtuality"? Furthermore, as the Heisenberg Uncertainty Principle posits, a particle does not leave its ambiguous state until observed by some form of external consciousness. If indeed there is an inextricable connection between consciousness and quantum physics, why shouldn't there also be the reverse connection between quantum physics underlying consciousness? Several contemporary theorists have posited precisely such a relationship (Herbert, 1993; Stapp, 2009).

As discussed in Chap. 3 on human psychology, one of the central philosophical and scientific questions occupying the greatest thinkers has been the source of consciousness and whether the brain and mind are dual or one and the same. Either school of thought can provide a logical basis for the human mind's ability and (perhaps) tendency to think virtually. The Vitalist/Dualist school believes in the dichotomous nature of material brain and incorporeal mind. If this is true, then we can view the "corporeal brain" as being the seat of our basic mental functions that deal with (sense, process and digest) the physical world. The mind would then be the seat of "virtual"—imaginative, creative—thought. For the Materialist/Monist school, there is also nothing to contradict human mental virtuality: just as the same universe of ours is comprised of corporeal elements with mass, but at the same time is full of "physics" (not necessarily physical) processes that cannot be explained in material terms, as well as comprising "entities" (strings, quarks, neutrinos), that have (almost) no corporeality, so too can one envision a human "brain" that also works on the basis of non-corporeal processes and produces "virtual" phenomena.

This, of course, might explain *why* the human mind tends to virtuality: it is merely mirroring the same phenomena found in Nature, perhaps even through some of the same processes. It does not explain *how* this might work in actual fact. No one can do so at this stage of our partial understanding of the quantum world, and its possible connection to the macro-"real" world. However, it does suggest a further "justification" for virtual thinking. In the past, when we understood the world to be completely Newtonian—with fully deterministic processes and materialistic in its exclusively physical, atomic structure—one could view any virtual thought as "irrelevant" i.e., without any counterpart in Nature. Indeed, one could perhaps even consider virtuality in that environment to be avoided as something "unnatural." In the modern era, though, inundated from all sides by mathematically grounded, *scientific* explanations of the virtual workings of the universe, there is a solid psychological foundation for mentally moving beyond the here and now, beyond corporeal reality.

Of course, I use the word "justification" advisedly—not in its normative, moral sense but rather meaning psychological normality. One need not fall into the naturalistic fallacy that "is" equals "ought." Rather, I am merely suggesting that seeing a clear "is" (i.e., the physical world is based at least in part on "non-physical" principles) can be enough to boost the confidence of those pre-disposed to virtual thinking; in other words, the type of internal cognition they are performing, and the "unrealistic" product of such imagination, is not necessarily alien to the external world. If our minds *create* virtual thoughts about virtual phenomena, this might well be because at the deepest level it *senses* and *reflects* the reality of the world's virtuality.

# Chapter 6
# Music, Literature and the Arts

## 6.1 Characteristics and Functions of the Artistic Endeavor

What constitutes "art," in the more general and generic sense of the term? It includes *artifacts* such as decorative objects; sculptures; paintings; tools (e.g., ceremonial sword), or other utilitarian objects such as ornate furniture, formed in ways that go beyond what is necessary for pure functionality; texts considered as objects (calligraphic scrolls, leather bound books); etc. Art also involves *performances* such as oral and written storytelling, as theater and novels; instrumental music; vocal singing; bodily movement, especially dance (Dutton, 2009, pp. 51–52).

What are the characteristics of "art"? What do they have that enables us to define something as being artistic? Davies (1991) argued that art must be *functional* (produce an aesthetic experience) or *procedural* (an artifact produced with the intent to be seen as art). Dutton offered twelve "cluster criteria," although admittedly each one is not exclusive to art per se: direct pleasure, skill and virtuosity, style, novelty and creativity, criticism, representation, special focus, expressive individuality, emotional saturation, intellectual challenge, traditions and institutions, and finally, imaginative experience (2009, pp. 52–59).

Obviously, "art" is an extremely wide-ranging human enterprise from several perspectives: the large number of types of art; the even larger number of traits found in art; and the breadth of art throughout history (time) and throughout the world (place).

Clearly, the arts are central to human experience. Our question is to what extent does art reflect or express virtuality? Langer (1953) argued that all the arts are virtual, going so far as to give each one a specific virtual name: narrative is virtual memory, drama is virtual history, film is virtual dream, dance is virtual gesture, music is virtual time, and so on.

---

The *Appendix: A Taxonomy of Virtuality*—listing by numbers 1–35, e.g. [V13] or [V34], some of the various types of Virtuality mentioned throughout this chapter—is freely available to the public online at https://link.springer.com/book/10.1007/978-981-16-6526-4.

We can begin to answer our question by delineating the three main functions of artistic expression (Dutton, 2009, p. 110):

1. Learn about our world, which is too vast for us to be able to view first-hand [V3; V9]. Paintings, photographs, novels, theater, cinema, radio, TV, et al., are what McLuhan (1965) called "extensions of man"—media that artistically represent[1] aspects of the world that we would not otherwise be able to obtain. Art also helps us *retain* information better than straightforward factual presentations: remembering poetry or prose set to music, or rhymed, or when the first letter of each consecutive sentence spells out a mnemonic. Visual and certainly aural cues are recalled better than textual information, so that people retained religious messages through church sculpture, paintings and architecture well beyond the weekly, oral sermon (in many cases, itself an artistic, storytelling art form).

2. Present the reality of our life in a new or different light [V5; V6]. Art can show us how others live (distant lands or alien cultures); make us see things that we miss in the humdrum of daily life; satirize, criticize, and even demonize aspects of our social or individual existence that we had not thought to be problematic; or conversely, socialize, mobilize or propagandize the populace into accepting cherished societal norms. In short, as Dutton suggested (speaking of stories), the arts "encourage us to explore the points of view, beliefs, motivations, and values of other human minds" (2009, p. 110).

3. Conjure up a "world" different from the one that we live in [V6; V26]. This can serve a change function, reformist or revolutionary, on the micro-individual and the macro-social levels—or can be reactionary in outcome. On the one hand, the arts can undermine the present status quo by showing how individual people and/or society can do things "better" (morally, functionally, economically). Here the arts (especially the fictional sort) offer a low risk, surrogate experience. We cannot live parallel lives and we are normally loath to destroy specific, traditional social institutions (not to mention longstanding societies) and rebuild them from scratch without prior "testing," but we can conjure up alternatives to what we individually are presently doing and/or the way we have collectively organized our social life—and through art try to work out the potential consequences of such alternatives before going ahead with significant change [V10]. Today, this can be undertaken through simulation (game theory etc.), at least at the macro-social level, but through most of history such an enterprise didn't have a (quasi-)scientific patina. Through dramatic theater, oral storytelling, novels, short fiction, and so on, different variations of common life themes, norms, institutions, and behavioral patterns, could be played with and judged for their utility and efficacy.

---

[1] The dividing line between factual *presentation* and fictional *representation* is not clear, for the *selection* of which facts to present will be an "artistic" decision. For instance, TV news shows will tend to prefer items with a stronger visual component. Newspaper writers/directors/editors will take other artistic aspects into consideration e.g., the audience's media expectations: an elite paper will have more text because of its readers' higher literacy level compared to a mass paper's readership wanting more visuals.

On the other hand, art can also serve to preserve and reinforce the status quo by enabling the populace to temporarily "escape" life's banal existence and withdraw into another world of the artist's creation. This is the Roman poet Juvenal's idea (1998)—"bread and circuses" (Satire 10.77-81)—in which the public gives up its authority to the leaders and contents itself with escapist entertainment.

When we sum up these artistic functions, only the first is primarily "reality-based" (notwithstanding the creativity in representing reality). The remaining functions, in ascending order, place increasing emphasis on virtual ways to express reality. Thus, the arts are heavily involved in *virtuality* overall—whether as an alternative to present reality [V6, 10] or imaginatively far beyond it [V26, 28, 30], or combinations thereof e.g., a fantastical novel that can be read for escapist pleasure but that also serves as a parable for the ills of contemporary society (e.g. *Gulliver's Travels, Animal Farm*); or a science fiction series or movie of a far-off world with problems similar to ours (*Star Trek, Planet of the Apes, Avatar*).

A second point: the proportional balance between artistic reality and virtuality does not stay the same, as the forms of expression undergo constant change. Even if we could clearly classify every artistic product as dealing in the "real" or the "virtual" it would be nigh impossible to count their number in any given society, or their relative quantitative ratio. Moreover, quantity does not equal quality or influence: one *Mona Lisa* can have greater impact than hundreds of private family portraits.

Trying to track the evolution of artistic expression is somewhat more doable, albeit a huge task. This chapter is not designed to be a comprehensive survey of virtual artistic expression through the ages. Rather, it will outline broad trends *as they relate to various elements of virtuality*. I turn first to some speculations regarding the historical and evolutionary origins of artistic expression, given the universality of the phenomenon among human societies.

## 6.2 The Art of Virtualizing the Arts: Biological Adaptations and Exaptations

How easy is a bush supposed a bear!
(Shakespeare, *1600*, Act V, i: 22)

On the face of it, the fact that there are no human societies without some form of art is somewhat surprising, for we do not normally view artistic endeavor as a fundamental need such as of food, shelter, and personal security. Clearly, the arts have always been in some way—and perhaps for several functions—an important factor in our evolutionary development, and perhaps humanity's physical survival as well.

The animal kingdom is replete with "deceptions" i.e., mis/representations of reality. These are expressed in four different ways (Camouflage, 2021):

*Cryptic camouflage*: blending into the environment to the point of imperceptibility [V19]. For instance, grasshoppers that look like a twig on a branch.

*Disruptive*: confusing the predator with hyper-stimuli that override the basic form or color of the animal in danger [V21]. Octopuses emit easily discernible ink that renders them hard to find.

*Mimicry*: copying the look (or scent or sound) of another (usually more powerful) animal [V6]. The Mimic Octopus can make itself to look like poisonous fish in the vicinity; recently, a beetle was discovered with an evolved body shape perfectly mimicking fire ants (and their pheromones), living in their midst unnoticed in order to eat the ants' young larvae! (Coghlan, 2017).

*Countershading*: removing several visual stimuli (e.g., depth perception) through reverse shading [V4]. Many animals have dark backs and light stomachs, counter-shaded to the immediate physical background, making it difficult for predators to see them.

It does not take much of a leap to surmise that early hominids saw this "natural" behavior in the animal kingdom and began to mimic or adapt such camouflage and deceptive practices themselves. Two examples: 2000 very ancient cave paintings in Lascaux, southern France, depicting hunters dressed as animals during a hunt; in South Africa, an "art workshop" from 100,000 years ago containing art implements and ingredients (e.g., ochre) for painting. These are evidence that early *homo sapiens* had the "conceptual ability to source, combine, and store substances that enhance technology or social practices [that] represents a benchmark in the evolution of complex human cognition" (Henshilwood et al., 2011, p. 219). Even Neanderthals showed a proclivity to symbolic art (Leder, 2021). One can speculate that at some point in humanity's early evolution the underlying concept—"what you see (or smell, hear etc.) is *not necessarily* what you get"—began to filter into other areas of life, somewhat less basic for survival.

This need not have happened consciously. As *homo sapiens* left Africa and migrated to colder climes, they would have sought thicker clothing. Animal pelts were the most logical choice; from there it was but a small jump to add an "animal head" to complete the hunter's animal camouflage. However, once dressed to kill (literally), the same procedure could easily evolve into "dressed to kill" figuratively, perhaps for social status reasons or for personal and group entertainment (costumes). Such dressing to present a "deceptive (virtual) appearance" [V5] was not restricted to clothing: at some point humans began to paint their bodies and use "cosmetics," perhaps copying colorful organs of animals around them, such as the bright, red backside of the baboon, or even to draw attention to their own (usually reproductive) organs e.g., red lipstick as a sign of a women's engorged labia (Ackerman, 1991, p. 114). Possibly the next artistic jump: from body art to external art (cave walls, rocks), using much the same media: vegetable oils and plant colors.

Similarly, "sculpture." Several hundreds of thousands of years ago hominids began to make rudimentary tools and weapons for finding, killing, and preparing food. Over time, hominid tool makers diversified into specialized objects. However, when not functionally employed a specialized instrument could serve as an aesthetic object: a knife used for sharpening a spear could also carve wood into an ornamental object after the hunt. Certainly, this would serve "religion" with carved statues and clay figurines as avatars for the gods and spirits [V6].[2]

Somewhat later the most important tool of all emerged: speech and language, evolving from approximately 200,000 BCE to perhaps as late as 40,000 years ago.[3] This enabled humankind to give full voice to its creativity, simulating not only discrete things but also through language signifying people, situations, actions, events—and through sound and later "music," mimicking the surrounding environment (Changizi, 2011). Here too there is little doubt that the primary urge and factor behind the development of speech and language was utilitarian functionality: staying alive ("that's a lion in the tall grass!"), hunting for food ("the tracks lead in that direction"), mating ("I like you"), and only later esteem ("how did you like that rhyme of mine?") and self-actualization (storytelling or wall painting for their own artistic sake).

In the terminology of evolutionary psychology, the interplay between early human evolution and the development of artistic endeavor illustrates the process of adaptations and by-products of adaptations (Pinker, 1997; Tooby & Cosmides, 2005; Boyd, 2009). The latter were not a product of natural selection (as were adaptations) but rather emerged because they somehow piggybacked on, or were produced by, primary adaptive traits. The ability to make tools was a primary adaptation of hominids; the later ability (and proclivity) to use those tools for artistic purposes (jewelry, ornamental decorations, statues) was a by-product of the evolutionary, toolmaking adaptation—an "exaptation." Similarly, spoken language probably served as a highly adaptive trait to transmit critical information to peers and progeny, from which the "storytelling" by-product—oral poems, written novels, theater—evolved later.

Conversely, the imaginative, artistic sense could have been a *primary* adaptation, considering the immediate *and* potential challenges facing the hunter-gatherer:

> ...humans live with and within large new libraries of representations that are not simply stored as true information. These are the new worlds of the might-be-true, the true-over-there, the once-was-true, the what-others-believe-is-true, the true-only-if-I-did-that, the not-true-here, the what-they-want-me-to-believe-is-true, the will-someday-be-true, the certainly-is-not-true, the what-he-told-me, the seems-true-on-the-basis-of-these claims, and on and on. Managing these new types of information adaptively required the evolution of a large set of specialized cognitive adaptations....
>
> Viewed against this background, fiction as "false" information no longer seems quite so strange or different from the other types of information humans are designed to actively represent and maintain.... [as it] consists of representations in a special format, *the narrative*, that are attended to, valued, preserved, and transmitted because the mind detects that such bundles of representations have a powerfully organizing effect on our neurocognitive

---

[2] Blascovich and Bailenson noted that "ancient sculptures seem to have been more closely tied to religion and the gods than the ancient cave paintings" (2011, p. 27).

[3] Very briefly, factors behind the development of speech and language run the gamut from neurological evolution (brain size and synaptic connections) to the development of the human thorax. Rudimentary speech (specific sounds denoting specific actions or situations) probably emerged close to 200,000 years ago; fully grammatical language evolved closer to 70,000 years ago (Robson, 2010).

adaptations, even though the representations are not literally true. (Tooby & Cosmides, 2001, pp. 20–21)

In short, the fictional virtual [V9; V28] became central to human survival and flourishing.

Not all ancient societies and cultures traversed the same artistic trajectory e.g., the Bible's Second Commandment prohibition against drawing pictures and making graven images—compared to the Greeks' creative emphasis on the visual (sculpture, architecture, and human body). Nevertheless, from the beginnings of ideogrammatic and then alphabetic writing around 3000 BCE onwards we can perceive a general expansion not only of "literature's" sundry manifestations (oral storytelling, epic poetry), but also artistic crafts: sculpture, ceramics, jewelry, painting, and so on.

Most, but not all, of these involved a fair measure of virtualization, the simulation of reality in changed form [V13, V22]—whether diminished [V4, V8], distorted [V5, V23], potentially real [V26, V27] or totally imaginary [V28, V35]. On the other hand, many used artistic subterfuge in attempting to mimic or duplicate reality precisely [V1, V6, V7, V8]: using perspective to give the illusion of depth (painting), physical embellishments to "duplicate" the sense of the original (theater), replicating Nature's sounds through music (Changizi, 2011). These examples illustrate how the arts incorporated a mixture of strong and weak virtuality, as measured by the natural materiality of the medium and the verisimilitude of the contents re/presented. Here is a sampling of the virtuality spectrum regarding media and performance arts:

*Weakest virtuality*: Employing real humans to physically present symbolic movement, as in modern dance; a three-dimensional film documentary of natural phenomena.

*Weak virtuality*: Employing real humans to symbolically relate natural, social phenomena e.g., oral storytelling—or to physically present a work of fiction such as dramatic theater; manipulating inanimate physical materials to symbolically present reality in three dimensions e.g., realistic statues.

*Mid-level virtuality*: Shaping natural and artificial materials to create non-natural, three-dimensional forms: jewelry, abstract sculpture; human voice projection to produce transient vocal music with lyrics (symbolic human language); 3D holographic reproduction of reality.

*Strong virtuality*: Using artificially created physical media to produce ephemeral sounds without parallel in the natural world e.g., instrumental music; composing words to create a fictional narrative, reflecting aspects of social reality; two-dimensional photographic reproduction (even stronger virtuality: with some digital manipulation).

*Strongest virtuality*: Digitally animating non-realistic phenomena in 2D e.g., a fictional video game; a surrealistic text on a digital screen e.g., a Borges' novel on an e-book.

In sum, "virtual arts" as we know them today—an almost purely aesthetic enterprise—did not begin with any artistic pretensions but rather closely followed the chronological evolution of Maslow's hierarchy of needs (1943): physiological (food, clothing), safety (weapons, tools), social (family, companionship), esteem (respect, achievement), self-actualization (creativity). The arts ultimately became an exercise of virtual *re*presentation among humans with time on their hands (figuratively and literally). One can speculate that humanity started indulging in weakly virtual art forms adhering to reality as closely as possible, and gradually transformed them into something more strongly virtual through deepening of powers of abstraction and creative imagination. Over hundreds of thousands of years (or tens of thousands depending on the specific art form), the gradual but seemingly inexorable expansion of human artistic creativity might well have led to an evolutionary feedback loop that ultimately inculcated art as an integral part of human life.

For example, artistic creativity (e.g., good storytelling) might have been appreciated by potential mates as a sign of cognitive intelligence—thus increasing the reproductive chances of that creative artist (Scalise Sugiyama, 1996), what Miller calls "aesthetic fitness" (2001). Over time, artistic ability would have become inbred into human nature through natural selection while lack of artistic ability (or appreciation of same) would have been gradually out-selected genetically. If such ability went hand in hand with superior toolmaking (dual functionality), such an evolutionary, positive feedback loop would be almost self-explanatory: many or some of the efficient and successful hunters (sharp spears etc.) would also have made better "artists"—producing more children (or enabling more of their well-fed progeny to reach reproductive age) than their non-artistically talented or inclined compatriots.

A tantalizing hint—and it is not more than that—of art being an integral part of human nature (or that it is at least a "natural" proclivity) comes from child's play. As Aristotle pointed out: "...it is an instinct of human beings from childhood to engage in mimesis" (1984, p. 1312b25). Children play make-believe without prompting, whether or not they have physical props—and this takes up most of their waking time when not involved in critical life functions (eating, bathing, etc.) (Opie & Opie, 1969). Such pretend play emerges in all children of normal development at around age eighteen months to two years, manifesting itself as the ability to think in terms of meta-representations—pretense divorced from the real world (tots do not expect their dolls to die if they fall out the window and hit the pavement), indulging in virtual play while distinguishing it from real play (Leslie, 1987). Continuing Aristotle's idea, Walton argues that make-believe continues through adulthood: "It [make-believe] continues... in our interaction with representational works of art (which of course itself begins in childhood)" (1990, p. 12).

Finally, as noted at the start, basic modes of artistic expression and aesthetic appreciation seem to be universal: "When these [absolute necessities] were provided, it was natural that other things which would adorn and enrich life should grow up by degrees" (Aristotle, 1984, p. 1329b30; see too: Scalise Sugiyama, 2003)—foreshadowing by more than two millennia Maslow's hierarchy of needs.

I end this introductory section with a caveat. When one coldly analyzes human artistic production and consumption patterns in *modern* times, we find in the popular media a preponderance of subjects and contents dealing with animal (or alien) attacks, fantasy empires, wild chase scenes, bloody revenge—all things that are rather "useless" in a world of institutionalized police, law, and jails (but highly relevant metaphorically to the Hobbesian State of Nature from whence our species emerged). Thus, in the contemporary age the arts do not *necessarily* serve a purely functional, societal purpose. Rather, the central point here is that the life of (imaginative) mind, of thinking virtually in artistic and aesthetic fashion, is natural to humanity and continues to play an important (if somewhat subtle and possibly indirect) function on the *micro-individual* level through several central forms of artistic expression, surveyed below.

## 6.3  Music[4]

How is music virtual? Singing by human voice is "real"—using the body to emit pleasing sounds. As long as the listener is present during the singing, s/he too is involved in a real, and not virtual experience—although it could be argued that even listening to sound is a virtual experience due to the subjectivity of sound perception (Grimshaw & Garner, 2015), and the invisibility of sound waves [V19]. Similarly, instrumental playing of sounds that occur in Nature (and the audience present during the performance); being emitted by some sort of human tool/technology (whether a real animal bone; or a somewhat artificial carved wood; or from completely artificial materials such as a plastic flute), does not make it any less "real," if the sounds are a relatively accurate rendition of something heard in the natural world (water flowing, birdsong). However, here too one could argue that such instrumental sounds are mere "representations" [V6]—however similar or close to identical. Even sounds from a natural object e.g., a bone, with holes made by humans, could be considered virtual in their aural, if not material, artificiality [V13].

Beyond this there are three different ways by which music can become virtual (listed here in increasingly virtual order). First, sounds that *are* not (and certainly, if they *cannot* be) found in Nature, without parallel in the "real" world [V13].

Second, and far more strongly virtual, is the spatial and/or chronological separation between musical producer and audience [V3], what we today call recorded music or sound *r*eproduction [V6]. Spatial separation increases virtuality, as when the audience listens by radio to a live musical performance; chronological divorce is even more strongly virtual, as in listening to a performance recorded earlier.[5]

Third and finally, from an objective, *technical* standpoint regarding reproduction and transmission—but not necessarily from the subjective, aural perspective of the

---

[4] As Wallin et al. (2000, pp. 6–7) argue: "The question what is music? has no agreed upon answer… Musical universals place the focus on what music *tends to be like* in order to be considered music." Nevertheless, some basic elements are common to almost all forms of music: pitch (with sub-elements: melody and harmony), dynamics, rhythm (with sub-elements: meter, tempo, and articulation), and sound elements such as texture and timbre.

[5] There is one additional form of "virtual music": musical notation (e.g., sheet music). Strictly speaking, that is not music at all but rather a *written* medium for a specific form of human communication—"ephemeral data of rhythm and pitch" [V25] (Hobart, 2018). It is worth noting (pun intended) that like the invention of "zero," musical notation also enabled the representation of silence i.e., a pause in the music [V19].

listener—digitally recorded music is more virtual than analogue.[6] Digital recording takes the original, *naturally* analogue sound waves and transforms them into binary bits that then divide and "*re*produce" the original sound into infinitesimally small "bites" of sound strung together, that the human ear cannot discern as "choppy" sound bites (our brain naturally "fills in" the "nano-pauses," faking us into hearing smoothly continuous sound).

Several music aspects can be listed with relative certainty (Wallin et al., 2000, pp. 8–11; Cottier, 2021, pp. 16–17): 1—music exists everywhere among human societies; some basic musical principles are close to universal (e.g. rhythmic patterns; octaves are aurally equivalent; etc.); 2—music originated a long time ago during the evolution of *homo sapiens* (perhaps even before); 3—music seems to be inextricably intertwined with human speech development (possibly predating speech, as Darwin thought) and human brain expansion; 4—music is connected to body movement in much the same way that verbal speech is tied to body language; 5—just as *homo sapiens'* language development had a strong *social* component (group solidarity etc.), music also probably served an important social function (Levitin, 2008); 6—musical instruments go back at least 35,000 years (Conard et al., 2009) and most probably to the beginnings of "modern" humanity around 50,000 years ago; 7—"there is no experimental evidence that any nonhumans show a preference for musical sounds" (Bloom, 2010, p. 123).[7] In short, it is extremely likely that music has been an integral element of the human condition from the very start of the human species and may even have been a significant factor in humanity's development.

However, during most of human history music was far more "real" than "virtual"—for two reasons. First, until the late nineteenth century, no technology existed for music recording (re-presenting music exactly as first performed); all musical performance was live and unmediated. Second, from time immemorial music was more heavily dependent on the human voice than on instruments that in most cases served as mere accompaniment to keep pitch or rhythm. An equally important factor underlying the centrality of voice was that music was usually performed/played within a religious context (normally ceremonial, but not exclusively), with the core part of the experience being prayer where voice had a central role to express the sacred message. For example, (King) David played the lyre but it was no accident that he was also credited with composing dozens of early Hebrew prayers (*mizmor le'david*: Songs of David) i.e., his instrumental playing was an adjunct of his vocal poesy/praying. Music among the ancient Hebrews/Israelites was played mainly during religious litanies i.e., ceremonial parades (Krahenbuhl, 2006), and during the Jerusalem Temple sacrifice service.

---

[6] All digital technologies are more virtual than analogue e.g., digital photography where multitudes of pixels ("picture bites") replace analogue's continuous picture, reproduced from a photographic (negative) plate.

[7] Birds and whales use sophisticated sound to communicate. Lacking speech language, they use "musical sound" to communicate. Humans, on the other hand, appreciate (and create) music for *non*-communicational purposes, although certain musical forms can send (or reinforce) a message, such as soundtrack music in movies.

With the Greeks one finds the first steps taken towards music as a secular, instrumental, and aesthetic phenomenon in its own right. Nevertheless, despite the use of instrumental music for theater performance, the vast majority of Greek music was vocal: *dithyramb* (choral song in honor of Dionysus); *enkomion* (praise for a person); *epinikion* (song for athletic or military victory); *erotikon* (love songs); *hymenaios* (wedding songs); *hymn* (praise for the gods); *hyporcheme* (song and dance during altar sacrifice); *paean* (song of praise); *phallika* (crude/lewd songs); *prosodion* (thanksgiving songs); *skolion* (banquet song); *threnos* (funeral song). All these were either sung by a single person (monodic lyric) and/or by a choir (choral lyric). Nevertheless, important steps were taken to greater musical virtuality through important theoretical and practical musicological advances paving the way for primarily instrumental music and the art as a more abstract or cerebral exercise [V12; V13].

The Catholic Church reverted to ancient form by sharply circumscribing instrumental music, viewing it as sensual and fomenting carnal (or non-spiritual) urges. Thus, Church music tended to be almost exclusively vocal (plainsong/chant morphing into Gregorian chant) at least until the later medieval period. Instrumental music didn't disappear—instruments were in relatively wide use, but mainly to accompany sporadic secular singing in the royal courts (medieval motets, canons, madrigals, chanson) and later the general public (minstrels performing music composed by troubadours) with flute, pan flute, recorder (all winds), lute, psaltery, zither, organ.

The one exception: dance music from the middle of the medieval period onwards—but here too instrumental music was the handmaiden of the main attraction: the dance. Such virtual, instrumental music represented only abstract sound [V23], but its *purpose* was tied to the corporeal human body in movement.

Vocal music's dominance began to decline around the late Renaissance period (1400–1600) into the Early Baroque (seventeenth century), due to three central factors. First, an evolutionary change from monophonic to polyphonic music, with a wider range of sounds, pitches etc., that the human voice couldn't easily provide. Second, technical advances enabled the creation of new or improved instruments e.g., clavier to piano. Third, the decline of the Catholic Church that constituted a bulwark against pure instrumental music. Secularism's ascendance, especially among the growing merchant class, educated people, and the nobility, created a growing audience for music divorced from the human voice. Instrumental music did not *replace* vocal music; an entirely new, popular art form—opera—emerged at around the same time! Rather, instrumental music joined other "high culture" musical art forms e.g., Bach wrote hundreds of voice cantatas (and dozens of Masses) but also composed an equally huge number of concerti, sonatas, and other purely instrumental pieces.

By the late eighteenth century onwards (Haydn, Mozart, Beethoven) instrumental music had gained primacy with large-scale orchestras playing major symphonies, concerti, tone poems and other modern genres. Nevertheless, on occasion, classical symphonic pieces adopted the human voice for extra effect e.g., Beethoven's Ninth Symphony, fourth movement (Ode to Joy).

For the audience, music had transformed from something palpably "real"—the human voice—to an increasingly, although hardly exclusively, abstract art, divorced from objective reality.[8] In a word, music in the early modern era had become virtual to a significant extent [V23]. "Classical" (i.e., highbrow) music had become mostly divorced from any social-environmental fetters, other than technical ones (hall acoustics and the like). Such music had no "subject" and not even a "narrative," other than the purely subjective pleasure it gave the audience, and occasionally a subjective "reminder" of a real-life experience [V11].[9]

The modern age witnessed several new musical forms combining instrumental music and voice: rock and roll and its predecessors (rhythm and blues, folk music) as well as compatriots and successors (country music, rap). Movie and television background music are intermediate points—abstract and ancillary to the medium but serving an important function in the "language" of those media as emotional reinforcement of the message. On the other hand, jazz has (mostly) taken abstract music to new heights, not to mention so-called "modern classical" that has even moved away from traditional harmonics (Schoenberg's twelve-tone scale). Thus, the contemporary period is marked by combinations of reality-grounding (voice, lyrics with real-world meaning) and virtuality-oriented music (pure emotional or cerebrally abstract sound)—and many combinatory forms in between. In this regard, one cannot posit artistic dominance in the present day of musical virtuality over reality.

Nevertheless, musical virtuality has become dominant in two different *technical* ways. First, as mentioned earlier, music recording; the amount of time spent listening to music while physically removed from the performance [V3] far exceeds attendance at live performances, a situation not even possible until the late nineteenth century.[10] Thus, in the most extreme sense, despite the ups and downs over the

---

[8] This, notwithstanding a few tone poems like Debussy's "Afternoon of a Faun" or Smetana's "The Moldau [Die Moldava]" that were heard as literal (or at least very representatively figurative) reflections of a deer lazing around, and a mighty river coursing through Prague, respectively.

[9] A major controversy still rages among musicologists and music historians as to whether purely instrumental music should still be understood to have a "narrative" or not. Kivy (2009, p. 120) surveys the arguments pro and con, concluding: "absolute music... [was a completely new type of] instrumental music without text, program, extra-musical title, bereft of either literary or representational content... an art of purely abstract but perhaps expressive sound." This is true as far it goes, and it reinforces the contention that musical virtuality expanded and deepened with this new form of music. However, as anyone faintly familiar with classical music history knows, one must add several caveats to Kivy's conclusion: (a) some "absolute music" was given real-life "meaning" by the composers themselves e.g., Beethoven's original intention to dedicate the Eroica (Third) Symphony to Napoleon's heroics; (b) some of it was combined with lyrics that had clear narrative meaning, e.g. Beethoven's Ode to Joy movement in the Ninth Symphony; (c) post-"classical," *Romantic* music in the nineteenth century clearly tied much of its musical content to real life themes (national identity, military heroism etc.); (d) much absolute music has been "appropriated" by later media and turned into accompaniment e.g., movie music. (I was at a concert recently where Mozart's Piano Concerto #21 was performed; the minute the Second Movement was played, I immediately nodded to myself, instinctively saying: "Elvira Madigan," a movie that I last saw some 40 years ago); (e) humans tend to "seek meaning" even in things that are not there (e.g., Rorschach Test), so that from the *subjective* perspective of the listening audience, what may have been "absolute music" for the composer may not be heard that way by listeners.

[10] An historical curiosity: Alexander Graham Bell at first thought that his invention, called the "telephone," would primarily be used to listen to live symphony hall concerts from the comfort of the listeners' homes!

past several centuries of voice versus instrument, musical virtuality—disembodied music [V25]—is a relatively new phenomenon that eclipsed embodied music by the mid-twentieth century.[11]

Second, musical sound production itself is becoming more virtual through the technical use of synthesizers that can create "unnatural" sounds, as well as "mashups" (overlay, dubbing) that combine sounds in ways that were not possible in the past (Metz, 2017). Indeed, synthetic, sound-generating devices will probably transplant (almost) all musical instruments in the near future, offering music that is virtual in four ways: not representative of natural sounds [V10], created by a non-natural instrument [V13], digitally recorded [V6], and with the studio "production" spatially and chronologically divorced from the listening audience [V17; V18].[12]

## 6.4   Virtual Storytelling: Voice, Written and Visual Narratives—Oral Epics, Literature, Journalism, Theater, TV and Cinema

A need to tell and hear stories is essential to the species Homo Sapiens – second in necessity apparently after nourishment and before love and shelter. Millions survive without love or home, almost none in silence; the opposite of silence leads quickly to narrative, and the sound of story is the dominant sound of our lives….

(Reynolds Price, 1978, p. 3)

One need not argue that storytelling is more fundamental than shelter and love, but inarguably it is quite important (and probably critical) to the proper functioning of humanity, individually and collectively, as can be discerned from the numerous aesthetic forms developed over time to express the storytelling urge. Storytelling, then, is the foundation underlying many forms of artistic expression, listed in the above subtitle. Despite the many technological and artistic differences between these "media," and despite the vast differences in historical time that separate them, they all have the same purpose: to tell a story, imaginative or real. Of course, some of these media are not employed solely for storytelling (e.g., biblical moral proscriptions, scientific and mathematical works, philosophical tomes). However, most of human talk, writing, fact relating, acting, and performing involves storytelling in one shape or another.

How is storytelling a virtual exercise? First, insofar as the story told departs from factual reality it becomes substantively virtual. In this sense, storytelling can run the gamut from the highly realistic (retelling with great exactitude and detail something that happened [V1])[13] to the completely virtual, totally imaginary narrative [V28,

---

[11] Radio was, and continues to be, a major contributor to the disembodied virtualization of music consumption, through *live* musical performances broadcast in the past, and in recorded format today.

[12] For a more extended survey of the virtuality of music, see Whitely and Rambarran (2016).

[13] There is almost no possibility in principle of recollecting and/or retelling a true event or ongoing story with *total* verisimilitude; human memory is not capable of "perceiving" everything that occurs

V35]. This pertains to fiction, and to narrative storytelling in print and the electronic media: documented (prose) history, documentary (aural and visual) history, and journalism. Again, such "virtuality" can be viewed objectively (how far is it from real truth?) as well as subjectively (how much does the audience think it is real?). As the 2020 U.S. presidential election results showed, this distinction is not "theoretical" but goes a long way in understanding the phenomena called post-truth, faction, truthiness, and alternative facts. When storytelling surprises—whether journalistic news, campaign oratory, or historical documentary—it goes against the grain of conventional wisdom (or even "fact"), being perceived as *subjectively* real despite its objectively virtual nature.

Second, as with vocal versus instrumental music, here too the greater the separation of the narrative expression from the human body and especially voice, the more it becomes (in)corporeally virtual [V25]. The advent of writing, then the invention of the alphabet and eventually the development of print, were all stages along the path of greater storytelling virtuality in the physical sense. Nevertheless, other forms of storytelling maintained physical presence (e.g., theater), although as time went on it offered greater virtuality through more sophisticated forms of "deception": stage sets, costumes, and makeup [V5, V6, V7]. However, even this traditional, visual, narrative form (from at least the time of the Greeks) eventually metamorphosed into something more physically virtual (i.e., distanced), as cinema, radio, and television became the dominant contemporary media for the art of narrative performance.

The wide range of narrative art forms in oral-based as well as literate societies strongly suggests that storytelling is innate and intrinsic to *homo sapiens*; its presence among all ancient and non-literate societies provides absolute proof. First, the existence of storytelling in primitive societies that still exist today provides a window into the practice of our forebears. Second, ancient paint art was somewhat narrative, probably used as a memory-enhancing device to enable the storytellers to hew to the original story as handed down over the generations. Third, in some of the earliest written stories the author states that it is based on an oral tradition; indeed, many oral myths that were written down continued to change or be reworked as time went along: "[T]he Greeks had no 'Book of Greek Myths'; they just kept tampering" (Mendelsohn, 2010, p. 74). Finally, modern methods of literary and linguistic analysis can identify oral traditions within ancient texts (Ingemark, 2007, p. 280).

Why do humans seem to be predisposed to storytelling? (No animal does this.) And what are the specific benefits that it provides?

Cognitive psychologists speculate that storytelling serves a prime social-survival skill in enabling us to hone our innate talent for "mind reading" our social partners.

---

at any one moment, and even if such sensory perception were possible we have to "make sense" of it all and thus "select" those details that most fit our understanding of what (is important to that which) occurred. In this sense, *all* storytelling must be partly virtual [V4] by the very nature of our limited sensory, perceptual, and intellectual apparatuses.

Humans have the almost exclusive[14] cognitive ability to hold a "theory of mind"—
we can interpret other people's mental state and comprehend their intent vis-a-vis
others e.g., "Peter said that Paul believed that Mary liked chocolate" (Cohen, 2010).
To accomplish this, we developed storytelling, the quintessence of mind-reading:
we read how internal thinking, and external behavior as a product of that thinking,
all work together in others within an almost infinite number of social situations and
interpersonal permutations.

Storytelling, then, is a major expression of virtuality—not only because the story
describes something that is not presently seen [V9], but also in that it enables us to "get
into the mind of others," substituting the protagonist in lieu of a real person [V6]. This
is useful because in our life it would be impossible to encounter every single type of
social situation and challenge. Instead, we either run imaginary "verbal simulations"
[V10, V28] of other possibilities, or we re-simulate situations that already occurred
in the past [V9, V11]. In either case, this enables us to "practice" in advance of
encountering most of the probable, social, "mind-reading" situations that we may
face in the future.

This is not merely something that we *must* do; probably because we are fairly
good at this, it is also something that we *enjoy* doing. First, we hone our mind-reading
skills through verbal simulation. Second, we enjoy playing the "mind-reading" game;
such virtual narratives surprise us with their novelty (man bites dog), and they can
also enlighten us with the storyteller's solutions, through plot twists and surprise
endings. Third, stories tend to have important moral messages that further social
cohesion. One of the conundrums of evolution is how *social* altruism developed if
the prime purpose of every organism is to ensure its *own* survival. The answer: genes
work not only for the individual's survival but also for the survival of closely related
genes ("relatives").[15] However, this form of altruism needs reinforcement; after all,
the genes in my body are 100% mine, and in my mother or my child only 50% mine,
and in my sister's body possibly a different mixture of mine even if we have the same
genetic parentage. Ergo, storytelling reinforces altruism—along with other socially
useful behavioral traits.[16]

It is worth noting how storytelling involves a very intricate dance between virtuality and reality, in this case between gene and meme. Narratives are cultural creations

---

[14] Higher primates probably have a rudimentary ability to "read other minds," guessing what their
peers are about to do based on their present actions. However, this is probably not a *conscious*
thought process, and certainly animals don't "think about thinking." See Corballis (2011) for a
comprehensive analysis of human "recursive thinking": theory of mind, mental time travel, and
embedding structures within structures.

[15] This generally accepted theory for the past half century is now under attack. For a balanced
survey of both scientific camps, see Zimmer (2010).

[16] It should be noted that this evolutionary development explanation is useful for literature but not
necessarily for music or most of the craft arts (sculpture, jewelry, body art, fashion clothing). There
are a few exceptions to this rule e.g., a single totem pole usually "relates" the historical story of
a specific family or tribe. Painting, too, on occasion can offer a narrative, mostly when there are
several scenes portrayed within the same product or enclosed space (e.g., triptych; cave wall art).
Nevertheless, even if they can grossly symbolize states of mind, they are no match for the power of
narrative storytelling inherent in literature, theater, and the like.

whose purpose is to help genes (our body) survive and thrive. One does not need to go as far as Blackmore (1999) who contends that memes (specific idea fragments) compete in a form of cultural natural selection process; suffice it to argue that memes are a uniquely human device for transmitting information aiding us in our personal survival and procreative enterprise.

How do we go about narrating? What precisely happens when we are involved in storytelling? I refer here not to the technical-sensory process of forming words and sentences but rather to the psychological aspect of where "we are" when we hear, read, see—*experience*—storytelling.

At the extremes, the two polar psychological states can be termed "distance" [V2] and "immersion" [V3; V21]. In the former, the reader is constantly aware for several reasons of being "here" and what transpires in the artistic product is "there." I shall use literature for illustrative purposes, but this is valid for any narrative form or medium.

First, the novel may not be written well, detracting from the reader's illusion of taking part in the story, or at least "being there" while it transpires. Second, the author can do things that force the reader "back to reality": the narrator writes self-reflexively about the story ("I probably could have said that better") or addresses the reader directly ("quite a strange happenstance, don't you think?"). For example, as Ryan (2010, p. 4) noted: eighteenth century literature had "a playful, intrusive narrative style that directed attention back and forth from the story told to the storytelling act."

However, in a master's hands this technique can enhance the theme, if not the feeling, of virtuality. For example, several of Shakespeare's plays consciously and self-reflexively play on the idea of life's virtuality. *The Tempest* shipwrecks people on a mysterious isle where everything is different because magic has taken over; the isle's owner suggests that everything is a mere mirage [V20; V31], for "we are such stuff as dreams are made on" (*The Tempest*, Act 4, Scene 1, lines 156–157). In *As You Like It*, folks escape to the Forest of Arden to change their social status, their names, even their sex [V6; V32], prompting the forest-philosopher Jaques to note that "all the world's a stage, and all the men and women merely players" (Act 4, Scene 7, lines 139–140). And then there's poor Hamlet who must work with actors who will play parts connected to those played by the real actors on the stage [V1; V6], as if he's a real person trapped inside a play (Bloom, 1998, pp. 383–431)! These might temporarily break the distancing spell of the play for the audience, but they also strengthen the idea of art and life as a virtual enterprise.

A third cause of distancing is when the narrative chronology is so jagged that the reader has a hard time figuring out *what* is going on, and *when* (although this might be aesthetically interesting, it undermines reader comprehension). This is a problem, especially for non-fiction (i.e., news), in the age of hypertext, where the flow of reading is constantly interrupted by the author's links that lead the reader to become sidetracked from the main narrative. Fourth and finally, when illogical things happen within the premise of the story.[17]

Such cessation of the narrative's logic weakens the virtual experience because the audience is made aware of standing apart from the story. Of course, the story can be virtual at different levels [V7, V8, V9, V28, V32, V35] but with too much "distance" the *experience* of the reader becomes non-virtual. To *subjectively* feel the fullness of the virtual experience, psychological immersion must take place.

Immersion as a psychological concept, the strongest feeling of "presence" [V3], has become a hot topic, especially regarding 3D and IMAX cinema, video games, and virtual reality. Literary immersion is no different: "[T]he complex mental activity that goes into the production of a vivid mental picture of a textual world. Since language does not offer input to the senses, all sensory data must be simulated by the imagination.... In the phenomenology of reading, immersion is the experience through which a fictional world acquires the presence of an autonomous, language-independent reality populated with live human beings" (Ryan, 2001, pp. 11; 14). In commonplace language, reading immersion is getting "lost in a book" (Nell, 1988); being transported elsewhere (Gerrig, 1993); mentally simulating another world (Walton, 1997).

This is almost never a case of *total* "transport." As Birkerts (2010) notes: "Where am I when I am reading a novel?... I am never without some awareness of the world around me—where I am sitting, what else might be going on.... I occupy a third state, one which somehow amalgamates two awarenesses...." Substitute "narrative" for "literary" in this paragraph, as well as "watching, hearing, experiencing" for "reading,"[18] and one can conclude that all good "storytelling" will subjectively (and temporarily) transport us from reality to semi-virtuality: we are psychologically "there" while remaining physically "here."[19]

Given the nature of the phenomenon, narrative immersion clearly existed in previous eras: book reading; listening to a spellbinding storyteller. Thus, the digital age does not offer a *qualitative* novum regarding literary immersion, as underscored in Ryan's book title *Narrative as Virtual Reality: Immersion and Interactivity in Literature and Electronic Media* (2001).

Nevertheless, something has occurred recently to significantly alter the dynamic of "distance" and "immersion": *interactivity*, or *prosumption*. Interactivity entails

---

[17] Science fiction is no different. Despite "aliens" and other foreign objects and phenomena, each story should have its own internal logic within the "strange" world created by the author. The fantasy genre can suffer a similar fate.

[18] Ryan goes on to expand the phenomenon beyond literature: "This fundamental *mimetic* concept of immersion... applies to novels, movies, drama, representational paintings, and those computer games that cast the user in the role of a character in the story, but not to philosophical works, music, and purely abstract games such as bridge, chess..." (2001, p. 15). Of course, the *degree* of immersion is not the same with every medium. For example, theaters are darkened to limit the many possible distractions, especially by other audience members.

[19] Ryan offers four levels of immersion: concentration, imaginative involvement, entrancement, and addiction. It is only with the latter that the reader is totally immersed, losing all sense of external reality (2001, p. 98). Moreover, immersion can have three objects of involvement: spatial (the setting), temporal (plot), and emotional (character) (2001, p. 121).

some sort of active engagement of the audience with the storyteller, whereby the former influences the latter's narrative.[20] True, this existed from time immemorial as storytellers worth their salt were attentive to the audience's mood, and if the latter were bored or otherwise distracted, the former would respond through raised voice, changed intonation, more humor, and the like (Ryan, 2001, p. 19). However, mass storytelling was invariably a unidirectional activity, and even more so when some of it ceased to be offered in live presentation (books).

This has undergone profound change in the recent past, with digital storytelling (Hartley & McWilliam, 2009) enabling the audience to become part of the storytelling production process, ergo "prosumption": producer and consumer simultaneously, or what some call the "wreader" Ryan (2001, p. 9). This is expressed in "citizen journalism," amateur YouTube video clips, social media rumor mongering and disinformation, musical mashups, and even on occasion in participatory-audience-authored literature (Saunders, 1999, pp. 117–118).

What does this mean for virtuality in storytelling? On the one hand, if two heads are better than one, then many imaginative minds can be superior to one creative mind. Thus, prosumption will tend to make the *substance* of storytelling more imaginative, novel, and unusual, than the traditional route of one "author" or several (e.g., theater and cinema have always involved team performances, including playwright or scriptwriter, stage or set director, prop/makeup/costume artists, and so on). Yet whether single or multiple-authored, the producers were mostly or all "professional," limiting the number who would participate in the process of artistic production. The age of prosumption and interactivity widens the net to include *all* audience members, thereby increasing the number (but not necessarily the proportion) of narratives with highly offbeat and occasionally unique characteristics.

On the other hand, interactivity significantly increases "distance" (between creators) while neutralizing audience immersion, at least in most forms of contemporary narrative. It is hard to be immersed in a novel when you are asked to self-reflexively think about how the plot might evolve. Thus, paradoxically, the cognitive effort demanded of interactivity in storytelling reduces our emotional immersion within the story framework. The one exception: the broad category of video games and artificial worlds in which the game's progress is solely up to the player, despite the underlying rules laid down by the original creators. However, this is the exception that proves the rule, as video games are less "storytelling" and more "life re-enacting" i.e., another "platform" for us to live out several hours of alternative life [V6; V10], much as we have the relative freedom to do in our own "real" lives.

This brings us to the last issue related to narrative and virtuality. Can one discern a "progression"/expansion of narrative virtuality (however jagged) through the ages,

---

[20] There are commentators who include a "weak" sense in their understanding of "interactivity" e.g., a reader who "engages" with the narrative and brings a personal interpretation to the text (Ryan, 2001, p. 17). I do not consider this to be "interactivity" in the main sense of the term. Nevertheless, Ryan does add two other levels of interactivity that are useful, what she calls "weak" and "strong." The former entails the reader choosing from several predefined alternatives (as in a video game), whereas the latter demands of the reader/surfer/player/etc. to participate in the physical production of the text (Ryan, 2001, p.17). It is mainly "strong interactivity" that I discuss here.

similar to music and religion? Or perhaps, as with interactivity, there's been a regression?

From a *substantive* (contents) standpoint, storytelling's historical thrust is not altogether towards more virtuality; fictional literature has been "weakened" in virtual terms by modern society's incredible technological advances. If in the past, narrators used their imagination to create non-existent "worlds" [V34] (and accompanying elements), modern science and technology with their amazing discoveries and inventions have circumscribed the possibilities of the "virtual imagined" [V28]. Moreover, the myriad imaginative ways of describing that fictional world have also been narrowed by ever-increasing, information-gathering technologies. For example, Borges' 1941 phantasmagorical short story *The Library of Babel* about a "universal library" would not have been written today in the "Age of Google," itself a near universal library.

In short, truth may not only be *stranger* than fiction, but also *stronger* than fiction. The more our real world "expands" the area of the possible, and the more the average reader has access to that expanding reality, the less hold virtual storytelling will have on consumers of narrative cultural products.

On the other hand, if real-life "truth" has supplanted large swaths of fiction, it has also opened new vistas for fictional treatment, for the more we know the more we know how much we don't know. It is within this *new* "don't know" that much fiction today continues to offer imaginative escape into strong (i.e., unfamiliar) virtuality [V35]. This explains a seeming paradox: we would expect science fiction to flourish in an age where science has not provided many answers—but historically quite the opposite occurred: it was only in the modern age, after modern science had already begun to divulge nature's secrets wholesale, that science fiction became a large-scale genre of popular culture. The more scientists discovered about our world, the more bizarre it became (e.g., quantum mechanics)—opening vast new tracts of this "undiscovered world" within which fictional narration and the literary imagination could flourish. Thus, it is far too early to eulogize storytelling as a central mode of virtualizing.

Regarding storytelling "expansion," there are two additional ways of looking at the question: a—time spent in such activity; b—the variety of means through which such activity has been expressed. In both, we can discern an expansion of virtuality, with one interesting caveat.

Although somewhat speculative, based on anthropological work of hunter-gatherer tribes living today, we can surmise that pre-agricultural human society had a large amount of leisure time, of which most was spent storytelling and gossiping[21] (Lieberman, 2021). Leisure time decreased radically for most of humanity with the

---

[21] Person-to-person storytelling, generally called "gossip," can be included as part of our "narrative" analysis. Assuming that human nature has not changed much since early human time, there obviously hasn't been much change in this regard. Nevertheless, the technological expansion of media—mail, telephone, and now social media—have enlarged the *means* by which we can receive more personal information from others, and also "tell stories" to others. Moreover, the enlargement of our social units over time has enabled us to encounter more people; the greater the *number of acquaintances*, the more opportunities for gossiping.

advent of agriculture that demanded (most of the year) non-stop daylight work. Widening trade and commerce in the early modern era, accelerating in the late nineteenth century with the reduction of manufacturing workhours, and workdays in the work-year, gradually returned leisure time and with it the opportunities to "consume" more storytelling culture.

Storytelling expansion was also driven by technology. Writing was not revolutionary (at first), given the very small number of literate people in ancient societies. The main leap forward occurred with the introduction of the printing press in Europe, enabling an increasing proportion of the population to consume "stories" (fictional or factual) in exponentially greater numbers than heretofore as a result of a sharp increase in literacy and increasing economic wherewithal (due to far cheaper prices of print) to buy reading material. Increasingly more people gained access to virtual worlds of others' creative imagination—leading to a further rise in number of authors who could now make a living from what heretofore had been a pastime for the rich or an avocation for the priestly class.[22]

An even greater leap occurred with the invention of electronic communication, especially cinema, radio, television (including video), and the internet. These further widened the number of people with access to storytelling, especially in the Third World where illiteracy ran high. Anyone can listen to the radio and watch television—including a very sizable part of the population that heretofore was completely dependent on others' help in accessing storytelling: pre-school children! True, parents could read books to youngsters and even tell them bedtime stories but the introduction of radio and especially TV changed the calculus immeasurably. This development was not necessarily for the better from an educational standpoint, but television certainly increased the amount of time youngsters were exposed to "professional" storytelling beyond their own imaginative play.

Moreover, the *quality* of virtuality intensified for the reader as compared to what the oral storytelling audience experienced—for reasons of greater disembodiment. As already noted, almost all storytelling before print (and certainly before writing on papyrus and parchment) demanded the mutual presence of speaker and audience. Conversely, most modern storytelling is consumed in "virtual space" [V17]—a meeting of minds, perhaps, but without the previously essential meeting of bodies [V2; V3] and not even the physical presence of (most) other audience members.

Distanced dis/embodiment also raises the question of whether unmediated oral narrative, or mediated written/performed/recorded storytelling, can be considered more virtual from the perspective of storytelling transmission *between generations and eras*. The answer is complex. On the one hand, writing and print hugely expanded such "vertical" or "historical" communication between those alive in the present and the deceased in previous generations. As Descartes put it: "…the perusal of all excellent books is, as it were, to interview with the noblest men of past ages" (1637, p. 6). This is an almost banal point, but from the virtuality perspective it reaffirms that

---

[22] As the title of a recent article suggests—"The medieval nuns' guide to virtual travel"—nuns spent a lot of their leisure time reading very numerous and popular pilgrim(age) diaries and journals, in highly immersive fashion (Merin, 2021).

humans are the only earthly creatures capable of such multi-generational communication (animals can learn from the previous generation, but only when the latter are alive, transmitting the "lesson" in real time, "face to face"). In sum, more modern forms of narration have immeasurably strengthened virtuality *quantitatively* as well.

However, one can also argue the reverse. Oral communicators could not *directly* transfer their story to future generations, but rather had to rely on each succeeding generation to do so through replication—open to distortion. Given enough generations, this led to extreme distortion, *a la* the children's whispering, mouth-to-ear game "broken telephone." Indeed, "distortion" ("entropy" in scientific parlance) was viewed in the ancient world as "re-creation." In a world devoid of "copyright,"[23] each successive storyteller was expected to add or embellish, and certainly no one took umbrage at the lack of duplication perfection. In essence, this meant that the narrative heard by any specific audience at any point in time was significantly "distorted" [V5] in relation to the original version. True, copying errors did occur during the reprinting of books, but such "revisions" were several orders of magnitude *lower* than orally retold narratives. The narrative you read and the film you view today, written by an author and produced by a director 80 years ago, is (almost always) the identical version of the story.

In short, there has been a huge numerical expansion, in audience size as well as production of creative narratives, regarding storytelling virtuality in modern, and especially contemporary, society. However, this has come at the "expense" of narrative fluidity over the generations, as greater replicative capabilities have concretized each story in its original state.

## 6.5   Painting, Sculpture and Other Visual Arts[24]

The visual arts include many types of expression: photography, sketching/cartooning, canvas painting, mosaics, friezes, collage, 3D animation, sculpture, and others. The common denominator here is the attempt to aesthetically (albeit artificially) "replicate" some aspect of reality [V7; V8; V10]. A central difference between them is dimensionality: from non-dimensional painting using only evanescent light [V25] (Rose, 2017), to absolute two-dimensionality (sketches on paper) and "expanded" 2D (thick oil painting) [V8], to hybrid "2D on 3D" (vase painting), through "enhanced" 2D (mosaics, friezes), all the way to "minimal" 3D (collage) [V7], and finally to

---

[23] Because it was frightfully time-consuming and expensive to make (rewrite) even one copy of a book, any such "copying" was the highest form of flattery and only aided in getting wider distribution of the original author's ideas/stories.

[24] Functional "crafts" are not included here because they are very primarily "material." Of course, a ceramic bowl can have graphics, or a picture, painted on (or in) the bowl—but that element is "art" whereas the bowl itself (wholly corporeal) is "craft." Nevertheless, almost all human craft has some limited amount of virtuality in their "artificiality" [V13] and occasionally, non-natural form [V5; V22].

complete three-dimensionality (sculpture) that at times approaches verisimilitude [V1].

From the standpoint of virtuality, all of these are differentiated in one important respect from music and storytelling. Whereas the latter two at least *originated* as completely non-virtual activities (use of voice and human body), the visual arts from the start contained a measure of virtuality in four different ways. First, painting and sculpture are virtual in that they either try to portray the *external* world in a form that must (by its artificial nature) depart from the real world [V4, V5, V7, V8, V13], and/or they express the artist's *interior* (i.e., emotional) world, subjective by nature [V28]. This regarding "classical" art; more modern "conceptual" art, and other art "isms" (Cubism, Dadaism, Surrealism, Expressionism, etc.), are obviously even more virtual in their purposeful avoidance of "objective" reality [V22; V23].

Second, for most of history, visual art has had to contend with the problem of representative accuracy: given the lack of technically advanced tools for high level artistic work, art (including sketching, drawing, and painting with different media) was incapable of representing reality in perfect fashion, even when the artist wished to do so.[25] Third, most ancient (and medieval) visual art was heavily religious in subject matter, so that the *objects* portrayed tended to be virtual as well, in light of their non-corporeal existence in the real world. Of course, *subjectively* the audience might not have considered this to be virtuality, as the audience's "visualization"[26] was based on a belief in the reality of the objects portrayed, such as angels, God, Satan, mythological figures, and the like.

Fourth, as Wertheim (1999, p. 87) notes, "earlier medieval [pre-Renaissance] art must be understood as essentially *conceptual*. Gothic and Byzantine art had been attempting to convey an immaterial conceptual order, but the new naturalistic art of Giotto's time was specifically attempting to convey the *visual order* seen by the eye." Post-Roman European art made little attempt to copy visual reality; artists painted objects as they felt their essential nature to be [V30], trying to capture the soul (i.e., the Ideal) of their subject matter. From their perspective, this was *non*-virtual art, but from a modern standpoint it was certainly virtual due to its lack of verisimilitude as we understand the term today—indeed, quite like "conceptual" art (discussed below).

None of this suggests that the visual arts, especially painting and sculpture, followed some linear progression from "primitive" (highly virtual) art to "sophisticated" (less virtual) art. Greek art (sculpture especially), in part relying on advances in geometry as well as the development of outstanding technical skills, tended to have a very high level of representational verisimilitude. Indeed, the Romans themselves viewed their own art as being inferior to Greek art. This "decline" in representational exactitude accelerated with the demise of the Roman Empire and did not regain the heights of Greek artistic glory until Europe rediscovered Hellenistic culture during

---

[25] I am not claiming that "representational" art is "superior" to its non-representational counterpart, but rather that the extent to which art diverges from representing reality renders it more *virtual*, whether the images' "non-representational" form is intentional or not.

[26] Visualization is "the role of viewers in generating what they see in images" (Burnett, 2004, p. 14).

the High Middle Ages, eventually surpassing the Greeks during the late Renaissance with the discovery of depth perspective [V7].

Moreover, the subject matter of Greek art tended to be more secular than religious, more oriented to the "real-world" than "other-worldly" in character, although the Greeks too portrayed their share of mythological deities and heroes. This obviously changed during the Christian era when most art—painting, sculpture, objects—was religious in nature, although most of these were quasi-historical characters portrayed in somewhat mythological (or divine) guise. Here too the Renaissance constituted a watershed, as increasingly secular themes, objects, and people were painted and otherwise represented in artistic fashion: still life, landscapes, common folk, royalty, historical events, and so on [V8]. This continued apace, as religious subject matter declined and secular themes ascended, employing the highest level of technical skill and representational verisimilitude (without any further "progression," given the incredibly high level of artistic *skill* of Michelangelo, Rembrandt et al., already at the start of the modern age).

Then, from the mid-nineteenth century onwards, *representational* art underwent a major revolution: no longer attempting a true representation of visual reality but rather personal expression of reality's "*essence*" as felt or understood by the artist—similar to medieval, religious art that sought the "inner essence" of the object. Art in the late modern period became "virtual" in distancing itself from representing the external, "objective" world; instead, by the second half of the nineteenth century (Impressionism and Expressionism—precisely as the words suggest), painting focused on expressing internal, subjective perceptions of the world, and in the twentieth century (Surrealism), of internal emotional feelings without any connection necessarily to the world-at-large or *against* the world-as-is (Dadaism).[27]

The height of late modern virtual art was reached by three different streams. The first was almost *sui generis* and a solo effort: Escher's art as "impossible reality." Here we have art that at first glance looks realistic (not real), but on second glance is physically impossible [V35]. The second was more influential: Abstract Expressionism, whereby visual art became completely disconnected from anything recognizable in the world (e.g., Jackson Pollock's "action painting") [V23]. Finally, starting in the 1950s–1960s Conceptual Art began to "dematerialize" art (Osborne, 2002) by removing the need for realistic representation, paradoxically placing the entire emphasis of "visual art" on ideas (concepts to be visually conceptualized) [V24] or "performance" [V10].

Parallel to these developments in "high art," the twentieth century also produced another art form that was highly popular with the general public: comics and cartooning. These were virtual in a different way: the characters tended to be anthropomorphized animals (Mickey Mouse; Bambi) or people with super-powers

---

[27] One can find a very few unusual examples in the very early modern period of such "virtual," non-representational art. For example, *The Ambassadors* painted by Hans Holbein the Younger in 1533 has a strange "anamorphic projection" (Tomas, 2004, p. 14): "There is a strange distorted form that occupies an area of its lower surface, as if the painter's brush had slipped over a wet paint.... Holbein has managed, in a singular manner, to combine and oppose antithetical spaces and metaphysical domains on one surface."

(Superman; Spiderman) [V5; V6]. This was not pure "visual art," but rather combined the visual with the textual—almost a throwback to ancient times where art and literature (not necessarily simultaneously) portrayed humanlike gods or godlike humans with superpowers (Hercules; Achilles).[28] Moreover, by the 1960s this visual/textual genre began to influence and infiltrate "high art" as well: Lichtenstein, Warhol et al. with their Pop Art. In any case and notwithstanding possible roots, comics became a worldwide phenomenon, simultaneously in distant lands: Japanese manga graphic novels, Asterix in France, etc.

Finally, one other type of significant art is doubly virtual: "counterfeit art." Ancient Roman artists copied Greek sculptures, but most probably sold them as duplicates [V1] and not as faked originals [V32]. Later, during the early modern period, "inauthenticity" in some cases became an integral part of the artistic process! Many famous painters had studios where apprentice artists would "fill in" the blanks or areas on the canvas where the master artist felt it needed only proficient hands to complete the work [V4; V6]. There are dozens of paintings today "by" masters, of which we are unsure whether painted in whole or in part by the master, or merely under that artist's supervision.

Perhaps even more vexing (artistically, not to mention morally) were forgeries of another's work by artists whom we today consider to be of the first rank [V6, V32] e.g., Michelangelo who sculpted a sleeping Cupid figure, treating it with acidic earth so that it would look ancient (Art forgery, 2021). Today, with the massive commodification of art, forgery has become big business to the extent that museums have been fooled, and several may even be complicit regarding the phenomenon. As a result, except for those pieces whose provenance can be completely proven, almost all "real art" today has a whiff of virtuality from the perspective of their "authenticity" [V15]—regardless of their other virtual aspects.[29]

Can one claim that contemporary art is more virtual than in previous eras? No conclusive bottom line can be offered because the visual arts today are so varied that one cannot make any "general" statement as to the "ruling" style, form, aesthetic approach, and so on. For almost every impenetrable artwork or super-minimalist sculpture installation, one can also point to a hyper-real painting or a wildly popular,

---

[28] It is not coincidental that the creators of Superman, Jerry Siegel and Joe Schuster, were Jewish-Americans (i.e., cultural outsiders, sons of refugees), and that the Superman story strikingly follows the story of Moses: born with the name Kal-El (in Hebrew: a light, or mini-God; perhaps the authors meant "Kol-El" which would be "Voice of God" in Hebrew); sent off as a baby in a small "basket"; adopted by a foreign family; turns into an extremely mild-mannered adult; becomes a fighter for justice against his intrinsic nature; and so on. For a survey of the connection between American comic book super-heroes and the Jewish past (and present), see Fleischer (2006), Fingeroth (2007).

[29] There is a parallel phenomenon between visual art forgeries and "fake jewelry." As with painting, "faux jewelry" (otherwise called imitation jewelry or costume jewelry) existed at least as far back as ancient Rome where glass jewelry was called *mandacio vitri* (lying glass). Nevertheless, only in the contemporary "Age of Virtuality" could a serious exhibition of such artificial jewelry take place—fake jewelry as an art form (Gomelsky, 2010)! In fact, the reason these pieces survived through the centuries is that their basic component sections and materials were not worth enough to bother separating and selling (or melting) in pieces. The multiple levels of virtuality here are difficult to disentangle...

kitschy landscape. Indeed, art has become so universally approachable—poster reproductions, museum exhibits, et al., from every historical period—that art is no longer something *apart* from, or accompanying, the real world but rather an integral *part of* our world (interestingly, "*art*" is the root of both words). Thus, the artistic *object* is no longer something we view as virtually distinct from the "real world," but at the same time it heightens our connection to the virtual by preserving and portraying for us worlds that have long disappeared, enabling us to retain more virtuality from the past.

## 6.6   Conclusion: The Unreality and Increasing Virtuality of Artistic Expression

To speak of the "arts" is to deal with an incredibly varied field of human endeavor, most of whose elements are found in all present and (to the extent technologically feasible back then) past societies. Another constant of the arts over time and place has been its dynamic flux. As seen in this all-too-schematic survey of the arts throughout human history, artistic evolutionary development is almost universal as well. True, the pace of artistic change has accelerated in the more recent past; among ancient human civilizations we can discern change *ex post facto* over hundreds of years, not the decades or even fewer that have marked the past century or so.

Throughout all this artistic fluidity and development, can any other generalization be advanced? Yes: one is universal; the other almost so. First, the emphasis of the arts on virtual experience:

> Our main leisure activity is, by a long shot, participating in experiences that we know are not real. When we are free to do whatever we want, we retreat to the imagination – to worlds created by others, as with books, movies, video games, and television… or to worlds we ourselves create, as when daydreaming and fantasizing…. Studies in England and the rest of Europe find a similar obsession with the unreal. (Bloom, 2010, p. 155)

Given what we know of the arts (and other leisure experiences) in the past, the same held true back then as well, albeit the nature of "unreality" was expressed artistically in different ways—especially, but not only, through religion.[30]

The second generalization, albeit somewhat less universal, is that artistic forms of expression have tended to move from the mildly virtual to the strongly virtual over time. Again, not in every art form, and not in any "ineluctable" straightforward progression, but viewed from a macro-historical perspective, one can certainly see evolutionary development in the direction of increasing virtuality.

Several factors were behind this. First, technological: the arts started out as something intrinsically tied to the human body: singing, dancing, storytelling. Over time, as human society became more technically sophisticated, several technologies were borrowed, adopted, and adapted for the realm of artistic endeavor. Media by their very

---

[30] Some ancient world artistic expression sounds almost modern. The Hellenic/Roman world "enjoyed" scary stories about ghosts, monsters, and other forms of haunting (Felton, 1998).

nature "disconnected" the artistic creator/producer/communicator from the receiver/ recipient; the arts became virtual in the sense of a physical break between artist and audience. Technology also strengthened the virtuality of art in another related sense—it enabled a chronological divorce too. For instance, if during our early story-telling era the audience had to belong to the same "generation" as the teller,[31] once writing evolved the storyteller could transmit the exact story to later generations as well. The same happened with music in the modern era of recorded sound and so on.[32]

Just as important was the revolutionary development of the human brain that increasingly was able and eager to think in more abstract terms. We have already seen such increased abstraction in the previous two chapters, so that increasing "abstraction" in the arts is not only *not* surprising but almost "naturally" concomitant to developments occurring in humanity's general cognitive and intellectual milieu. If the Hebrews could conceive of an invisible, universal Deity [V19; V33], and the Greeks could talk about Ideal Forms and conjure up unseeable, minuscule "atoms" [V19], then it should not surprise us that the arts at some point would also develop ways of "re"presenting the world in imaginary forms that bear little relation to what our bare senses enable us to perceive as "reality" [V5; V22; V23; V28]. In some of the arts, technology enabled this to occur with greater ease. For example, the human voice alone could not develop a full-fledged, *non-vocal,* musical art form (what we call "classical music") that made "unnatural sounds." The technology of musical instrumentation did enable us to do this; however, without the cognitive idea that it is worthwhile producing something not found in nature, technology by itself would not have led to the lyre, violin, and synthesizer.

Notably, as we have added more virtual forms of expression to the arts, we have not abandoned the less virtual types. Far from it: today we still flock to live vocal singing performances, love going to standup comedy shows, appreciate realistic sculpture, enjoy dance recitals, pile makeup (and tattoos) on our bodies, and so on. Nevertheless, for many different reasons to be pursued later in the book, the *amount* of time we spend today consuming the arts in their virtual forms far outstrips the

---

[31] Again, although stories were handed down from each generation to the next, each successive generation embellished or otherwise added to/subtracted from the original story, not to mention offered different intonation, emphasis etc. (all the techniques of rhetoric and oratory) so that the next generation could be said to be telling and hearing a "different" work of art. Not so with a piece of written literature that maintains its exact character from generation to generation (notwith-standing newer print editions that might introduce a few typographical errors or occasional editorial censorship).

[32] A similar, parallel evolutionary process can be seen regarding music and painting—from sacred to secular, from *re*-presentation-*al* to virtual/abstract—except the changeover occurs earlier in music than in art. This is mostly due to technological factors; it obviously was easier to produce better musical instruments than to develop better artistic reproduction technologies (the camera). Further-more, the emergence of instrumental music and the technical development of instruments were both a product of ideational changes in society—desacralization and secularization. Non-representational art, on the other hand, developed not so much in response to *ideational* changes in society but rather to the *technological* threat of the camera. Thus, musical virtuality was a product of changes in the *mind*; artistic virtuality (painting) was largely a product of changes in *matter*.

time spent on non-virtual art forms. This is not at all surprising, for as Dutton (2009, p. 104) glosses Kant: "when we appreciate the beautiful, we respond only to what he calls the *presentation* of the object—how it looks or sounds to us in the imagination, how it comes to us in, again, the theater of the mind." The virtualized arts, therefore, retain their aesthetic hold on us despite their physical divorce from the producing artist, or from the physical/"real" form in which each art was originally expressed millennia ago.

# Chapter 7
# Economics

> The real has died... an extermination... of the real of production
> and the real of signification.
> (Baudrillard, 1994, p. 7)

## 7.1 Introduction: The Historical Eras of Economic Activity

On the face of it, there is no area of life that is more intrinsically "real" than economics: trade, commerce, money, etc. Economics deals with the corporeal world: material goods and human-to-human services. Finding large doses of virtuality here, therefore, reinforces the general thesis of virtuality permeating most areas of life. For if virtuality is to be found in this most "concrete" field of human endeavor, where wouldn't it exist?

*Homo Economicus* did not start off with any significant form of virtuality. Mankind's economic systems can be divided into five consecutive eras: hunting/gathering (pre-history to approximately 8000 BCE), agriculture (8000 BCE to 1750 CE), manufacturing (1750–1950), service (1950–1990), information (1990–present). Each newer era is based on the largest cohort of workers in that period. For instance, during the Agricultural Era many people still subsisted on hunting and gathering, and a smaller number dealt in manufacture (i.e., artisans) and trade, religion, government, or education (service), but most workers toiled in the fields.

The first three eras were heavily marked by "material" goods. There was nothing virtual about the hunting and gathering life; our ancestors simply took what they could from the external physical world. In the next stage, farmers worked the land and grew food and other useful natural products (e.g., hemp for clothing). During the third stage, industrial workers also largely produced palpably corporeal goods. True, in the latter case many of these goods were no longer "natural" i.e., things found in

---

The *Appendix: A Taxonomy of Virtuality*—listing by numbers 1–35, e.g. [V13] or [V34], some of the various types of Virtuality mentioned throughout this chapter—is freely available to the public online at https://link.springer.com/book/10.1007/978-981-16-6526-4.

© The Author(s), under exclusive license to Springer Nature Singapore Pte Ltd. 2021
S. N. Lehman-Wilzig, *Virtuality and Humanity*,
https://doi.org/10.1007/978-981-16-6526-4_7

Nature, so that they were weakly virtual in their artificiality [V13], but *materially* they were and are as "real" as animals, berries, and grown food.

It is only in the twentieth century that the *brunt* of economic activity becomes virtual, although of course not all of it. Very roughly speaking, the Service Economy is divided into two main categories: physical and symbolic. The former category involves almost everyone who works face-to-face with other people: educators, soldiers/police, taxi/bus/train drivers, technicians, salespeople, (most) bank and government clerks, physicians and nurses, caretakers, and so on. On the other hand, a much smaller group at first—but growing into dominance by the end of the twentieth century (ergo: the Information Age)—includes workers such as journalists and news editors, TV and movie actors, financial analysts, auditors and tax consultants, government planners, graphic designers, book authors and editors, marketing and advertising professionals, meteorologists and environmental planners, scientists, librarians, air traffic controllers, software programmers and webmasters, and so on.

The fact that the essence of these professionals' work is manipulation of symbols—textual, numerical, aural, or visual [V24]—does not mean that such activity is completely virtual. First, almost all these workers employ material objects (laboratory, computer, scientific equipment etc.). Second, many work side-by-side *with* other professionals e.g., the journalist works with the editor, the research professor also stands in front of classroom students. However, when performing the core of their work (news writing, library research) they are working mostly *for* other people and not *with* them. The converse is also true—most physical service workers whose main job is to directly, and in some ways physically, help other people, also have a measure of virtuality in their work: teachers grade exams and papers, security workers file reports, technicians read and learn the instructions, etc. Thus, virtual activity is part of almost every worker's job in the contemporary world; the differences are a matter of (sometimes great) degree.

## 7.2  The History of Money: Virtualization of Economic Exchange

The type of economic era depended on social size. When the sole social unit was very small—up to 50 people, no more than the extended family during the very long hunting/gathering era—there was no need for anything other than a system of "Communal Sharing" (Fiske, 1992) of the limited bounty. What was mine was yours, and vice versa. Around 10–12,000 years ago humans began to "tame nature" i.e., agriculture, and this enabled the social unit to expand in size because now there was enough food within a relatively small territory to support larger numbers of people.

This was revolutionary from several related standpoints. First, people began to divide the land, so that the "communal" economy now gave way to a more "individualized" one with each farming family able to grow a different crop (however limited the possibilities), leading to "Equity Matching" (Fiske, 1992): you give me goat's milk and I'll give you wheat. This was still a completely non-virtual form of economy. However, a second consequence of such private domestication of the land (and animals) was that on occasion, when the environmental elements were fortuitous, each farm family—or society as a whole—could actually create a *surplus* of food. This surplus ultimately led to social and economic stratification. Some families managed to produce more grain whereas others grew and accumulated less.

What to do with all that surplus? With no sophisticated means of storage—especially for perishable goods—the solution was to trade surplus foodstuffs for non-perishable materials (perfume, spices, fine wood, metals). This was still Equity Matching, but an advance from bartering immediately useful goods. Now individuals could (for example) store the wood and build an extra house for their children.

A third consequence of the agricultural revolution is directly related to virtuality: the concept of private property. True, property itself back then was still completely "real," but the idea that it could belong to "me" and not to "us" added a layer of social virtuality: material goods could now be in someone's legal "possession" without being in their actual, personal possession! The full consequence for economic virtuality wouldn't be felt until the modern, capitalist era, but even in the mists of early, domesticated human history such a development was of great "philosophical" import as it separated the individual (and socio-political) body from a direct, physical relationship with what it "owned."

Another consequence of the growing social unit and introduction of surplus was that if not every person had to be involved in "survival"-type economic activity, it became possible for some to start "indulging" in non-economically useful activities e.g., artistic activity (poet, painter, jewelry maker), or economically useful technology creation. The latter sent human society into a spiraling virtuous circle: more technology made economic activity more efficient, freeing more workers from economic survival activity (food production) into thinking (education, "science," tools).

At some point, these technologies and the ever-expanding size of the social unit (city-state, "nation," and finally empire; see Chap. 8) led to another transformation in forms of exchange. Socio-politically, a higher level of stratification was reached; economically, this led to a system of "Authority Ranking" whereby higher-ranking people get better (and more) things than their subordinates, based on taxes or confiscation (Fiske, 1992, p. 700). Taxation is at least 5000 years old (Stahl, 2009, p. 80). At this point, of course, society and the economy had become highly differentiated between haves and have-nots, with the former accumulating more sophisticated "goods" (for an easier life and to display higher social status) that would retain their value over time.

Moreover, the evolution of city-states and larger socio-political entities, coupled with better means of transportation (roads, ships) enabled "international" trade between economies distanced from each other. However, such trade-at-a-distance [V3] deepened previous difficulties and produced new challenges: spoilage, volume, and weight. If a merchant sent a ton of grain on fifty camels to a distant trading partner, it was obviously cheaper or more efficient to sell all the grain and forty camels, and then return with only ten camels carrying some brass, gold, and silk—instead of returning with all fifty camels each carrying a relatively light load. From

2500 BCE, when Egypt and Mesopotamia began using gold and silver bullion as currency (Stahl, 2009, p. 80), and especially from 1600 BCE onwards when relative peace was the rule in the Near East for centuries, gold and silver bars became the norm for international trade (Quiggin, 1949, pp. 271–272). The value of the metal was still commensurate with its "intrinsic worth" i.e., the cost of mining and shaping it. Here "money" was still a one-to-one substitute for the original goods [V6].

Then, around 610 BCE in Lydia the first coins came into existence (Holt, 2021). These retained their intrinsic worth but now with the government's imprint a trader no longer had to weigh the metal—its value was clearly stamped on the coin. This currency form is called "representative money" [V22], although from a virtuality standpoint it maintained its "substitute/similarity" nature [V6].

This was the basis for "Market Pricing" (Fiske, 1992). Once differentiation of labor reached the point where few could self-sufficiently supply themselves with all necessary life goods, the system became overly unwieldy. There were too many suppliers of too many types of goods to have everyone barter with everyone—not to mention that I may have pears to sell today but the producers of wool clothing might only have such goods next month after shearing season is over. I can't wait until next month to sell my pears—and the woolen merchant is hungry today! The former town "market" of material goods slowly gave way to the monetary market economy with its Market Pricing and legal coinage.

Here we find for the first time a break with money's actual value, its place taken by the idea of quasi-symbolic money [V24] in the form of promissory notes between traders, and loans from a "bank." In both cases, one side symbolically promises another to repay real money at a *future* date, whether to the seller directly or to a third party (the bank) that proffers the means to purchase the goods now. Indeed, as early as 1750 BCE temple priests in Mesopotamia were issuing loans to local residents (Stahl, 2009, p. 80). This took money to another i.e., temporal, level of virtuality—*potential future* payment [V26] for *present* (real) goods.

Nevertheless, throughout the ancient era one thing remained constant: the value of "monetary currency" was always tied to the underlying value of the material from which the currency was made. A silver coin was worth its weight in silver and not some arbitrary sum that the authorities or anyone else gave it, so that money remained "real."[1] When the debtor repaid, it was with something (bullion or goods) of similar intrinsic value.

This situation continued for many centuries until China's Song Dynasty (tenth century CE) issued the world's first paper money: "jiaozi" [V12]. The effort was disastrous; without proper understanding of underlying economics the authorities continued to over-print these bills, leading to hyper-inflation. The Chinese ultimately discarded this new type of money.

Nevertheless, "fiat money" can be a viable form of currency. The government decides the value of the (intrinsically worthless) paper, and as long as its political

---

[1] This does not mean that the currency could not be debased. During times of financial crisis, the rulers might put less silver or gold in the coin, but soon enough that meant the coin was worth less and "inflation" ensued.

authority holds sway—and it prints additional currency roughly in line with the growth of the country's Gross National Product, and not much more than that—the value of the currency is maintained. However, paper money has two levels of virtuality.

To understand this, let's return to paper money's history. It next appeared in the American colonies as "bills of credit" issued by several provincial (pre-U.S.) governments to people it owed money. These were paper notes used to pay taxes. Continental Europe also employed fiat currency e.g., paper currency issued in Vienna during the Napoleonic era against gold ducat reserves. A somewhat later example was the "United States Note" (popularly called the "greenback") issued by the Federal Government during the Civil War. Such "promissory notes" (promising to exchange the paper for "real" currency) also were issued by companies e.g., the British South Sea Company.

They were all "backed" by either bullion or at least coins (specie), so that the holder of paper currency could (in theory and in principle) trade it in for something with real intrinsic value, although this became less likely (and possible) in the early twentieth century with governments overprinting paper money without enough bullion in the vaults to back up the paper. In short, all forms of paper (or cheap metal coin) currency with some real worth backing can be called "*substitutive* fiat money" [V6]—in that sense, only partly virtual.

The second type of currency has been in use since 1971 when President Nixon unilaterally abrogated the 1944 Bretton Woods fixed currency regime backed by gold. In cutting the dollar loose from gold—or any other commodity of value—the world entered the age of "official" virtual money, as the paper currency and cheap metal coins no longer had much if any intrinsic value. This can be called "*symbolic* fiat money" [V24]: "The value of our money…resides not in the ink and paper of the note that is carried in our wallets and purses. The note is just… a virtual stand-in for something else. The real value of our money is… itself something entirely apparitional and constitutes what is… a virtual reality" (Gunkel, 2010, p. 136).[2]

Non-governmental bank checks constituted ancient precedents, first found in ancient Rome: *praescriptiones*. Different forms of the same idea were later used (chronologically) in Persia ("chaks"), by the Moslem Abbasid Caliphate ("sakk"), by the Knights Templar for Christian pilgrims to the Holy Land (probably the world's first "traveler's checks"!), by the Venetians for international trade ("bills of exchange"), within England by private bankers ("drawn notes"), and then Bank of England issued pre-printed "cheques." Contemporary virtual money variations are certified checks, money orders, payroll checks, and warrants.

The next virtual money stage commenced in the late 1940s: the modern "credit card" (Gerson & Woolsey, 2009), not even pretending to be "money" but rather a means of identification to electronically transfer money "elsewhere" [V17] from

---

[2] Just how confused the virtual "money" situation can become was made clear when a Warhol silkscreen of two hundred one-dollar bills was sold at auction for over $43 million dollars (Panero, 2009). This is an example of an artistic artifact that symbolizes money [V8, V24] being worth far more in economic terms than the original "real-virtual" (symbolic fiat) money itself!

buyer to seller [V22].[3] Money is now almost completely virtual—the only physical thing is plastic in the buyer's hands, itself materially worthless.

Electronic transfers [V25] almost complete the virtualization of money, with two forms: banks' back-office clearinghouse and customers' internet banking. Such transactions reached approximately $5 trillion globally in 2020 (Digital Payments Worldwide, 2020). In another decade or two the advanced world will most probably no longer have any corporeal money whatsoever, becoming a totally virtual phenomenon, albeit retaining its very "real" function in the modern economy. An even greater virtualization trend is cryptocurrency, not backed by any government, using digital blockchain technology recording each transaction's entire history of (What is cryptocurrency? 2018).

Finally, the ultimate incarnation of monetary virtuality are derivatives (options, futures, swaps, forward)—financial contracts with no asset value, but constituting a "bet" on a hoped-for *future* price [V26] of any specific underlying asset: a material good, stock share and even foreign currency. As Arnoldi explains, "through the technology of derivatives, uncertainty and risk can be rendered virtual and be traded" (2004, p. 23).

Financial derivatives combine the non-physicality of electronic money with "goods and services" that have no corporeal existence at present or altogether [V25]. They do serve an economic purpose (albeit controversial). For instance, interest rate futures: the homeowner pays a slightly higher interest rate (compared to a variable rate), guaranteeing that the mortgage will hold steady at the initial agreed-upon rate regardless of future interest rate fluctuations. This example highlights how derivatives are virtual [V6, V24]—interest rates are merely reflections of the market's agreed price for loaning money, a shadow (interest rate) of shadow money, to borrow Plato's metaphor.

One cannot overestimate the centrality of financial derivatives to the contemporary world economy: "Less than 5% of daily transactions can be traced to commercial operations in goods and services: 95% are speculative capital movements" (Jacque, 2010, p. 6). Simply put, the majority (by far) of financial transactions in the world today involve the virtual future rather than the real present!

To summarize money's increasing virtualization: "Finance, the beating heart of the global economy, is unquestionably one of the most characteristic activities associated with the growth of virtualization. Money, the basis of finance, has desynchronized and delocalized labor, commercial transactions, and consumption on a large scale, all of which have traditionally operated with the same units of time and place" (Levy, 1998, p. 68).

"While it is tempting to see modern capitalism as an all-encompassing break with the past of human society, the virtuality of the circulation of money and the very transcendence and abstraction of money show that the ways in which we separate the religious and the sacred from the secular and the profane do not help us to achieve better understandings of the disjunctures and differences that constitute social life" (van der Veer, 2012, p. 191).

---

[3] An American Express ad even portrayed its card as a "virtual credit space" (Schulz, 1993).

## 7.3 Virtualizing Economic Organization and Property

A parallel development took place regarding the organization of economic activity, as well as ownership. These two elements combined to form the revolutionary virtuality development around the start of the Industrial Revolution: the "corporation."

During the hunter/gatherer stage, there was little formal organization other than a rough division of labor between the sexes (males hunting; females gathering[4]). Similarly, the agricultural era: individual families tended to their plot, or the "group" worked the land together. However, with increased socio-economic, and especially political, stratification, the upper/ruling class with large-scale farms and lands needed to organize economic production. All sorts of organizational systems were developed: outright mass slavery, indentured servitude, tenant farmers paying their landlord a percentage of their produce; and later, salaried work.

Here virtuality began to rear its head, as the larger organization—and stratified work hierarchy—distanced lower workers from "managers" [proto-V2]. True, those workers had contact with more fellow workers than their hunter/gatherer (or even small farm agriculturalist) predecessors; but from our vantage point this was the kernel of modern, large-scale worker virtuality within huge economic organizations.

Trade marked another step in the direction of virtual economics, distancing producer/seller from buyer, formerly trading in local markets and bazaars. With an enlarged socio-political order, the economy widened its territorial scope, creating merchant intermediaries [V3]; economic activity started becoming socially (inter-personally) virtual.

The next step along this road (literally): the introduction of international trade, resulting from the same increased political and economic, territorial scope. Here producers never interacted with consumers, hundreds if not thousands of miles apart. The paradox is that international merchant traders were the only ones whose activity was "real" through contact with producers and consumers, in addition to physically transporting goods—whereas the producers/sellers and buyers/consumers no longer had a physical relationship with each other [V2].

From an economic organizational standpoint, the average worker felt no steady progression over the millennia in the direction of virtualized organization. True, one can discern a general trend (albeit not completely consistent or clear) in international trade and the continued distancing of producers from consumers. Some empires in the ancient world had large scale international trade, but these tended to be *within* the territories under empire rule; by the medieval period, trade began to pick up *between* empires and civilizations e.g., Moslem and Chinese, Christian and Moslem, and to a limited extent between the Christian world and the Far East (the

---

[4] This division has now been questioned by the discovery of a burial site where a female was laid to rest with an extensive hunting kit of stone tools—leading to a second look at previously studied burials throughout the Americas that revealed between 30 and 50% of large animal hunters could have been female (Wei-Haas, 2020).

Silk Road). This received an added boost post-1492, with European powers "discovering" and conquering the New World, thereby widening international trade to encompass almost the entire globe ("Australia" joined in the eighteenth century).

In the modern age, such trade was first carried out by (or with official support of), "national" governments.[5] However, for historical reasons—government inefficiency, waste, corruption, incompetence, as well as the rise of a very large and wealthy middle class—the center of gravity for macro-economic organization shifted from government-chartered institutions to the modern, public corporation. In terms of virtuality, this was a revolutionary step for two reasons.

First, it severed the traditional relationship between person and property. True, until the modern era not every "product" had a "rightful owner" e.g., primitive societies had little concept of private property; until post-Gutenberg there was no legal "copyright." For thousands of years, though, ownership did exist, inevitably assigned to a specific person, family, or the ruler. The corporation, on the other hand, is jointly owned by stockholders who have no palpable connection with their "property" other than a piece of paper attesting to their infinitesimally small [V19], part "ownership."

Second, just as electronic money transfers replaced most human monetary interchange, stock trading has also become increasingly virtualized. Instead of paper-based stock certificates being traded manually, automated exchanges have recently taken over, leading to a completely virtual system with humans no longer actually trading or even setting the buy/sell price [V2; V25], now performed by algorithms [V6] (Bowley, 2011). In short, not only is the corporation virtual, as well as the stock certificate attesting to ownership, but now the actual exchange of ownership is disconnected from human hands and increasingly from human decision-making.

Third, the corporation is a wholly artificial entity [V13], a legal "fiction" to enable masses of people to financially establish a large-scale economic institution—given legal "rights" almost on a par with real-life people[6] e.g., the U.S. Supreme Court ruling that limits on *corporate* funding of election campaign advertisements are unconstitutional (*Citizens United v. Federal Election Commission* 2010). This "corporation-as-person" is not necessarily illogical as it is the sum of its owners who do enjoy such rights. One could argue that as the corporation limits stockowner financial liability it should have limited rights too i.e., such an imbalance gives the corporation more "relative" rights than the individuals owning it!

Modern corporations have increased two more related levels of virtuality. First, the national corporation has become an international, even a multinational, entity.

---

[5] In Chap. 8 I discuss nation-state virtuality. For now, there existed a rough symmetry between government virtuality (an "imaginary" construct connecting people who never see each other) and of virtual economic organization (traders forming a network to bridge vast distances between economic actors invisible to each other).

[6] In the 1886 case *Santa Clara County v. Southern Pacific Railroad*, 118 U.S. 394, the United States Supreme Court "suggested" that corporations could be recognized as persons for purposes of the 14th Amendment. "Suggested" and not "ruled," because the comment came in the oral preface to the official ruling, and the Court Reporter (an officer of the court) took it upon himself to include it as an integral part of the ruling's summary. Nevertheless, subsequent court rulings did entrench the idea in law, with occasional limitations on very narrowly defined issues.

*International* in selling its goods and services outside the borders of the country in which it is incorporated; *multinational* in decentralizing and diffusing its sundry core functions around the world. These have taken "action-at-a-distance" [V2] to a new level, with full-time employees all around the world. This "glocalization" weakens the centrality of "central headquarters" [V17] but also strengthens non-virtual activity at the local level (Lohr, 2010b). Overall, such an organizational arrangement has a huge amount of mediated communication, involving much virtual "presence" [V2; V3].

Second, multinational corporate workers are completely distanced from each other physically. In previous eras international trade separated producer from consumer; the contemporary age has also distanced joint producers within the same corporation—work-at-a-distance [V2, V3, V18[7]].

Wouldn't the sheer size of the corporation (or any large institution for that matter) undercut efficient consultation and decision-making, as the managers are physically cut off one from the other, not to mention from the regular workers who are closest to the customers, clients, or suppliers? And if the need for coordination has now become international in scope, doesn't that render the problem almost insuperable?

The answer lies in a seemingly historical coincidence: the rise of the modern corporation occurred at the same time as the emergence of modern electronic communication. However, it was no coincidence that the corporation appeared alongside the telegraph, telephone, and the modern (penny press) newspaper that relied on "instantaneous" information.[8] Virtual communication [V2, V3, V8] in large organizations was adopted to overcome the communication-at-a-distance problem between workers, and also between laborers and their managers. Conversely, some new communications technologies were invented to meet an economic and/or corporate need. For instance, without fast communication over long distances, early American companies could not increase in size along with territorial expansion because of the difficulty of working without market information from afar—ergo the telegraph.[9]

---

[7] Globalization has also added a *temporal* element of virtuality to the mixture. If X in New York works with Y in Beijing, the notion of "real" time becomes blurred for both.

[8] The influence of the telegraph and telephone on the newspaper was not only a matter of news transmission *speed*. The *presentation* of news changed drastically as well. Until the mid-nineteenth century, newspaper articles tended to be extremely longwinded, almost literary in style—a function of literary authors doubling as "journalists" (think of Charles Dickens' "journalistic detail"; his books were first serialized in the newspapers, and he was paid by the word). However, once news reports were sent by telegraph to the editorial office (paying the telegraph operating companies for each word), they became much shorter and to the point i.e., the 5 W's style came into vogue—putting What, Who, Where, When, and Why in the opening paragraphs. Phone calls were also quite expensive, so that the introduction of the phone did nothing to change this other than to reintroduce the term "deadline" (first used in the Civil War to denote an imaginary line that prisoners were not to cross)—the late hour when reporters found that the editorial office "closed" next day's newspaper and thus the news office phone became a dead line...

[9] Of course, the telegraph also met a political-national need: integrating the new western states into the Union—a case of expanded *political* virtuality.

In short, virtuality in an important sphere of life (communication) heavily influenced the development of virtuality in another critical sphere (economic organization), and vice versa. This was not the case until the modern age; beforehand, it was improved, non-virtual *transportation* (ships, overland road networks) that enabled economic commerce and trade to flourish—with communication (letters and bills of trade) piggybacking on such improved transport.

As work-at-a-distance continues to expand (with an extra push by Covid-19), a final stage of virtuality is emerging: the "virtual corporation." As noted above, by its artificial [V13] and incorporeal nature [V25] the corporation is *ipso facto* "virtual" by definition.[10] Taking the modern communication logic of "action-at-a-distance" [V2] to its extreme, corporations dealing with information need not have much physical presence, nor many full-time employees [V6]. Such radically downsized, but widely diffused, companies constitute extreme organizational virtuality, with three elements: 1—a network of independent organizations grouped under a unique identity, visible as a single organization but comprising different units [V15]; 2—mixing different partners, each focusing on a core competency [V6]; 3—using compunication technologies intensively to distribute and process information in real time, for rapid decision-making and timely operations coordination [V2, V17] (Simon et al., 2006, p. 298).

The term "virtual corporation" is a catch-all of two different configurations: virtual enterprise and virtual corporation. Virtual enterprises involve one-time or temporary cooperation between different entities lacking formal hierarchy, to accomplish a goal or carry out a project (Virtual organisations..., 2008). Virtual *corporations* are permanent ("incorporated") entities operating in mostly virtual fashion with all functions "outsourced" or "sub-contracted" to workers who are *not* employees but rather employed by other companies or self-employed. The core is a small "headquarters" staff (not necessarily quartered in any single place) that develops corporate strategy—what to produce, at what price, to sell or rent to which sort of customer—and then farms out most of the actual work to others: R&D, manufacturing, marketing, finance, legal, and so on.

Each of these two company types is virtual differently. The virtual *enterprise* is strongly virtual due to its impermanence [V18]; yet the workers are part of a permanent economic entity that on this occasion is working with other companies. Conversely, virtual *corporations* have limited numbers of co-workers, all of whom spend most of their working hours with employees outside their own corporation. Formally and organizationally, there is less virtuality here, but from the workers' socio-psychological perspective their labor is carried out inter-personally almost completely virtually.

There are several other virtual aspects of contemporary economic activity. First, in the compunications revolution even traditional corporations enable their employees to abandon the centralized office for work at home or "on the road" i.e., telework [V2],

---

[10] The term for creating the "incorporeal" company is "incorporation." On the other hand, another word for an economic company is "firm." The paradoxes of the English language....

through highly mobile devices. Today's newsroom, for example, has become a physical "mausoleum" with a few editors producing an ever-changing, dynamic product created by dozens of distant journalists in constant "touch" with their headquarters.

Second, numerous corporations have no physical contact with their retail *customers*—not to mention their *suppliers* – and even without any physical contact with the product itself [V2, V4, V6]. The modern-age virtualized distancing of seller and buyer started with the introduction of mail order, first by Benjamin Franklin in 1744, and more systematically by Hammacher Schlemmer in 1848. Montgomery Ward (1872) and then Sears (1888) started the monthly mail-order catalogue business that has turned into a $2 trillion global business in the twenty-first century (now mostly digital). Another distance innovation arrived in the twentieth century: direct marketing, whether by post or by telephone (telemarketing).

Shopping is gradually becoming virtual from another perspective—the growing use of "augmented reality"[11] to enable the shopper to digitally "try out" the product from home (Papagiannis, 2020). There is no technological reason that with future haptics (touch technology), shopping for almost any product (including food) can't be done by "doing there without being there" [V2].

Shopping, buying and even secondhand trading are all done today on the internet through "sites" (itself an interesting terminological development from physical to virtual) [V2; V3], in what has come to be called B2C (Business to Customer). Overall, global 2019 online B2C revenues reached $3.53 *trillion* (Retail e-commerce sales worldwide…, 2020)—and that's only half as large as online *wholesale* commerce (B2B: Business to Business), usually accomplished without any direct physical contact between the two sides, until physical delivery of the "product." As the digital age continues apace, more such "products" (e-books, streaming music, etc.) have no "materiality" [V25] so that even physical contact with the "delivery man" and the "product" itself will slowly disappear in the future—and if it is a corporeal product, a drone will deliver it [V6] (Palmer, 2020).

Finally, economic crime is also evolving and expanding virtuality. In pre-modern times, such crime was exclusively material (oxen, slaves, crops etc.). Indeed, the Bible rarely refers to *stealing* money, even though there are lots of commandments regarding loans and monetary transactions. As representative money and ultimately fiat money expanded, monetary theft and other virtual economic crimes dominated because it was so much easier to "carry" money, especially paper, than bulkier property such as furniture, animals, and real estate.

The digital age represents a further watershed in virtual crime. White collar crime—today generally a matter of stealing *electronic* money [V25]—far exceeds property theft of material goods. It is extremely rare to read about someone having even a few million dollars' worth of real property being stolen, other than art theft (paintings themselves being "virtual," at least representationally). Tax evasion has become easier as well, with offshore tax havens [V2] offered by virtual governments to virtual corporations; hyper-fast electronic stock trading has led to an upsurge in

---

[11] Augmented reality is discussed at greater length in Chap. 13.

illegal stock market manipulation and insider trading [V32]; and of course, the opportunities for simple embezzlement have skyrocketed as well, when the money to be
taken is comprised of computer bits [V25] that can be manipulated and hidden by
employees from within the company.

Moreover, newer forms of economic crime have evolved in the digital economic
ecology. For instance: identity theft, phishing, hacking, and cracking,[12] malware,
ransomware, viruses, denial-of-service (DOS) attacks. Overall, it is estimated that
cybercrime will cost the world $10.5 *trillion* annually by 2025 (Morgan, 2020).

There is a newer type of digital crime of even greater virtuality—"criminal"
activity within computer-generated, cyberspace virtual worlds [V10; V17; V25]
(Lastowka & Hunter, 2006). This involves completely virtual "property"—avatars or
objects that act as stand-ins for real people [V6; V28]. Is X liable for civil damages if
his avatar "destroys" the painstakingly created virtual house of Y? Should an avatar
be punished if it harms other avatars (if so, how?); or should X be punished in, and
by, the real world? In real-world criminal law, is stealing virtual money like real
money theft—when the former can only purchase virtual artifacts?

These are no longer "academic" or hypothetical questions. With millions of people
spending huge amounts of time "gaming" virtually, and with increasing "cross-
border" interaction between virtual worlds and the real world, "virtual law"—or
law for virtual worlds—has become real (Heck et al., 2021).

## 7.4   The Economics of Digital Virtuality

Given digitality's lack of materiality and its inherent incorporeality, the "economics
of digital virtuality" will differ from traditional, material economics, especially
concerning when comparing the "value" of material vs. digital products. Whereas one
had to purchase each analog vinyl record to hear the music, a digitally produced album
theoretically can be copied an unlimited number of times without purchase (and
very minimal storage cost) i.e., digital technology enables infinite reproducibility at
(almost) no cost—never possible in a *physical*, analog economy. To ameliorate this
economic "problem," the audio industry has developed subscription (or ad-based)
streaming—another layer of virtuality with music arriving "incorporeally" [V25]
from a distance [V2].

The ethereal [V25] nature of digital materials also affects the *locus* of consumption. Humans have always daydreamed and fantasized, so that (as noted above) it
is not surprising that many are now drawn to "living" part of their life in a "virtual
world" or playing "computer/videogame worlds." Here the avatar's "act of consumption," and the "product" itself, are completely virtual [V2, V3, V6, V8, V10, V17,

---

[12] "Hacking" is breaking into an internet site for the challenge and thrill of it, without intending to
cause harm; "cracking" is the same act but with the express purpose of injury to the site: defacement,
theft, blackmail, replacement of content with other content, etc.

V22, V25, V28]. The phenomenon is so widespread that by the end of 2019 there were 2.5 billion "gamers"—a third of the entire world's population! (Narula, 2019).

What is especially fascinating about virtual worlds is its permeability with the real world—in both directions. Real-world companies and governments have started advertising their "real" products through virtual billboards and opening "embassies" in virtual worlds. Conversely, players/avatars in virtual worlds perform "economic activities" and then sell their virtual products in the real world for real dollar money (Heeks, 2010).[13] This "economic development" is not totally novel considering that humankind has assigned real economic value (through copyright and patents) for centuries to "symbolic work" such as books, music, art—mostly consumed imaginatively in our heads: "in the arena of synthetic worlds, the allegedly 'virtual' is blending so smoothly into the allegedly 'real' as to make the distinction increasingly difficult to see. There is nothing revolutionary in this, though. It is merely a recognition that these things were always as real as anything else in the human culturesphere [sic]" (Castronova, 2005, p. 148; see too Lehdonvirta, 2014).

Thus, our ability to present and manipulate virtual products in an external-to-the-brain, virtual environment, is a function of technological progress and not necessarily, intrinsically different from what we were doing in the pre-digital age. Indeed, one could argue that "virtual worlds" in our heads (daydreaming etc.) are *more strongly virtual* than playing in a digitally produced virtual world, where we are in physical contact with a computer mouse (or another interface), reacting to what we see on the screen or other platform.

Given all this, the latest two developments should not be all that surprising. The first are NFTs—Non-Fungible Tokens: "a digital asset that represents real-world objects like art, music, in-game items and videos… bought and sold online, frequently with cryptocurrency" (Conti & Schmidt, 2021). Here we find legal "ownership" of a digital artifact without the owner taking actual possession. Indeed, the artifact is still "seen" and "used" by anyone who wishes to, without gaining the owner's approval (in that sense, different from copyright). The most notorious NFT is the $69,000,000 paid for a Beeples collage, digital artwork (Kastrenakes, 2021)!

Second, and a harbinger of future legal complexities involving creativity by artificially intelligent (AI) entities (algorithms, etc.) is the autonomous production of program code by AI [V13]: "The software engineers of the future will, themselves, be software" (Automating Programming, 2021). Who owns such code? Can AI have "economic rights and obligations" as do corporations, themselves a legal fiction? An Australian court has already ruled that an AI algorithm can be considered an "inventor" for the purpose of receiving a patent! (AI can invent…, 2021). Still, the overall issue is far from settled. To paraphrase Churchill: a conundrum wrapped in a riddle inside an enigma…

---

[13] This is not as clear cut as it may seem. The creators/owners of the virtual world claim that everything "manufactured" within it is rightfully theirs—much as a letter to the editor at a newspaper becomes the sole property of the paper, especially when published. Others disagree. See Castronova (2005, pp. 156–158) for a discussion of "property rights" in virtual worlds. See too Dibbell (2006), describing the author's year-long experience buying and selling virtual property in the highly popular *Ultima Online* game, an activity that turned into a full-time occupation.

One final area of "economic virtuality" that goes by several names combines digitality, virtual "ownership," and virtual "organization": collective production, creative commons, wiki-work, open source. These involve a joint effort of self-selected "strangers" to each other, willing (usually from a distance) to work cooperatively and incrementally to digitally[14] produce (gratis) something that is of value economically. Wikipedia and other wiki projects, the Linux program (computer operating systems kernel), HTML and HTTP code (underlying the entire web), recommendation systems regarding products and sellers (Yelp) or news (Reddit), are examples of such voluntary digital employment.

This is the logical endpoint of economic *organizational* development: first hierarchically structured, personal interface constructs; then greater territorial and communication diffusion (e.g., multi-national companies and virtual corporations); and finally open-source, wiki and the like bring organizational activity to an end point of virtual expression, best described as *anarchic cooperation within a partly organizational vacuum* [V15]. These do have a small modicum of "direction" or at least "supervision," but not necessarily strictly hierarchical, as many cooperative projects rely on the mutually agreed reputation of elite contributors in determining who supervises, edits, regulates.

Similarly, from an *ownership* standpoint this constitutes the end of the capitalist road. Not only isn't there a specific, legal owner (or group of owners) of the "product" but the producers explicitly *prevent* anyone from taking future ownership of the goods produced by the group's cooperative effort. For example, the GNU General Public License (GNU GPL) grants the recipients of any such program right of use without "purchase" or "rent" payment, but each successive contributor must include the GNU GPL in the "upgraded" program so that this upgrade too cannot become private property. This is now colloquially called "copyleft," constituting a return to the time (pre-Gutenberg) when any symbolic creation (art, book, etc.) could be copied at will without having to pay "royalties" to the original creator. Copyleft might be an idea born of ideological zeal (the credo: "software is meant to be free") but undoubtedly it is philosophically in tune with the infinite replicability and free transmission challenge of digital technology. Ownership of *material* goods does continue to exist, but the growing open source/wiki "economy" reverts to the original, virtual state of ownership conceptually and in practice.

Digital technologies, along with the internet and other wireless media, will continue to widen in scope, reducing the cost of communication and transmission of symbolic products (text, pictures, sounds) almost to zero. These parallel and reinforce the underlying forces of the "Information Economy," so that one can safely project the continued expansion of free collaborative/cooperative work as a major, if perhaps not dominant, force in the twenty-first century's neo- or post-capitalist economy. Only time will tell whether "property" as we have come to know it, and

---

[14] It is striking that as opposed to the "good neighbor" example (not uncommon in the pre-modern past), wiki-work is performed at a distance *by people who have never met each other.* Thus, the wiki phenomenon is not only somewhat revolutionary from an economic standpoint, but also from a *social* perspective as well.

relatively permanent corporations and other socio-economic institutions as we have come to work with, will manage to survive the virtual collaboration and organization revolution.

Indeed, several of the above economically virtual activities raise an even bigger question—and for governments, a significant headache: how to include such virtual economics in its official statistics, and its tax regime? As Castronova (2005, p. 255) cautions: "If transactions, production, and consumption increasingly occur in synthetic worlds, the corresponding measures of economic activity on Earth will fall. Central Statistical Offices do not, to my knowledge, collect any data about economic activity inside the membrane [virtual world]. As a result, even though economic activity may be going up, the statistics will say it is going down."[15] However, the problem is far larger than that as it pertains not only to virtual worlds but to a huge chunk of economic activity in the real world, increasingly based on virtual content and transactions:

> Many services… cannot be readily counted or tracked. Suppose an investment bank in London has its 'back office' in Mumbai. When a trader in London purchases a bond, thousands of bytes of data flow from the UK to India, where the bank's employees obtain confirmation, arrange payment, and maintain records accessible to their colleagues around the world. The data themselves have no value that can be recorded in India's import statistics. No government records how many bytes flow in either direction, nor the amount of British or Indian labour required to perform the work. So far as government statisticians are concerned, there has been no transaction, so no trade has occurred.
>
> The extent of trade can be even harder to measure when it comes to a Facebook page. The annoying advertising banner that flashes on a user's screen might have been posted by a server physically located in another country, and the fact that the user clicks on the ad might be recorded by a server in a different country still. These separate transactions both involve advertising services delivered across borders, but neither leaves a trace in any country's trade statistics. (Levinson, 2021)

This is important not only from the taxation standpoint, but also regarding economic policymaking. Among government concerns is the question of who wins and who loses during macro-economic disruption, but without clear statistical data (or *any* data regarding a specific industry or profession) [V4; V5; V15; V16] it is impossible to create reasonable and effective policy regarding international tariffs, industrial subsidies, vocational retraining, etc. This is another example of the problems facing capitalism in the virtualized, Information Age.

## 7.5 Material Virtuality

Despite all the above phenomena, we remain corporeal beings with critical needs of a very material nature: food, clothing, housing, and myriad other goods (cars, furniture, appliances). While these are less dominant in "dollar" terms, they certainly continue

---

[15] For an analysis of economic activity in virtual worlds, see Lehdonvirta and Castronova (2014).

to be a vital part of our lives. Nevertheless, there's a gradual virtualization of the goods economy too—in consumption and production.

Regarding consumption, contemporary society seems to have a penchant for the "fake" [V6; V32]: counterfeit labels, imitation brands, art replicas, fake masterpieces, faux furs, and other indistinguishable substitutes of "politically incorrect" originals (e.g., stuffed animals). Indeed, fake products have become so "mainstream" that a "rephotograph" (a virtual picture of a virtual image) in 2005 was auctioned for $1 million in 2005! (*Untitled—Cowboy*, from a Marlboro Man cigarette advertisement). Some fake "masterpieces" have even been resold ten times to different collectors, while university art departments use them as teaching tools (Esterow, 2021).

To an even greater extent, "image camouflage" extends to our own bodies. To stay slim, we eat artificially sweetened foods; spend fortunes Botoxing wrinkles, enlarging breasts, artificially pumping up muscles with steroids (and in the Far East, "westernizing" eye shape); dye our hair or cover baldness with toupees and hair weaves; not to mention snake oil solutions for body "handicaps" e.g., penis enlargement, and sexually attractive olfactory pheromones. The "service" industry is also replete with fakery: psychics and tea leaf readers, Elvis impersonators, lip-synching singers—the list goes on....

Another form of "fakery" has also become increasingly abundant: conspiracy theories, aka Voodoo Histories (Aaronovitch, 2009)—from the antisemitic and totally fictitious *Protocols of the Elders of Zion*, all the way to the recent (completely debunked) 2020 U.S. Presidential "election fraud" theories. The contemporary age has produced neologisms such as "truthiness," "alternative facts," "fake news" and "post-truth" [V32]. Although these are not necessarily *economic* forms of fakery, they are tied to the modern capitalist system in helping sell print and electronic media products to an increasingly eager and (partly) gullible mass audience.

The other side of the coin from consumption is *manufacture*—having largely become an automated and robotic enterprise. Robots, of course, have physical materiality but from a human standpoint they constitute another important divide between labor and its fruits—the latter remain the same whereas the former has largely left the human fold. In almost every significant manufacturing industry today (and several service ones too), one finds assembly lines "manned" by robots [V6] (a few humans still do "supervise," but even here remotely by computer more often than not) [V13]. Thus, workers are beyond being "alienated" from their work; in the more advanced economies they are physically divorced from the manufacturing process [V2]. Somewhat ameliorating (de-virtualizing) such alienation is 3D (additive) printing [V7]. Here "manufacturing" can be undertaken at the consumer's home, programmed by others from afar.

Finally, the ultimate form of manufacture is still awhile away but in in sight: nanotechnology, whereby in principle *anything* can be manufactured *anywhere* with *any material* within a Fabrication Box (Gershenfeld, 2005). Future nanotechnology processes will use cheap raw materials (sand, water), transmuting them at the atomic level into other materials, including biological ones (Marquit, 2010)—and from there building anything, just as a child builds objects from Lego blocks [V1, V13].

This is not necessarily an intensification of virtuality. One could claim that such "alchemy" is virtual because the entire manufacturing process takes place out of view within the "black box" of nanotech "magic." However, here again manufacture "returns home." Certainly, this is a higher level of virtuality compared to our forebears; the artisan had an intimate, hands-on connection to the product. Automation, 3D printing, and nano-technological processes all physically separate producer and product, but these newer technologies also restore some connection (actual presence) in creating the goods that we produce. Thus, as Information Age economic elements continue to become increasingly virtual, the Industrial Age—*piggybacking on Info Age technologies*—will be making a physical comeback into our personal lives.

## 7.6 Conclusion: Economic Theory and Economic Virtuality

Technological, organizational and socio-economic change over the millennia, and especially during the past few centuries, has pushed mankind's economy in a more virtual direction from several perspectives: the notion of private property; the relationship between workers and their work; communication between workers themselves and with customers or suppliers; the territorial scope and nature of economic organizations along with the spatial "distancing" of managers from workers; the increasing economic value of non-material "goods" and services; a financial system based on monetary virtuality; among other significant changes. Thus, it is not farfetched to claim that "[t]he United States [and by extension the world] has an economy that in major respects is based on virtuality" (Pfaff, 2008).

It is not altogether coincidental that thinkers who gave thought to the whys and ways of economic activity through the ages, also have something to contribute to the discussion of economic virtuality, even if only tangentially. I shall conclude this chapter with a very brief analysis of the two competing, economic ideological frameworks in the modern era: capitalism and Marxist socialism, understanding full well that there are many variations of each, in addition to those who look for an intermediate or hybrid "third way" (usually called social welfare or social democracy).

The economic framework underlying the entire capitalist edifice is the *invisible hand*: "by directing that industry in such a manner as its produce may be of the greatest value, he intends only his own gain, and he is in this, as in many other cases, led by an *invisible hand* to promote an end which was no part of his intention.... By pursuing his own interest, he frequently promotes that of the society more effectually than when he really intends to promote it" (Smith, 1776, Book IV, Chap. 2, parag. 9). From the perspective of this virtuality, the "invisible hand" constitutes an interesting virtualization of economics [V19]. This is not to say that on the micro-individual level economic activity is somewhat abstract or amorphous; for each individual such activity is very real. However, from the *macro*-economic perspective, Smith's theory envisages a framework without a "designer" or "coordinator"—the system works virtually by itself without ongoing, close supervision.

Marxist Socialism (also called Communism) has two central tenets relevant to our subject: 1—a society in which citizens should contribute according to their ability and receive according to their need[16]; 2—historical materialism.

Marx's "each according to ability and according to need" necessitates some form of central planning. In terms of our earlier discussion regarding the evolution of economic organization, this is a "regression" (in an historical sense, not necessarily in a moral sense) back to a hierarchical system where constant interpersonal supervision is carried out (slavery, indentured servitude, vassalage).

Historical materialism is even more obviously connected to "real" phenomena, for it views everything as being based on *materiality*—even such ostensibly "virtual" things as ideology that in the Marxian conception is merely an ideational expression of concrete factors (the social, political, and technological environment) and the palpable self-interest of all the economic actors within the system.

Thus, as food for thought I offer the following conjecture: the failure of Marxism[17] in the contemporary era is in part due to the mismatch between the Marxian ideology's underlying, non-virtual nature—and the virtual character, indeed *zeitgeist*, of the world's contemporary economy. Of course, Smith had no notion of "virtual" goods as opposed to Marx's purely materialistic outlook. Rather, by not having an overarching philosophical theory of "im/materiality," Smith rendered his approach "goods-neutral" i.e., any sort of goods and service would fit into his economic framework—whether manufactured products as was the dominant type in his age, or service/information evolving two centuries later.

Nevertheless, this chapter's economic survey is more "Marxian" than Smithian in one important respect. Marx's *dialectical* approach to economic history offers a good way to understand the process of major changes in the historical evolution of economic practice and phenomena—organization, money, type of product.

---

[16] This principle is most fully described later in Marx's career (1938), well after his analyses of the inherent flaws of Capitalism. Indeed, one of the strange aspects of Marx's writings is that he spent far more time addressing the factors supporting, and in his estimation also ultimately destroying, capitalism, than he did on the future Communist system that would take its place. This can be explained by the fact that Marx was at base a social scientist (and in his prime years only tangentially a "philosopher"), so that he was far more interested in the concrete here and now than in the far-off future. Moreover, despite the underlying (material) "determinism" in his system, he did not feel that it was possible to predict with any accuracy what would emerge from the rubble of capitalism. That explains the point just noted above: why there were so many different manifestations of Communism in the twentieth century—everyone could read into the term what they wanted, lacking much specificity on the part of Marx himself. In short, the "virtual" as future potential [V26] was not a central part of his very "real," here-and-now agenda.

[17] Marx's "failure" might also have been a result of his success! His trenchant analysis of capitalism's serious flaws (labor exploitation; work alienation; proletarian revolution) pushed leaders of the industrial world to introduce such palliatives as social security, unemployment insurance, universal health care, that removed the most egregious flaws of the capitalist system, thereby short-circuiting the proletarian revolution. It is not coincidental that the first two governments to implement such social welfare policies were *conservative*: Disraeli in England and Bismarck in Germany.

This doesn't mean that every new "synthesis" is a mark of "progress," but it does suggest the existence of an underlying historical force that is driving human economic endeavor from being a wholly mate*real* (sic) undertaking to an increasingly virtual enterprise.

# Chapter 8
# Community and Nationhood, Government, War

## 8.1 Introduction

Human beings are social animals. Interaction with other people is critical for their development and continued survival—to supply basic material needs and for psychological well-being. However, this doesn't tell us much about *what kind* of social framework we need: its nature, number of members, etc.

Through most of human history, micro-social intercourse was not very virtual, if at all: people interacted physically with other people. Even with the advent of writing and a few other minor forms of long-distance communication (drums, torches), there was no interaction other than "face-to-face" for the vast majority of humans. Around 10–12,000 years ago, with the introduction of agriculture and settled communities, "society" took on an increasingly virtual character. Since then, everyone has not only lived in the "real" world of physical proximity and social intercourse, but also in the "virtual" social world of "non-physical" entities, of mental construction.

## 8.2 Political Society's Development and Expansion

From time immemorial, *homo sapiens* lived in a "clan" social entity—an extended family—that numbered around fifty people, the optimal size for some division of labor, for security, and for sustenance, in what was generally a tenuous, hunter-gatherer existence. The clan, of course, was built on the family—not at all virtual, for in a group of fifty souls everyone knows and interacts with everyone else.

The gradual invention of agriculture was revolutionary from a survival standpoint (a steadier source of food) and socially as well. The two were inextricably connected

---

The *Appendix: A Taxonomy of Virtuality*—listing by numbers 1–35, e.g. [V13] or [V34]. Some of the various types of Virtuality mentioned throughout this chapter—is freely available to the public online at https://link.springer.com/book/10.1007/978-981-16-6526-4.

S. N. Lehman-Wilzig, *Virtuality and Humanity*,
https://doi.org/10.1007/978-981-16-6526-4_8

149

as the "conquest of Nature" (growing food on arable land) enabled more people to dwell in contiguity, and thus also to enlarge the social unit from clan to village, and then later to town, city, city-state, nation, and ultimately to empire.

However, this huge enlargement of the social unit's territory and inhabitants entailed a process of social "distancing"—not merely a "geographic" phenomenon (different residents of the social unit now lived kilometers apart) but with social ramifications as well, for if I don't see my neighbor daily, or even much at all, then in what way can the residents of said social unit consider themselves to be part of a common social enterprise?

Here the concept of "peoplehood" emerged.[1] In its first stages it certainly was *not* some artificial social construct but rather an expansion of the idea of family, taken to exponential proportions: not six to eight people of the nuclear family; not the fifty plus-minus of the clan; not even the several hundred of an extended tribe; rather, thousands and even tens of thousands of descendants from some real, common ancestor. Whether historically accurate or not, the Hebrews' biblical census in the desert (*Numbers*, I, 1–3) is an example of the concept as understood by ancient peoples: they were counted by tribes, with "exact" patrilineal descent measured over several generations. Nevertheless, such a social unit is based on some form of virtuality: the remembered past [V11], as it incorporates for the first time a distanced social *identity*. Whereas the extended genetic relationship might be real, there is no actual interaction between distant members putatively belonging to the same "people" [V25].

Over time, with the increasing number of people found within that unit, the familial feeling of "belonging" to the same social entity gradually weakened within a territorially defined city, a psychologically minded ethnos, a politically fashioned empire, or a culturally oriented civilization. In place of daily face-to-face contact that bound people together in the smaller clan, these larger social units had to find some substitute to maintain a measure of social cohesion. They were numerous and varied, but with a common denominator: social constructions based on some quasi/semi/largely fictitious "unity idea." The level and type of virtuality differed significantly depending on the unity-idea variant: ideational system such as distorted historical memory [V5; V11],[2] common religious belief [V25], pseudo-scientific racial (or pseudo-historical caste) purity [V31; V32], amorphous cultural affinity [V15], ideological solidarity [V33], and imagined nationhood [V28] (the latter discussed in greater detail below). Others were more connected to the material, real world and thus less virtual: commonality (if not communality) of economic status e.g., peasantry, aristocracy; or even common servitude to brute, dynastic power [V2].

What is interesting about this (incomplete) list is the fact that most unity-ideas enabled almost anyone to become a member of the social unit. Indeed, the greater

---

[1] Pre-agricultural societies can also have a sense of belonging to a social entity larger than the "clan." African herder groups today still feel kinship to an "ethnic" *tribe* larger than the traditional, limited "clan." However, as nomads they have few technologies nor a writing system; their overall virtuality experience remains very circumscribed.

[2] Gellner (1983a) noted that only through a shared amnesia/collective forgetfulness can a nation emerge.

such openness, the more such a social unit is distanced from the biologically based origins of human social formation. Of the list above, only racial/caste purity raises almost insuperable barriers to social inclusion, resonating with, and perhaps emanating from, earlier familial origins. In short, even in the ancient world most socio-political communities were porous [V15] i.e., not based on the real-world, genetic-generational principle of the still-earlier social unit of the family and extended clan [V6; V12].

Nevertheless, until the late modern era one cannot speak of any "natural progression" from one unity-idea to another; there isn't any discernible transition from lesser to greater socio-political virtuality. True, from approximately 500 BCE to 1500 CE some sort of "progression" to greater social virtuality can be discerned. At first the Greek city-states based their social identity on territorial contiguity as well cultural affinity, and then the Hellenic civilization expanded outwards leaving behind concrete geography as a determining factor in social identity. The Romans went through the same process—from the city of Rome to Roman Empire, followed by the community of Christians whose social cohesion (loosely defined) was theological; Christianity didn't demand the potential "member" have common blood or history, but simply accept its belief system in order to join the Holy Roman (i.e., Christian) Empire. Similarly, Islam with its concept of *umma* (extended nation). Indeed, this is why both religions succeeded so well in expanding their membership—the barriers to social inclusion were almost nonexistent.[3]

However, this two thousand year trend to stronger virtuality in social identity came to a grinding halt in the late fifteenth century when the Spanish (Catholic) Inquisition reintroduced "race" into the social community equation through its policy of *limpieza de sangre* ("purity of blood") in order to exclude "New Christians"—former Jews who converted to Christianity (some voluntarily, others under duress)—from holding high positions and in general to distinguish them from "pure-blood" Spanish. Of course, the idea of racial purity in this case (as in almost all cases) was nonsense [V31], for the Spanish Catholics themselves had been conquered hundreds of years earlier by the Moors and significant "miscegenation" had taken place over the centuries between these three religion-communities.

Thus, in the early modern era "race" became once again a "unity-idea" (forgive the irony), a social identity reversion from virtuality back to putative corporeality. Of course, the idea of "race" was not the only, or even the main, reason for the rise of Nationalism. There are several historical factors for nationalism becoming the dominant socio-political unit in the past two centuries (see below), but "race" is clearly the most "regressive" one historically (and most people today would add, morally too).[4]

---

[3] The third major Western religion, Judaism, did not expand for two reasons. First, it demanded more than theological "belief"; conversion depended on accepting a huge body of commandments that had to be carried out *in practice*. Second, and more germane to the present discussion: the Jewish People always viewed their nation in religious *and* ethnic terms. It is the only Western religion to claim that it started at the same time of ethnic nation-creation: Mount Sinai.

[4] Any discussion of the *biological* basis of "race" (within the human species) would lead us too far astray. In any case, the question is not so much "whether" race exists but rather "to what extent?"

## 8.3   The Virtuality of Nationalism

"[N]ationalism is a theory of political legitimacy, which requires that ethnic boundaries should not cut across political ones, and, in particular, that ethnic boundaries within a given state…should not separate the power-holders from the rest" (Gellner, 1983b, p. 1).

The concepts of nationhood, nation-state, and nationalism[5] are relatively modern. For most of history the usual political units—from small tribes to political dynasties and empires—did not contain nationalist principles, and rarely contained discrete "linguistic and cultural boundaries" (Gellner, 1965, p. 152), the *sine qua non* of "nationhood."[6]

From our present-day perspective the origins of Nationalism can be found in the Late Middle Ages—at first with the very gradual consolidation of political sovereignty in the hands of kings at the expense of the feudal aristocracy, and then with the concerted effort to eliminate local vernacular, replacing it with a "national tongue" (Eisenstein, 1983, pp. 82–83).[7] This latter phenomenon could only occur after the appearance of the printing press and the spread of newspapers that encouraged ever-widening public literacy, thereby destroying the already weakened hold of Latin (Anderson, 1991, pp. 18; 38; 44–45). The printing press also enabled the king to duplicate royal images and portraits in large numbers, thereby establishing monarchical "charisma" in visual fashion among the still mostly illiterate masses (Eisenstein, 1983, p. 97). It was only then, with strong central power and a budding bureaucracy, coupled with a trans-local "national" awareness of what was happening

---

and "how important is it in explaining differences between people and groups?" For example, on the one hand no one denies that there are superficial differences in skin color between Native Africans, Europeans, and Sino-peoples—or that a few diseases strike one "race" disproportionately (or even exclusively) compared to others e.g., Tay-Sachs among Ashkenazi Jews or sickle cell anemia among Africans. On the other hand, the genetic differences between the "races" are minute. Indeed, there are more and greater differences between different people *within* each "race" than there are *between* the "races" collectively. For instance, there are no significant differences in average height of Africans and Europeans, but there are huge differences between tall and short fair-skinned Europeans, and the same is true within "yellow" Sino-peoples, black Africans and so on. Thus, one can posit that in most important aspects, "race" is not a real but rather a virtual concept [V15; V20; V31; V32].

[5] To avoid confusion, I differentiate between three related terms: "nation" or "nationhood" (the socio-psychological feeling of meta-cultural commonality), "nation-state" (the political institution expressing nationhood), and "nationalism" (the ideology underlying the primacy of nationhood and importance of having a nation-state).

[6] In the modern world, there are only three examples of countries that do not have these two criteria: Switzerland, Belgium, and Canada. For Gellner, the latter two are exceptions that prove the rule—countries that continue to suffer from serious binational stress. The former is also an exception in that it is a confederation of semi-sovereign cantons, so that Switzerland is not a "nation-state" in the generally accepted sense of the term.

[7] Actually, the centralizing ruler had to remove two opposing linguistic barriers: the local vernacular (most Europeans in the late Middle Ages could not understand compatriots living a hundred or so kilometers away) and continent-wide Latin. The latter was the "lingua franca" of the scholarly and elite classes in all of Europe, the main cultural glue holding together the Holy Roman Empire.

in other places under the same state's rule, that the necessary and sufficient conditions existed for the rise of Nationalism as a significant force for socio-political solidarity.

This admittedly highly schematic historical review demonstrates that nationhood is an "imagined political community" [V28] (Anderson, 1991)—"*imagined* because the members of even the smallest nation will never know most of their fellow-members, meet them, or even hear of them, yet in the minds of each lives the image of their communion" [V19] (1991, p. 6; original emphasis). Gellner (1965, p. 169) is even blunter: "Nationalism... invents nations where they do not exist" [V31; V32]. Obviously, their approaches differ. Anderson assesses the phenomenon as being of intermediate-strength virtuality i.e., making a (national) mountain out of a (weakly social but existing) molehill; Gellner views nationhood as being strongly virtual—an outright "deception and self-deception" (1983b, p. 57), a product of its proponents' overheated imagination and its supporters' blind acquiescence in the fable.

It is possible that they are both correct—depending on the "nationhood" in question. Those countries with a relatively long history of gradual, social, cultural, and linguistic collective development (for instance, Germany and France), are closer to Anderson's model; others with a far shorter and more artificial, collective experience, such as former colonies in Africa and the Middle East, artificially put together regardless of tribal/ethnic mix, are exemplars of Gellner's view. In any case, it is true that a few contemporary nation-states can be said to have a more "real" than "imagined" collective past (the United States; Japan), but the vast majority of nation-states in the contemporary world are virtual to a large degree [V13].

It is no coincidence that the Nation-State (and its attendant ideology of Nationalism) became the defining political paradigm in the eighteenth century, around the time that religion began to lose its hold on the modern world, as Scientific Rationalism undercut the mystical and metaphysical explanations provided by Religion. For the masses, however, science could not provide the strong emotions once held by religious believers. This was Nationalism's contribution in its nineteenth century, Romantic form. Nationalism, then, was (is) not so much an *ideology* as it is a *secular theology*: "With the ebbing of religious belief.... What then was required was a secular transformation of fatality into continuity, contingency into meaning... few things were (are) better suited to this end than an idea of nation" (Anderson, 1991, p. 11). In short, religious virtuality was transmuted into the virtuality of nationhood.[8]

It bears mentioning that this intense national emotion produced other virtual elements of nationhood in the place of religious artifacts [V6]. For instance, music and the arts. What is the national anthem if not a stand-in for a glorious *Stabat Mater*? National (sculpture) monuments—or the national flag—can be understood as secular-national substitutes for omnipresent Jesus-on-the-cross figurines festooning every home wall in the past. In some countries that don't clearly demarcate religion from state, former religious symbols have been transmuted into emblems of

---

[8] This explains in part why most modern nation-states insist on the separation of religion and state. Religion constitutes the main competition to nationalism for the hearts and minds of the public.

the nation-state e.g., the State of Israel's *menorah* (the Jerusalem Temple candelabrum). Even "sainthood" has its nationalist parallel: the Medal of Honor for military heroism, and the Presidential Medal of Freedom for those making "an especially meritorious contribution to the security or national interests of the United States..." Overall, then, "Nationalism is the functional equivalent for industrial societies of the world-religions of agro-literate polities, and of animistic cults of pre-agrarian bands, nomads and tribes" (O'Leary, 1997, p. 199). Indeed, recent research in social psychology has discovered the continued connection between these two areas of life *within the mind of individuals*: whenever the perceived level of the government's controlling authority goes down, belief in God rises; and vice versa—when people start having doubts about God's involvement in the world, they tend to lend greater support to government legitimacy and power (Day et al., 2011).

Such a reverse nexus between the nation-state and religion brings us back to an important element alluded to earlier: the fact that the nation-state is a collectively imagined concept with several, embedded elements of virtuality, does not mean that from an *emotional* standpoint, the citizenry's feelings of national patriotism are in any way "false." Quite the opposite—emotional attachment to the "Motherland/State" is very palpable, with significant real-world ramifications as witnessed by the willingness of countless young men and women to "defend their country" even if it means leaving their flesh and blood loved ones behind and dying for this "higher cause" [V33]: "The language used to mobilize citizens to go to war—sacrifice, honor, country—is a virtual rhetoric" (Ignatieff, 2001, p. 201). This suggests that under certain conditions, virtual constructs can spawn human emotions that are even stronger than those engendered by real, live people.

This can be discerned in the "cousins" of nationalism (perhaps "illegitimate children" would be more precise) from the recent past: Fascism, Nazism, and other similarly extreme, xenophobic ideologies motivating adherents to perform the worst atrocities as a result of intense, emotional manipulation (e.g., *Deutschland über alles*).

Nationalism does not normally *negate* the existence and legitimacy of other nations or peoples—within or external to the nation-state. Rather, what Gellner derisively calls "reasonable nationalism" (1983b, p. 2) uses its nationhood to define its values and goals, while enabling non-national groups to live within the nation-state with some measure of cultural, religious, and even occasionally linguistic freedom. Thus, nationalism does not necessarily demand purity of nationhood within its midst [V4]. When such freedom for minorities is not forthcoming, then and only then does Nationalism become Fascism, a sort of metaphysical, ideal-type [V30; V33] e.g., race-based Nazism attempting to "cleanse" Germany and Europe of Jews (and gypsies, homosexuals, etc.). However, there is nothing inevitable about nationalism turning into fascism. Indeed, one of the first nation-states in modern history was England, its national development occurring parallel to its slow but steady transition to liberal democracy. Even today, some of the more "nationalistic" countries in the world are also the most liberal—the United States and France constituting two exemplars.[9]

From the standpoint of historical mythology, the main difference between Nationalism and Fascism is the degree of myth inherent in their "imagined past." Where liberal nation-states *revise* or somewhat *reinterpret* history to "imagine" the "community" (using Anderson's terminology), Fascist ideology is much more an inventive *fabrication* of the nation's past. It is almost pure mythology, taken as the gospel truth—compared to nationalism's relative openness to historical critique and revision. Thus, nationalism incorporates mild-to-medium strength virtuality in its perspective on nationhood [V4; V5; V11]; fascism's narrative is much "stronger" virtuality [V32; V33].

Why is nationalism such a powerful phenomenon in the modern world? After all, Nationalism has not spawned very many serious, political philosophers/thinkers or even intellectual ideologues—as have Liberalism, Conservatism, Communism and the like. As noted at the beginning of this chapter, humans are social beings to provide for their physical wants and satisfy psychological needs. Darwinian evolution explains this duality: an individual's survival is dependent on social interaction and cooperation, undergirding the evolution of "survival of the most social." During hundreds of thousands of years when *homo habilis* evolved into *homo sapiens* in the metaphorical "jungle," those proto-humans with a proclivity to sociability would have had a greater chance of surviving than the "individualists" (similar to the social utility of religion; Chap. 4). It is for this basic reason of evolutionary psychology that human history (based on the archeological record) starts with clans, tribes and so on, and not with the individual.

What determines the *form* of such sociability? The same factor that originated sociability in the first place: economics. As seen in Chap. 7, humankind successfully raised the ante over eons i.e., "economic survival" depended on larger and more complex economic systems of production—from hunting/gathering, all the way to mechanical, industrial-scale manufacture, and finally to service and information provision. As a general rule, the more complex the economic system, the more extensive the political system.[10]

Thus, the very large-scale economies of the Industrial Revolution necessitated a new type of polity—the nation-state: "Its economy depends on mobility and communication between individuals, at a level which can only be achieved if those individuals have been socialized into a high culture, and indeed into the same high culture" (Gellner, 1983b, p. 140). Un/fortunately (depending on one's perspective), "an industrial high culture is no longer linked... to a faith and a church. Its maintenance seems

---

[9] The nation-state today does suffer from mass immigration and multicultural stress rendering the concept of "nationhood" more amorphous and harder to define—and spawning anti-immigrant "nativism."

[10] Of course, occasionally the causality was reversed or circular: larger political systems created the proper environment, or initiated the demand for, higher and newer levels of economic production. For example, the early modern European monarchs needed greater tax revenues for their state consolidation efforts, so they encouraged the development of science and technology, while also seeking new resource and trade sources such as found in the New World and the Far East.

to require the resources of a state co-extensive with society…. A growth-bound economy dependent on *cognitive renovation*…" (Gellner, 1983b, pp. 141–142; my emphasis).

In short, the novel situation of religious decline along with mass education and growing higher education among an expanding elite, coupled with market demands for greater socio-economic integration, all led to the emergence of a new type of "high culture" that we call "nationalism," buttressing the nation-state. Again, it is not the natural, social "nation" that produces the nation-state but rather the artificially constructed nation-state [V13] that creates ("imagines") the strong sense of nationhood [V28] among the populace.[11] Its great popularity, then, is based on two related elements: the human socio-biological need to "belong" to something larger than oneself; the success of the nation-state in raising—for the first time in human history!—the vast majority of the population well above the minimum baseline for mere survival, and many/most citizens into a lifestyle of significant material comfort. Paradoxically, then, the virtuality of the nationalism concept maintains its strong hold in the contemporary age precisely because it provides the population with a real-world, socio-political environment for significant, material benefit.

## 8.4 Other Unifying (Virtual) Isms

Despite the modern-age supremacy of Nationalism, it was hardly the only socio-political "ism" created and used as a social super-glue. Certainly, for the hundred-plus years from the mid-nineteenth century until the waning twentieth century, *economics*-based, internationalist Marxism was a major competitor, notwithstanding the tyrannical political structure in which it was subsequently packaged by Stalin, Mao and others.

Despite (or perhaps, *because of*) the coercive way Marxist states tried to transform "class" into the main factor underpinning social solidarity, it did not succeed.[12] On the face of it, the lack of success is somewhat surprising because economic status seemingly is something more concrete and easily perceived by the class member than a more amorphous, virtual "nationalism." However, in the modern (especially capitalist) world, class is in fact no less virtual than nationalism given the relative ease of upward economic mobility (downward too, but most workers ascended the economic ladder into the middle class). In other words, personal—and even more so, generational—socio-economic identity became indeterminate [V15]; one could be born into the proletariat but end up in the bourgeoisie or even among the nouveau riche. Nationalism, despite its ostensible inflexibility—or perhaps *because* it offered a

---

[11] Anthony Smith (2000) argued that while nationalism is a modern concept, it draws heavily from the pre-existing history of specific groups that try to create a sense of common identity out of such historical memory [V11]. Nevertheless, he agreed with Anderson and Gellner that much of this shared history tends to be a distorted, or flawed, interpretation of the real past [V4; V5].

[12] In Sect. 7.6 I provided an alternate explanation for Communism's failure.

more fixed identity—proved to be far more durable than international or even national socio-economic solidarity, as can be seen in the collapse of the Soviet Union and breakup into fifteen nation-states, not to mention the steady numerical rise of U.N. membership.

International Communism wasn't the first ideational attempt at creating some sort of unity among humanity as a whole. In 1313, Dante wrote *On Monarchy*, calling for a universal and secular empire as the only hope for mankind to express its full potential. In 1625 Grotius extended this idea in *On the Law of War and Peace* (1814), the first serious mention of "international law." In the late eighteenth century Kant suggested the need for a universal government arguing that "the greatest problem for the human race, to the solution of which Nature drives man, is the achievement of a *universal* civic society which administers law among men" (1784, Fifth Thesis). Still later, others like St. Simon raised further proposals for a world government. All these were considered utopian with no possibility of materializing [V28; V30]—until the twentieth century.

In one sense, these philosophers' recommendations constituted virtual sociality carried to its ultimate extreme, incorporating billions of people who have no chance of ever meeting or knowing each other [V2; V27], and getting them to feel part of one "society." On the other hand, such social universalism was at least based on every human being's biological similarity.

What changed? Increasing globalization and world war. Starting with the first transatlantic (telegraph) cable in 1858 and continuing through the twentieth century with the internationalization of telephony, radio, television (cable and satellite), and of course the internet, the practical means for worldwide, inter-national communications emerged for the first time—the necessary (albeit, not sufficient) condition for any sort of global identity. This was a replay, in a much wider context, of print revolution's underpinning of nationalism through an ever-widening circle of readers sharing the same "news" and common national experience.

The two world wars, with their unprecedented loss of life and international scope, added a sense of urgency to forming an international regime. The resulting "solution": the League of Nations and then the United Nations. Although they did not aspire to becoming a true "World Government," in a practical sense they did lay the potential foundation for such an entity in the far future.

However, formal regimes of civic agglomeration don't survive unless based on some ideational-psychological component that the members can believe in as the "real" reason for staying together—just as family, religion, ethnos did in the past. Is there evidence for this in the international arena today? Yes, no, and maybe.

*Yes*: The various trends of globalization are found everywhere: expanding trade, finance, and multinational manufacturing; migration; communications; international treaties to resolve global problems (disarmament, ozone depletion, species extinction; pandemic warning system); huge growth of international, non-governmental organizations (NGOs); quasi-governmental, international institutions such as the IMF, World Bank & the International Court of Justice (Hague); informal but regularized trans-governmental meetings such as the G7 and G20. All these are socio-politico-economic frameworks, potentially the beginning of a post-modern, "citizen of the world" mentality among a growing segment of the population in the "First World." These same institutions have a tremendous influence on ordinary people's lives; "internationalism" is no longer a figment of philosophers' fecund imagination but rather a growing (albeit still limited) reality in practical terms.

*No*: Nationalism is actually increasing in strength among those countries that have only recently joined (or wish to join) the fraternity of nations. In many countries that gained independence in the twentieth century, nationalist feelings run strong— an "adopted" unity concept taken from Western colonial powers (Anderson, 1991, p. 118).[13] Ironically, many of these new countries now suffer from internal, irredentist movements demanding secession for their own "ethno-nationalist" reasons (former Yugoslavia splitting into several countries), religion (Christian southern Sudan split from the Moslem north), and language (potentially: Quebec, Belgium). Even patently international events such as the Olympics, designed to bring nations closer together, seem to increase nation-patriotic feelings rather than lessen them.

*Maybe*: Are these two trends—globalism and nationalism—contradictory? Not necessarily. First, as just noted, practical globalism already co-exists with psychological nationalism for real-world reasons. Second, the former exists almost exclusively in the more economically advanced countries that have been formal "nations" for at least a century if not longer. The latter expresses itself today most forcefully in the newer nations. However, even in the former group there's another trend that is stronger than globalism: "Continental Regionalism."[14]

The European Union is the most obvious example. After centuries of massive bloodshed, many European nations finally decided to try the regional/continental path, while simultaneously supporting the establishment of an international body (the U.N.). Despite ups and a few downs (e.g., Brexit), the steady deepening of the European Union seems to be inexorably leading most of those nations in a direction towards a continent-wide form of government *and common, socio-cultural identity*,[15] overlaid on their respective local-national identities. It is too early to tell whether similar proto-regimes in other regions—NAFTA, SEATO, OAU, Arab League—will lead to Continental Regionalism, but this cannot be dismissed.

What are we to make of this complex and ostensibly contradictory situation where contemporary socio-political units are moving in a national, regional, *and* international direction? I believe that they all represent much the same phenomenon, albeit

---

[13] Such "Colonial Nationalism," to use Anderson's expression, was introduced to bind the colonies more closely to the Crown or "mother country" but the cultural, historical, and linguistic gaps proved to be too great. The colonies detached "Nationalism" from the "Colonial," demanding to go their own national way.

[14] "Continental Regionalism" usually occurs within a macro-region of a specific continent, albeit not necessarily with all countries within that continent or macro-region joining the new political entity.

[15] Mammone (2011) noted that despite the seeming "ultra-nationalism" of extreme right-wing parties in Europe today—the French Front National party, Italian Northern League, Danish People's Party—they are less "local-nationalist" and more pan-European, white racist (or at least monoculturalist), adding a critique of "Regional" level pan-European politics to former, narrow, national concerns.

each at a different stage of development. The socio-psychological process of an individual's identity formation can help in understanding how and why.

A child has no fully realized self-identity; s/he is a member of a family and/or the child of Mom/Dad. The teenage years are marked by a struggle to understand and determine the "real me," somewhat rebelling against (or at least behaviorally distancing themselves from) those who represent the older, familial identity. However, until the process of solidifying this new identity is complete, the young adult will usually not enter an official, permanent relationship with another (even if biologically ready for procreation). Only when young adults have enough self-confidence in their new identity will they be ready to give up some freedom and even "dilute" the newfound self-identity in the name of a higher social construct, the new family unit.

A social collectivity is not exactly the same as an individual, but the psychology involved is similar.[16] Modern nations started out as "children"—either as small political units (city-states, cantons, principalities) in need of protection by other larger forces (e.g., the Church) or as colonies of a conquering empire. Rebellions ensued and ultimately most of these "teenage" social units became countries and/or coalesced into culturally homogeneous political units with a relatively clear national identity. Ultimately a number of such nations—after decades or even centuries of independence and stable nationhood—began a process of "marriage" to other contiguous national units with some cultural affinity. Thus, where any specific country lies on the national/regional/international spectrum is highly dependent on political chronology. As a general rule, the more "national" seniority a country has, the further along it will be on this spectrum.

All this suggests that regardless of the size and "national" affiliation of the socio-political units within the future international political order, the average human being will live within a unit far larger than anything that s/he can deal with on a socio-physical level. From the micro-perspective of citizens, the macro-structure is largely virtual—whether the political structure's scale is the large nation-state, a huge continental-regional union, or a humongous global government. There is little chance in the foreseeable future of a return to an exclusively non-virtual, human-level, framework.

This doesn't mean that there's no place for local neighborhood involvement, city government, or any other small-form, human scale political activity or sense of loyalty. Such face-to-face, "clan"-scale sociability continues to exist and flourish—probably fundamental to human existence. However, this does not undercut or contradict the growth of super-national, political constructs. Human beings are quite capable of multiple levels of social relationships and political loyalty. Indeed, that's *glocalization* (Roudometof, 2005), based on the need for a local, "face-to-face community" identity *precisely because* the more official political frameworks are so large, distant,

---

[16] Anderson (2011, p. 205) suggested that the psychology is virtually *identical*: "As with modern persons, so it is with nations. Awareness of being embedded in secular, serial time, with all its implications of continuity, yet of 'forgetting' the experience of this continuity—product of the ruptures of the late eighteenth century—engenders the need for a narrative of 'identity'."

and virtual from the individual's standpoint—although "virtual" and "long-distance" [V2; V3] are not necessarily antithetical to qualitative, interpersonal communication. In any case, we can have our cake and eat it too: psychological satisfaction through social relationships on the local, human scale, as well as the sense of security and economic benefits resulting from membership in much larger, virtual, socio-political entities: nation, region, and perhaps in the future, global humanity as a unified whole.

The above discussion has taken place from the individual outwards and "upwards"—large political constructs as virtual from the average citizen's standpoint. From a top-down perspective, we can also find indirect evidence that the nation-state itself views the matter somewhat virtually—particularly evident in the concept of "citizenship," formal membership in the nation-state. If citizenship was based strictly on geographical residency, one would not be able to continue holding citizenship while living outside the country's borders [V2; V17]. Certainly, if citizenship was defined as having a shared, ethnic-cultural history, one couldn't become a naturalized citizen [V11]. Indeed, if being a citizen meant offering complete loyalty to a nation-state, then one would not find several countries allowing dual citizenship [V15]—metaphorically, "virtually" the same thing as light being simultaneously a photon and a wave.

If the contemporary nation-state enables different types of "virtual citizenship," so too it has begun to view its own "identity" vis-a-vis some of the larger political entities mentioned above. All nations ask to become "members" of the United Nations. On specific problems and issues, the vast majority of nations have voluntarily conceded some sovereignty (freedom of action) by binding themselves to international treaties and their implementing institutions, such as the World Trade Organization enforcing economic rules of trade, Nuclear Non-Proliferation Treaty with its institutional arm the International Atomic Energy Agency, and so on.[17] Indeed, recently a strong case has been made that we are moving from a situation of state-centered law to "humanity law" (Teitel, 2011). This constitutes a paradigm shift in international relations and law in which the protection of persons and peoples increasingly takes precedence over national law and policy—both within states and between them. Here too one can see the beginnings of the oxymoron: "shared (or incomplete) sovereignty" [V4].[18] In short, even the bedrock nation-state is showing signs of increased flexibility and ambiguity [V15] regarding its hard-core identity and authority.

---

[17] Most European nations have gone much further than this, even abolishing their own national currency, border (passport) control, and the sovereignty of their own laws if such contradict EU legislation.

[18] Most recently, yet another type of quasi-multinational institution has begun to emerge: "incipient 'polylateralism,' in which the agents of global civil society and subnational governments are taking up the slack of faltering nation-states... NGOs, businesses or subnational governments who have little or no place in the formal multilateralism of nation-states" (Gardels, 2021) e.g., Doctors Without Borders. Polylateralism does not aspire to replace the nation-state but "merely" functions as a supplement to national action when the problem cannot or is not solved by any specific country or group of countries.

## 8.5   The Virtualization of Government

Whatever the size and the form of the (post-clan) state framework, there is always a need for an institutionalized governmental structure implementing the state's ideals and goals. Here too there are virtual elements, despite the real-life impact on the citizenry of the government's policies.

First and perhaps foremost is the size of the government. The average citizen gets to physically deal (or even communicate) with a very small portion of a government's workers: in the distant past because the system didn't demand public accountability from the authoritarian leaders and their functionaries; in modern times, especially among democracies, elected and civil service officials are held accountable, but their numbers have increased precipitously, so that it's impossible to have contact with them all (not that anyone would want to). Thus, from a *physical*, social interaction standpoint, "government" is only partly real as most of its workers labor "out of sight, out of mind" [V19]. Lately, even the small amount of previous, normal contact has been significantly reduced by virtually communicating with the government apparatus (telephone, internet) [V2].

Second, most citizens never have to deal with many departments of government— rendering major parts of the government apparatus beyond the immediate, daily life of the average citizen. Of course, this does not mean that we are not interested in what many of these government institutions are doing. We are, but here too almost the only information available regarding them is mediated [V3], either by the press or through social media.

Third, the (s)election process of those running the government is strongly virtual. Past monarchs (by whatever title) were "selected" by "Divine Right" or through some other ostensibly metaphysical process [V33]. Indeed, in ancient civilizations he (almost always male) was actually He—a semi-god, with metaphysical powers. In practice, the process was non-virtual: biological inheritance of the "throne," or military conquest. Thus, as a general rule the actual selection procedure was based on *concrete* actions or factors; however, the ideational justification for such rule tended to be *metaphysical*.

In the modern age, the advent of democracy meant that the selection process was based on greater transparency: election campaigns. However, the banal procedure of voting, completely distanced from the candidates for office, means that even the central act of democracy is mildly virtual [V2]—we don't all sit together in a meeting hall raising our hands or voice in affirmation of a candidate.

Governmental decision-making also slowly became more transparent, but again in mediated fashion [V3]. Moreover, this form of government maintains significant virtuality in another sense: the system is one of *representative* democracy [V6]— the leaders make decisions as substitutes for the sovereign public. Compare this with direct democracy in the ancient Greek *agora* where decisions were rendered by all citizens together, although most people living in democratic Athens did not have a vote: women, slaves, foreigners—a somewhat incomplete [V4], if not illusory democracy [V31; V32].

Modern democratic practice reveals other virtual aspects. There are two main philosophical camps regarding the question of representation. The Burkean approach calls for the electorate to choose candidates that are most suited to do what is best for the whole country, Rousseau's "General Will."[19] Conversely, the more accepted approach demands that elected representatives (e.g., Senators, Members of Parliament) make decisions based on the "platform" upon which they were elected.

However, both approaches are founded on a real-life fallacy [V32]. Burke's approach, in essence incorporating a modern-day incarnation of Plato's philosopher-king, assumes that one can objectively discover what is "ideal" for the nation [V30], but that assumes a value consensus among the citizenry—something that doesn't exist. Even the most "objective" decisions are based on interests and/or ideology. The opposite approach suffers from a different practical problem: representatives tend to focus more on the *upcoming* election than on the previous one, based on the increasing use of public opinion polls showing the leaders where the political winds were blowing. In this environment, decisions are not made based on the "contract" connecting previous campaign promises with present political action, but rather on speculation regarding future election outcomes as represented in a small *sample* of citizens [V6]. In terms of virtuality, contemporary democracies have taken to running constant pre-election *simulations* [V10], and even these are usually not carried out in person[20] (as occurs at the ballot box) but rather from a distance: phone, mail etc.

Regarding the election campaign itself, in the past several decades elections in advanced democracies have moved away from the *mass* meeting/indoor assembly/outdoor event to a much more mediated type of campaign in which the electorate receives most of its information through television/press news, analysis, as well as campaign ads [V3]. Even more recently, digital new media (SMS, email, social networks, party and candidate internet sites) have further distanced the electorate from physical contact with the candidates [V2], even if perhaps getting them more psychologically and actively involved (although such activity tends to be mostly virtual as well e.g., donations through the internet).

Elections are virtual in still another sense: they no longer are about the "true" candidate [V5; V32]. Rather, the image projected, and the message conveyed by the candidates for office, are now almost always determined by "election consultants" and "media advisors." The public cannot get to know the "real" candidate because s/he is constrained from being natural (in behavior or in policy proposals) lest being blunt, forthright, and open will alienate too many voters. Instead, we are presented with candidate-*actors* whose appearances are scripted and staged, manipulating their image to produce a "palatable" leadership identity for public consumption [V13; V15].

---

[19] Admittedly, the juxtaposition of conservative Burke and radically democratic Rousseau is strange. However, given the ambiguity of the General Will concept, at least on this specific point they are close to agreement regarding what leaders should aspire to in decision-making.

[20] The major exception is the focus group: face-to-face discussion by a selected number of people for in-depth understanding of political opinions. However, given its very small size the focus group suffers from a different form of virtuality: it cannot be, and does not even attempt to be, "representative" of the public at large [V4, V5].

In short, the sophistication (or at least heavy use) of public opinion polls and marketing techniques make a mockery of any possibility for Burkean democracy as well as the more traditional, pure-representative form. The public cannot judge who is truly "best" suited to lead nor is the citizenry offered an unalloyed policy platform, given that most candidates shy away from saying anything too "obligatory" on controversial issues.

In sum, government activity has real-life consequences, of course. However, the *experience* of government for the average citizen contains several layers of virtuality. First, the lack of physical interaction with most workers and even government departments. Second, the nature of contemporary media campaigns that place policy obfuscation and invented personal image above a true appearance of the candidates. Third and finally, the decision-making process in which legislators and other elected leaders (along with their respective technocrats who devise the policy options) determine what to do based more on their estimation of potential consequences in a virtual, future election than the actual promises of the previous campaign that brought them into office.

It should be noted that recent proposals and e-democracy experiments utilizing digital technologies have been put forward to ameliorate several of the above problems (Perez et al., 2018). However, even if they succeed, this will not lessen the virtuality of the democratic experience, given that they call for deliberation and decision-making from afar without physical contact [V2; V3]!

## 8.6 The Virtualization of Warfare

War would seem to be as non-virtual a socio-political phenomenon as we could imagine. Killing one's human enemy is an act of *corporeality* par excellence; ditto the use of armaments and munitions, the most material of man-made objects (which is why a synonym for armaments is "materiel"). Such was the situation for much of the history of human warfare.

In what way, then, can war be said to have a degree of virtuality? First, "distance." When war involved hand-to-hand combat—or at least killing within eyesight—it was completely "real" from the standpoint of the combatants. However, over time armies began to develop weapons that could be delivered from afar, removing the warrior from the opponent [V2]. At that point, although the *results* of war had real, physical consequences, the subjective involvement of the warrior doing the killing became virtual [V19]. Indeed, paradoxically in most cases such at-a-distance weapons only *increased* the real-life (actually, real-*death*) carnage because it was psychologically easier to "pull the trigger" the more soldiers were physically removed from the destruction inflicted on their military (and especially, civilian) victims. Thus, although objectively warfare damage became more extreme over time, subjectively for the combatants fighting turned increasingly (although almost never *exclusively*) virtual.

The weaponry that enabled this greater distancing and virtualization included the following, roughly in chronological order of their appearance until the late modern period: slingshot; bow and arrow; crossbow; catapult; four-spiked caltrop (non-explosive land mine); and most important of all—gunpowder that led to the development of guns; cannons; bombards; rifles; land mines; rapid-fire machine guns.

Nevertheless, almost all these early modern weapons still demanded visual contact between the combatants (except the land mine). The complete disconnection, physical *and visual*, between fighters—"indirect fire"—commenced during the mid-eighteenth century (150 mm field howitzer) and was significantly "improved" in the nineteenth century, attacking the enemy well beyond eyesight range [V2]. Another "virtualizing" invention was the tank, removing the soldier altogether from any unmediated visuality of his surroundings [V19], not to mention ever-increasing firepower through weapons such as land-based bazookas and higher ascent, short-range missiles (Katyushas) that could even send a projectile beyond telescopic range. The third distancing element, introduced in World War I was the most powerful of all: the fighter plane, placing the pilot too far up in the air to see any but the most general outlines of the target below, if at all.

Post-World War II, the rocket engine and long-range missiles further increased virtual warfare (again, from the attacker's perspective). During the past few decades, distancing weaponry has taken over the brunt of the fighting in many respects, with the soldier contributing some judgment in timing (when to fire) and against whom. This gradual evolution to "pushbutton warfare" [V3] among the most advanced armies has become the standard against an enemy force well divorced from any physical contiguity, through employment of longer-range missiles—cruise, intercontinental ballistic, air-dropped bombs—and unmanned, armed vehicles (e.g., drones).

In the contemporary age, we have added several virtual elements (beyond "distance") to warfare. First, the development of super-weapons—especially nuclear bombs—for the precise purpose of *not using them* but rather to "merely" threaten the other side with mass destruction [V26]. The balance of fear based on Mutually Assured Destruction (MAD) almost guarantees that neither side will employ such massively devastating weaponry, holding up since the first and only use of nuclear bombs against Japan in 1945. This is no longer *subjectively* "virtualized" warfare from a distance but rather "virtual" weaponry in the stronger sense of virtual—the weapons only exist to ensure that neither side employs them [V27]! That's why the term "Cold War" was used to describe the non-war from 1945 to 1989 between the Soviet Union and the United States: "cold" in the sense of the threat of nuclear war without its actual "hot" use.

A second type of virtual warfare is cyberwar, in which the battle takes place within the virtual cyberworld of bytes [V17]. There are three general goals in cyberwar (or cyber-terror, violence directed at civilians): 1—hampering or destroying the enemy's physical, civil and industrial infrastructure (power systems, data centers, communication systems) by attacking the software that enables these critical enterprises to work efficiently; 2—neutralizing internet sites, servers, and other elements of the enemy's cyber-infrastructure (such as massive attack causing Denial of

Service); or 3—stealing digitally-preserved, corporate, government and military secrets, and/or planting false information within these files, documents and other types of information sources.

The third category is obviously more virtual than the first (with the second somewhere in the middle) in that the intended attack recipient of the former is situated in the material world. In all cases, however, the weapons of attack are virtual/cybernetic. Two famous examples: the 2007 pro-Russian hack attack against many government internet sites in Estonia; Iran's alleged nuclear weapons program being seriously hampered in 2010 by the Stuxnet worm that caused significant damage to thousands of centrifuges that went haywire as a result.

A fourth element of virtual war involves training.[21] As the technology of warfare becomes more sophisticated the preparation of soldiers is also slowly moving away from "boot camp" reality to simulated virtuality [V10]. For reasons of greater verisimilitude, lowering costs, and protecting life and limb, many soldiers today spend large amounts of training time in simulations. Paradoxically, computer simulations of battle are more realistic than physical training exercises. We are slowly moving towards preparing for total virtual warfare (whether cyber or merely at a distance from the enemy), exclusively through virtual simulations of the "battlefield."

Fifth, whereas combatants have always tried to camouflage themselves [V5], cutting-edge physics has brought us into the Age of Invisibility [V19], with the so-called Stealth Bomber only the first of many invisible warfare elements to come: radar evading ships (Johnson, 2011) and invisibility cloaks for soldiers (Axe, 2008).[22]

Sixth and finally, we are at the cusp of perhaps the greatest revolution in the long history of warfare: the substitution of "robots" in place of human soldiers. The term here is used in the broader sense of Robotic, Artificially Intelligent (RAI), autonomous objects[23]—not necessarily a piece of metal with a head, arms and legs to replace soldiers [V1],[24] but rather all forms of weaponry (and other military artifacts) that are employed either completely or at least significantly without human intervention [V6]: unmanned aerial vehicles (UAVs) used mainly for surveillance; land-based reconnaissance robots; tiny, nanotechnology-manufactured, flying "swarm missiles"; automatically-piloted jet planes and tanks; robotic soldiers. Thus, as RAI removes more and more human soldiers from the battlefield and even from much logistical support [V4], warfare will be conducted from a distance [V2] and

---

[21] Military training through "games" has a very long history, well before the digital age (e.g., miniatures, board games etc.). See: Smith (2010).

[22] "Invisibility" will be dealt with at greater length in Chap. 13.

[23] Artificial intelligence also comes in a non-material form in computer programs, algorithms, and the like. However, in this paragraph I am mainly referring to AI embedded in *corporeal objects* that will continue to constitute the main means of warring in the future, notwithstanding the discussion of cyberwar above.

[24] Research has shown that the more the robots resemble humans, the more the human controller of that robot will cause it take violent action. This is because we anthropomorphize such humanlike robots, so that human controllers tend to psychologically remove responsibility from themselves and place it on the "quasi-human" soldier in the field (Koerner, 2011).

through non-human action [V13], even without any human guidance in the field i.e., "autonomous warfare" [V6] (De Vynck, 2021).

Where does all this lead? All the virtual elements mentioned here suggest some disturbing conclusions but also provide cause for optimism. The bad news, as Ignatieff concluded (2000), it is now so much easier for a nation to go to war: the number of soldiers killed and injured relative to the damage inflicted on the other side is the lowest in history. Moreover, the continued distancing of the fighters from their victims renders warfare almost psychologically painless—a trend reinforced with the expansion of autonomous RAI warfare. Even worse, weapons are becoming increasingly "democratized," enabling terrorists and guerilla fighters to wreak much more havoc than in the past: "once the price barrier to war is reduced, as it may be if molecular assemblers [nanotech] are able to manufacture many weapons anywhere, the potential may rise greatly for disaffected individuals, survivalist sects, and religious extremists to wreak havoc" (Mulhall, 2002, p. 278).

On the positive side, other important elements of virtuality work as a strong counterbalance. First, the media are far more successful in bringing the ravages of war to everyone's living room [V3]; it has become harder for governments to maintain wide public support for the war effort under the steady drumbeat of graphic pictures displaying war's horrors. Although citizens and increasing numbers of soldiers might be physically distanced from the battlefield, warfare impinges psychologically on the wider, civilian public more than ever. Second, cyberwar that causes destruction well beyond the field of battle (e.g., disabled power plants) means that future warfare will also *materially* affect civilian quality of life, the effect much more than psychological. This too should reduce public support for entering and conducting war. Third, if the trend to greater RAI in warfare eventually (almost) abolishes human intervention, there is little reason for each side to try and kill the enemy's population, civilian or military. Victory will go to the "army" that destroys the enemy's RAI physical infrastructure and technological capabilities [V6]—massive destruction, but relatively bloodless.

Fourth, non-military, nano/molecular technologies hold the long-term promise for ameliorating several of the main scarcity causes of war: energy could be plentiful with solar panels (and possibly atomic *fusion*) of remarkable efficiency, food could be copious with agro-genetic engineering, mineral resources could be abundant with nano-transmutations at the atomic and chemical levels of matter; and so on. Although this sounds somewhat utopian today, war could become virtual in the future in the sense of being something remembered in the human past [V11] but no longer occurring in practice [V27]. Indeed, two studies have shown that violence and war have decreased markedly in the contemporary era compared to all previous historical periods (Pinker, 2011; Goldstein, 2011).

Fifth and finally, a factor returning us to the issue of the nation-state and larger super-national, political constructs. Future warfare technologies such as nano-nuclear arms, genetically mutated biological weapons, and other esoteric forms (weather modification; virtually invisible swarms of nano-missiles) could become so destructive and indefensible that there will be no recourse other than to establish wider-ranging disarmament regimes, if not some sort of world army or global police force

(Mulhall, 2002, p. 280). This would constitute one additional—perhaps decisive—factor leading to the end of the *sovereign* nation-state as we have known it for the past few centuries.

In sum, current trends are headed in a direction whereby future warfare could be more virtual in several important respects while simultaneously having even greater real, physical impact on our lives. Paradoxically, such a potentially devastating influence might, in the final analysis, force humanity to abolish warfare altogether—turning this age-old curse of humanity into a virtual thing of the past [V11].

# Chapter 9
# Communication To/With/By the Masses

## 9.1 Preface: Content, Modes, and Media of Communication

Communication involves three core elements. The first is virtual *content*, dealt with at length in Chaps. 4–8. All these fields of endeavor contain some form of information transfer i.e., communication. Thus, the present chapter will *not* focus on the *content* of communication, but rather on the other two core elements: *modes* and *media* of communication that enable all other fields of endeavor to function.

*Mode* relates to the *ways* in which we "express" the contents that we wish to communicate: orally, textually, symbolically, linguistically, etc. *Media* are the technological *tools and objects* through which we transmit those messages e.g., tablet etchings, print on paper, telephone, radio, television, and so on. In short, I can communicate in quite non-virtual fashion: gesturing to a person in my presence to scratch my back (physical action without symbolic meaning, by way of a non-linguistic mode, without using any intervening medium). Or I can communicate in completely virtual fashion: writing a story about an alien civilization that I send to an e-book (the contents are virtual, the linguistic mode uses symbols/letters, and the medium, e-ink or digital bytes, has a large measure of ephemeral virtuality). Most forms of human communication are to be found somewhere in between these two poles.

## 9.2 Earliest Origins

Aristotle argued that what made us human above all else is "logos," speech (Berns, 1976). Thus, whereas animals communicate with each other orally (song, grunts),

---

The *Appendix: A Taxonomy of Virtuality*—listing by numbers 1-35, e.g. [V13] or [V34], some of the various types of Virtuality mentioned throughout this chapter—is freely available to the public online at https://link.springer.com/book/10.1007/978-981-16-6526-4

© The Author(s), under exclusive license to Springer Nature Singapore Pte Ltd. 2021     169
S. N. Lehman-Wilzig, *Virtuality and Humanity*,
https://doi.org/10.1007/978-981-16-6526-4_9

chemically (scents), and gesturally, they do so almost exclusively "face-to-face."[1] Moreover, Aristotle's "logos" denotes *rational* speech—the ability to communicate thoughts and ideas that deal with matters beyond the concrete need for survival and reproduction.[2]

If purposeful communication is fundamental to being human, and as seen in previous chapters, virtual experience is also intrinsic to humanity, it is not surprising to find that communication and virtuality are inextricably connected. Indeed, the other fields of virtualized human endeavor surveyed in this Part II of the book would not be possible (or would be severely circumscribed) were it not for the ability of humans to communicate virtual content by virtual modes and through virtualizing media.

Of course, this does not mean that all human communication has a measure of virtuality. Until the advent of symbolic writing approximately 5000 years ago almost all communication for most humans was *unmediated* i.e., non-virtual: oral speech and body language. Even in our contemporary, hyper-mediated era, we still devote a good portion of our communication time in face-to-face, verbal and nonverbal interaction.

Similarly, from the perspective of humanity at large, past and present, the over-whelming preponderance of communication *modality* has been, and continues to be non-virtual: "language is so overwhelmingly oral that of all the many thousands of languages—possibly tens of thousands—spoken in the course of human history only around 106 have ever been committed to writing to a degree sufficient to have produced literature, and most have never been written at all. Of the approximately 3000 languages spoken that exist today only some 78 have a literature" (Ong, 1982, p. 7 based on Edmonson, 1971, pp. 323, 332).[3]

Thus, any discussion of virtual communication must keep proportionality in mind. From a media standpoint, over the entire course of human communication history, proximate physicality has been heavily preponderant over distant virtuality [V2]. On the other hand, it is precisely because of the monopoly of face-to-face, oral communication in the past that the progression in the modern age to various media that "virtualize" communication (written; electronic visual and aural) has become

---

[1] There are some quite rudimentary exceptions to this rule. For example, dogs will urinate on a spot to communicate that "this is my area." Whales can "sing" to other whales thousands of kilometers away, although with today's international shipping ambient noise, this has declined to "only" about 50 km (Roman, 2008). Thus, human communication is *not uniquely* "virtual" from a distance perspective [V3], but the variety and richness of human non-face-to-face communication clearly sets it apart qualitatively from very limited animal "virtual" communication.

[2] Some evolutionary development psychologists claim that *everything* humans do is based on survival and reproduction: "A complete understanding of human social and cognitive development, and any associated sex differences, requires an understanding of human evolution and the associated epigenetic processes that guide the development of evolved social and cognitive phenotypes" (Geary & Bjorklund, 2000, p. 63). For the influence of evolution on human communication, see: Hauser, 1996.

[3] More recent estimates place the number of world languages at around 6000 (not 3000 as Ong thought), but of these approximately 3000 will most probably be extinct by the end of the twenty-first century.

so noteworthy. Similarly, the modes of communication have expanded, although not to the same degree: symbolic text (writing) and some sophisticated forms of visual symbolism (such as art and graphics).

As opposed to animals who communicate mainly through body "language" and a limited array of sounds (or songs) and chemicals, humans have gradually expanded communication beyond instinctive and natural animalistic modes. We now know from archeological and anthropological studies that even in the primitive hunting-gathering stage of *homo sapiens*, and to a lesser extent, even among their predecessors, *homo habilis* and so on, "we" have been using some form of symbolic language for over two million years, around the time our ancestors started making stone tools. Based on *homo sapien* fossil morphology, we also know that around 150,000 years ago these ancestors began to articulate sounds similar to what we speak today. Speech "language" as we understand the term (full blown grammar etc.) emerged approximately 50,000 years ago (Anitei, 2007). What is interesting for our purposes in this all too sketchy chronology is that symbolic, abstract thought *preceded* human language (but not necessarily speech sounds) and was not a product of language. The "modern" human brain developed symbolic and abstract thought [V23, V24], in turn enabling or "pushing" language development.

Nevertheless, the entire fossil record shows relative stasis in technology and "culture" for hundreds of thousands of years—until around 50,000 BCE when *homo sapiens sapiens* (defined as anatomically modern humans) finally begin "talking" language. While there is still some controversy as to whether 50,000 BCE constituted a "creative explosion" (Klein, 2000) or was a further step along a more continuous, socio-evolutionary path, many new things began appearing around this time: new types of stone, bone and ivory utensils and tools (harpoon heads, leaf-shaped points of flint, projectiles, scrapers), communal graves with accompanying ceremonial artifacts, symbolic art (cave paintings), artificial "housing" structures, and mass migrations to Australia, South Asia and Europe (Johanson, 2001). It is highly unlikely that all these things could have occurred without the virtuous circle appearance and development of "modern language": language begat culture, that begat more refined language and culture, that begat more sophisticated technology, and onwards. In any case, human *language* communication was the catalyst for almost everything we today call "modern."

How can language, especially the speech variety, be considered "virtual"? Obviously not in the transmission—brain to vocal cords to air to ear to brain. However, if one goes a bit further back in history and compares formal language to what preceded it, its increasing "artificiality" [V13], divorce from Nature (but not human nature), becomes clear. We have no recordings of pre-*homo sapiens sapiens'* "speech sounds" but they must have been based (at least in part) on onomatopoeia, mimicking of nature, itself mildly virtual (V1, V6). Our language still contains residues e.g., sizzle, cuckoo, boom, chirp and so on. Over time, these natural sounds began to metamorphose into sounds with less connection to the natural world, and ultimately took on "meaning" in only a symbolic way [V24], totally devoid of any environmental object, sound, or action.

We tend to have trouble thinking of language as "artificial" precisely because we have been doing it for so long (as a species) that it has literally become second nature. As Chomsky (1957) argued, we are born with an innate *ability* to learn language with what he termed a "language acquisition device" in our brain (he did not claim that language is innate and that there exists a unitary "universal grammar").[4] However, this should not hide the fact that the sounds and words that make up all human language have no direct source in the natural world; indeed, the fact that one (human) species can come up with over 6000 different languages is proof enough of the arbitrary and artificial expression of our most ancient and central method of communication. If it were not so, each of us would find it relatively easy to figure out the meaning of any language foreign to us. We can't because languages are symbolic creations of humans.

Indeed, this progression from natural sounds to symbolic speech is precisely what we find in the development of writing—the next major communication revolution in human history. At first, writing looked like drawing; in fact, it *was* drawing. To communicate events and ideas, the earliest "writers" in the Middle East put down pictures and later pictographs (figures not fully fleshed out but still representing the original object) that were instantly recognizable by all other "readers" [V8].

For reasons of speed, as well as the universal trend to information entropy (disorder), certain pictures over time began to represent ("re-present") an idea or concept that was loosely connected to the original object (e.g., five perpendicular lines = "hand") that we call an ideograph [V4; V5]. The final transformation was for certain ideographs to become even more stylized [V12] as each one represented only one specific sound—the beginning of what we call the "alphabet." Thus, for instance, the fourth Hebrew letter "*daleth*" (similar to the fourth Greek letter "*delta*") was shaped as a simple triangle in Paleo-Hebrew. However, the word *daleth* literally means "door"; and the door to the main form of abode back then—the tent!—was shaped as a triangle. In this case, not only did the shape of the letter preserve its original source but its name too. In most cases, though, the alphabetical letters of most languages have lost their provenance in both name and shape, having no connection with real world representation—thereby becoming pure symbolic artifacts of communication (V3, V4, V23, V24); daleth today is written ד.

There is another basis of comparison between the character of spoken language and writing—the *medium* of communication was certainly not virtual at all. The human throat, tongue and lips on the one hand, and clay, stone, sticks, knives on the other hand, were as corporeal as can be. This banal point is important because a gradually increasing degree of virtualization occurred in the *media* of communication, other than speech—the same sort of general (albeit not inevitable) progression to greater virtuality we have already seen in previous chapters regarding several other areas of life.

---

[4] Chomsky's *innatist* approach is not the only theory, although it is dominant. The two main competing theories are: a—*behaviorist* that claims language to be mostly learned through repetition and conditioning; b—*hypothesis testing*, arguing that children learn syntactical rules by postulating hypotheses and then testing them (not necessarily out loud).

## 9.3 Writing and the Virtualization of Other Spheres of Life

Before we continue along the story of mode and media progression, it is worth noting an important point that will be more fully fleshed out in principle in Part III: virtuality in one area of life may have important ramifications on (the development of virtuality in) other areas of life. In our case here, the advent of writing as a mode, and print as a medium (in their several incarnations), was crucial to the development of increasingly virtualized religion, government, and commerce.

The influence of writing on Judaism and Christianity needs little elaboration. The Old and New Testaments are universally considered to be classics of world literature (The Bible is the all-time best-selling book) that have immeasurably advanced the scope of these religions' teachings. Given that these two religions (and many centuries later, Islam as well) are the fount of monotheism which itself has had many important direct influences (e.g., moral "natural law") and indirect ones too (e.g., universality of scientific law), the connection between writing as a mode of mass communication and the spread of modern-Western civilization is clear.

The virtuality influence is quite direct as well. As noted in Chap. 4, ancient world religions started out pantheistic and pagan with heavy emphasis on tributary worship of concrete idols and human or animal sacrifice. Over the millennia, Judaism and then Christianity displaced this religious approach with monotheism that departed significantly from any corporeal view of the Gods and ultimately also from temple sacrifice. In place of pagan idol worship, these monotheistic religions introduced prayer to an unseen God, and at least Judaism forbade the use of visual portrayals of the divinity (e.g., the Second Commandment). It is no coincidence that the Decalogue, the very first instance of *written* law for monotheistic Judaism should command the abolition of *concrete, visually realistic* representations of the Almighty. The same abstract mindset that placed the virtual "Word" at the core of its religion, also simultaneously (largely) removed the non-virtual, "real," concrete manifestation of Divinity.

This is not to say that writing influenced all religions in the same fashion nor that the spread of these religions is due exclusively to the written word. Obviously, Christian proselytizing in its initial stages was mostly oral, as the pagan world was largely illiterate. On the other hand, whereas the written word enabled Christianity to consolidate itself theologically and institutionally along the lines of a central, unifying power (the encyclicals of the Church and the Pope), Hebrew and Aramaic texts enabled the decentralized Jewish people to remain spread out among many diasporas and yet still maintain some unity within growing religious diversity.

Indeed, overall the written word had been far more integral to Judaism (and Islam) than to Catholicism. The Church claimed a monopoly on biblical interpretation so that other than the relatively few educated monks and church scholars who were literate (and most of them used their skill merely to transcribe older texts, not to write new ones), the common Catholic believers received their gospel *orally* from preachers and local priests. This was a continuation of early Judaism, in which the people made a pilgrimage thrice yearly to Jerusalem, where they would hear the

Priests read out loud or declaim the Torah in public. However, paradoxically (and perhaps not so coincidentally), around the time of Jesus's appearance and initial oral efforts of nascent Christianity, the Jewish approach changed course and began a process of writing up the religio-legal (*Halakha*) discussions that were taking place. In lieu of Temple sacrifice and worship, Judaism became a very scholarly-oriented religion.[5] Henceforth, the Jewish people as a whole emphasized literacy in order to be able to understand God's word—leading to a massive and intellectually impressive religious law literature that evolved over the post-Temple period (e.g., *Mishnah, Gemarah* (together: *Talmud*), thousands of Q&A letters sent back and forth to rabbis across the world [V3], who also authored dozens of Jewish law compendia). Much the same occurred among Moslems after Islam's irruption in the seventh century and conquest of half the known world back then.

When the printing press media revolution arrived in Europe in the mid-fifteenth century (to be discussed below), the written word once again had a major impact on the ability of Protestantism to propagate itself. This was reinforced by Protestantism's core idea that literacy is a religious imperative for Christian believers to read the holy texts by themselves and come up with their own personal understanding of what the texts mean [V6]. Thus, Protestantism could not have come into existence without the written-printed word and also had little *raison d'etre* without each believer having the personal capability to read, and the physical ability to obtain, the biblical text.

Writing was equally important for the development of government and its bureaucracy; it was absolutely *required* for any government attempting to form an extended empire. Without written laws from the "emperor" that provincial rulers could use to govern the hinterlands, without written instructions to army generals in the distant periphery, and without other forms of written proclamations (holidays, taxes), it would have been exceedingly difficult to maintain control over vast expanses for any length of time.

Finally, in the economic realm we find a close interaction: not only did writing influence commerce, but as seen in Chap. 7 writing itself emerged out of the invention of tokens and then abstract numbers, tools of commerce.

After internalizing the new cognitive skill (number signage and manipulation) the human mind was ready for the next level of abstraction: the invention of writing. Whereas the tokens and their abstract numerals were limited to representing concrete, "real" goods, writing went further and abstracted immaterial sounds of speech. The first writing objects were phonetic, syllabic signs that recorded individuals' names. From there, writing symbols (whether ideographic or alphabetic) eventually represented other nouns and then ever more abstract concepts (Schmandt-Besserat, 2009).

Another economic example of cross-field influence can be seen in one of the (in)direct effects of the communication medium we call "print." Writing was always

---

[5] The rabbis instituted a religio-*political* revolution, establishing that henceforth it would not be the actual, *literal* text of the Bible that is determinative of religious practice, but rather the Rabbis' *interpretation* [V6] of that biblical text that would define how Judaism was to be conducted. As we'll shortly see, this is an interesting harbinger of the Protestant Revolution 1500 years later!

a solitary affair—one work produced at a time. The invention of the printing press by Gutenberg ultimately led to one of the other most significant "inventions" of the modern age: the assembly line and industrial mass production. Actually, the printing press itself was the first mass production "assembly line" of replaceable parts (McLuhan, 1962, p. 125; Ong, 1982, p. 118).

Altogether, then, we can see the interplay of virtuality across several areas of life. As agricultural surplus increased, there was a greater need to count and record commercial transactions. This led to more sophisticated and abstract counting systems that eventually morphed into writing systems. In turn, the invention of writing enabled the economically (and therefore militarily) stronger states that were expanding geographically beyond their "real" social boundaries into "virtual (distanced) territory" to maintain their empires through improved communication and expanding international commerce. Ultimately, this new modal invention called "writing" and its accompanying, abstract cognitive mindset, led the people of the Near East to turn that abstract approach in other directions as well: monotheistic religion, philosophy, and the rational investigation of our world (science).

*This "cognitive transference" from one domain to another is a core idea of the present book.* It is not that the "idea" of virtuality spread into various areas of life (certainly the ancients did not think about virtuality per se), but rather that the "experience" of virtuality—cognitive or socio-cultural—in one sphere of life tends to open up the individual to virtual experiences in other spheres, even if qualitatively different from the first. Indeed, it is hard to think of three spheres further apart qualitatively than religion (quite spiritual), commerce (highly material), and writing (completely symbolic). And yet, by digging a little deeper one can discern the virtuality commonality: religion deals with unseen deities [V19]; international commerce entails making deals with unseen traders [V3] and leaving one's goods out of sight in return for a piece of cuneiform [V12, V24]; writing describes unseen people, objects, long past events, and so on [V9].

Before returning to the development of human communication, one further example of such "cognitive transference" is in order—precisely because it can be seen so clearly in the archeological record, as well as adding to the virtuality interaction of another area discussed in Chap. 6: narrative art. Most art in the pre-literate era (going back several millennia, if not more) was stylized: more geometric shape and design than an attempt at realism. Moreover, there is no "running story"—each figure painted on a vessel is a self-enclosed world unto itself, although they might be part of a larger domain (demi-gods; types of grain). Then, somewhere and sometime in the Near East from 3500 to 2900 BCE, during what is called the proto-literate period, pottery painting disappeared completely—only to reappear in completely different guise. What monumental occurrence took place during this 600-year period that could explain the disappearance of pottery art—and then its rebirth as something quite different? The invention of writing.

Once Mesopotamian society began to develop writing in its early stages,[6] it seems to have devoted all its cultural efforts in the direction of this new medium. Later (2900 BCE onwards), with literacy now established as a full-fledged mode of communication, painting was "rediscovered" but in an altogether new form(at): "narrative art." Artists now took their cue from the newer medium, use art to tell stories, thus packing onto the object (urn, jar, vase) as much information as possible. Why would this happen? As a noted scholar on the origins of writing explains:

> All this may not have happened consciously. There was probably no single 'adapter' who announced: 'We shall henceforth borrow grammatical rules from cuneiform writing to tell stories in images.' Very likely the meaningful and efficient organization of signs on tablets simply suggested to vase painters a meaningful and efficient way to arrange images.
>
> In any case, it was a revolutionary development in the history of art. Although preliterate painters could produce beautiful images... they could not tell complex stories involving multiple figures. To tell such stories, painters required rules to organize the images – and they found these rules in writing. (Schmandt-Besserat, 2004, p. 55)

This is precisely what McLuhan (1964; the title of Chap. 1) was referring to in his famous aphorism "the medium is the message." It is not so much the content of each medium (or mode) that is important but rather the way the medium/mode works on, and occasionally changes, the way we think about the world and present it to ourselves and to others. Writing had several revolutionary characteristics—it was linear, it was relatively easy to learn (once we got past the ideographs into letters), and it enabled infinite amounts of linguistic flexibility.[7] Most germane for our purposes, it was purely symbolic [V24] and thereby provided its early practitioners with a mentality particularly suited to appreciating and expanding virtuality into many other spheres of life.

## 9.4   Media Virtuality: Orality/Literacy; Time/Distance

The symbols underlying what we call "writing" do not, and cannot, exist in and of themselves. They must appear on some physical artifact in order to perform their communicative function. To put this in modern terminology: letters, words, sentences are the software/mode whereas the material object upon which they appear is the hardware/medium. Together they form the "medium" we call writing.

Is writing and reading a fundamentally different communication experience than talking and listening? Also, given that writing's media "hardware" is made of atomic matter, can such hardware also make a difference regarding the virtual experience of human symbolic communication? Yes and yes. Written communication is a far

---

[6] The latest archeological evidence shows that writing was invented *independently* on at least four historical occasions: Mesopotamia around 3200–2900 BCE; Egypt around the same time; China in 1200 BCE; and in Meso-America sometime during the first millennium BCE (Mullen, 2010).

[7] The English language today has approximately three million words—all formed from a mere 26 letters. The number of grammatically correct sentences that can be formed is innumerable [V16].

greater virtual experience than oral communication; the actual *form* and *underlying characteristics* of the medium's materiality can significantly influence the *nature* of the virtual experience.

The central characteristic of oral communication—utterly banal, yet of surpassing importance—is that until the late nineteenth century all oral communication took place exclusively in the physical presence of both speaker and listener. To the modern mind, surrounded by radio, recorded music, audio lectures and books, etc., this point is hard to fathom, but to anyone before the mid-nineteenth century it was a simple truism. You could not (and would not) talk without an audience, and you could not hear another person without their being in physical proximity to you. "Text" and context were inextricably tied to each other. The lack of any material "medium" meant orality had no communicative virtuality (it did have some virtuality regarding *content*)—and insofar as speech was the main mode of communication throughout most of human history, for most people, for almost their entire lifetime, this banal point takes on central importance in any understanding of human communication and reality/virtuality.

Writing as a general mode, and more recently print and all forms of electronic communication, changed this situation radically. As Ong (1982, p. 171) pointed out: "the reader is normally absent when the writer writes and the writer is normally absent when the reader reads, whereas in oral communication speaker and hearer are present to one another." In other words, *media* of communication separate the sender and receiver of the communicative content, thereby rendering non-oral communication virtual. Text and context are now separate entities.

In what ways virtual? First, such disconnection is expressed temporally (time: V18) and territorially (space: V2; V17). Second, oral communication tends to be continuous; the speaker does not usually stop to enable the receiver to cogitate over the contents (excellent lecturers do this, but they are particularly skilled in pedagogy). Thus, audience members receive and internalize the message as presented; they aren't afforded the opportunity to think about what is being said (which is why propaganda is normally more effective in rousing speeches than in print). Moreover, as Ong argued, "[i]n an oral culture, to think through something in non-formulaic... terms even if it were possible, would be a waste of time, for such thought, once worked through, could never be recovered with any effectiveness, as it could be with the aid of writing. It would not be abiding knowledge, but simply a passing thought..." (1982, p. 35). With writing and reading (handwritten or printed), readers take charge of the pace. They can stop at any point and carry out several mental exercises, including an imaginary argument with the author of informational/argumentative material, or imagining the scene/character in a work of fiction. In other words (literally), the reader has a far greater opportunity to "add" virtual material to the reading process than possible in most forms of oral communication. Moreover, that "addition" will have permanency once the reader does what the writer did: putting it down in writing.

There is also a psychological component: in many and perhaps most cases involving the communication of substantive contents, the speaker—teacher/lecturer, priest, political or military leader, etc.—is considered by both sides to be more authoritative than the listener. The listener will therefore find it difficult psychologically to challenge the "expert" speaker (especially in the presence of other audience members), or even to ask a potentially "embarrassing" question. In the case of *reading* the same contents from the same authority, such a challenge is far easier—not to mention that the reader can turn to other written works to find answers to the bothersome questions.[8] In other words, not only is there a greater chance of virtual internal "dialogue" *within* the reader's mind, but there will also be more opportunity to expand that dialogue *between* the reader and other authors [V3]. Of particular note is the revolutionary impact this separation had on the field of education where the teacher/lecturer no longer had a monopoly on student learning; they could now be replaced, circumvented, challenged, or at least supplemented by the wisdom of others (Eisenstein, 1983, p. 35). This is related to the influence of print on abstract, logical thinking and world horizon expansion, as discussed further below.

I turn now to the second question: the influence of the specific medium on virtuality. Innis (1950) pointed out that different physical media will have a huge differential impact on their respective type of communication (and society), especially regarding time and space. Time-binding media tend to be "heavy" and durable (e.g. clay tablets, leather-bound parchment), and thus will last for generations. However, they are not easily transportable (and some e.g., wall hieroglyphics, are completely immovable), so that they cannot support a society and polity that seek control and power over wide expanses of space. One cannot run an empire based on the communication medium of clay or stone. However, this does enable a community's continuity over a long time, with high levels of stability and traditionalism as the rules of conduct (in any sphere: economic, political, religious, social) will not easily change.

On the other hand, much "lighter" media—wax tablets, papyrus documents, paper-based newspapers, radio, television—can be transported or communicated over great distances, lending themselves to spatially horizontal communication. This supports societies spanning large territories—political, economic, or cultural empires and civilizations—based on a more secular and materialist world view due to their dynamism and expansive interaction with other cultures. Unfortunately, such light media also degrade quickly. They tend to be entropic i.e., prone to relatively high information decay, so that they will not adequately support communication over generations because of their ephemerality.

What does all this mean in terms of each society's experiencing communication virtuality? As already illustrated in previous chapters, virtuality can be expressed in terms of memory [V11] or in terms of distance [V2]. Thus, a society utilizing "heavy" media will tend to be oriented historically i.e., its virtual experience will involve thinking about, and com*memor*ating the historical past. The Jewish people are a classic example: their main medium until the modern age, and to a large extent still today, has been the heavy parchment scroll (with wooden rollers adding to the weight) on which the Torah is written and kept accurate through weekly, public reading. This is the core of the Jewish people's historical memory (Yerushalmi,

---

[8] 9 The origin of the word "author" is Middle English when authors hand wrote their own book and therefore were considered to be an *author*ity on their subject.

2005), perhaps the main element in its ability to survive three millennia of serious vicissitudes. Traditional Jews continue to live more in the past than in the present, their prayers suffused with detailed collective memory based on the Bible (*Torah*), the *Talmud* (written down from the sixth century CE), and the hand-copied/printed Prayer Book.

Conversely, societies using "light" media will concentrate on the here and now—but from a virtual distance, such as the vast Roman empire, or Andersen's (1991) "imagined community" (large nation-state), discussed in Chap. 8. It can even traverse other virtual, social constructs such as national borders and religious communities in a proto-Regional sort of way. For instance, the early modern period (fifteenth through eighteenth centuries) saw the rise of the "Republic of Letters"—a community of scholars throughout Europe, the Middle East and the Americas who sent letters and books to each other through the post, "supported by physical technology that allowed transmission of the correspondence which bound it, and by techniques of virtuality that allowed the realization of this community in the minds of its citizens" [V2; V3] (Callisen & Adkins, 2011, p. 61). In such examples, the media will bind people from afar, who may not have much of a shared tradition, by setting up a virtual social construct.

American history and society offer an instructive example. In order to "connect" the increasingly great distance between America's east coast and the new nation's westward expanding borders, the telegraph was invented in the early nineteenth century. It was first widely implemented in the U.S., despite Britain, Germany and France being far more industrially advanced at the time. Nevertheless, it is only in the twentieth century—after the wounds of the civil war healed to some extent, and with the advent of radio and then television—that America could truly be called a unified nation socio-psychologically and conceptually.[9] On the other hand, given the centrality of television, the quintessentially "light" medium (and now also the internet) in the average American's life, it is probably not all that coincidental that modern American society is extremely present-oriented; historical ignorance, or at least indifference, is a given for most of American society.

## 9.5  Maps: The Virtuality of Visual Print

Before turning to print and the modern era of mass communication, we should round out the artistic picture in Chap. 6 with one additional form of visual content incorporating a few different layers of virtuality: maps.[10]

---

[9] The newspaper, a light-heavy medium (discardable, but easily archived in libraries), was exclusively local through much of American history; its central role was to maintain communal feeling in large, increasingly multi-ethnic cities.

[10] The following schematic history of map-making is comprehensively referenced in Edson (2001): "Bibliographic Essay: History of Cartography."

Maps have existed since at least Greek and Roman times. Although there are almost no extant maps from the ancient world, we know that they used maps based on several ancient authors and medieval copies—hardly surprising given the vast territorial extent of the Roman Empire and the need to understand who was where. In the second century CE, Ptolemy wrote *Cartography*, explaining how to make maps, including the idea of using longitude and latitude lines. The Chinese too have a map-making tradition going back at least 2000 years.

During the Renaissance, Europeans made numerous maps of their cities, as well as of the larger Mediterranean Basin. When the nation-state emerged in the early modern period, kings began to use maps to nurture national identity, as well as for military campaigns. Meanwhile, the Gutenberg and the Protestant "revolutions" augured a different type: historical maps, especially for the biblical period e.g., Saint Paul's peregrinations, the Israelite kingdoms, etc.

Once overseas colonialism commenced there was a great need for mapping the oceans and the New World. Interestingly enough, the explosion of mapping the Americas ultimately led European nations to map their own countries in greater detail—at first in traditional geographical fashion, and in the nineteenth century in thematic fashion. This latter was a true revolution in that maps were no longer conceived as only representing "space" but rather of other aspects of "reality," be they mineral resources, population density, topography, weather, and so on.

Notwithstanding such attempts at conveying reality graphically, print or flat maps are virtual from several different perspectives, literally and figuratively. First, and most obviously, almost all have been two-dimensional [V8] (except for the ubiquitous desk globe). The lack of a third dimension renders the map *ipso facto* partly virtual [V4; V8]. Even worse, it is inherently impossible to retain all elements of "truth" when representing a three-dimensional object in two-dimensions. A classic example: the Mercator map has had a huge influence on mankind's (distorted) view of the geo-political world (Kaiser & Wood, 2001). In 1596 Mercator drew his world map with straight longitude and latitude lines to enable better navigation. However, this enlarged the territorial size of countries closer to the poles (e.g., Canada, the U.S., Russia, Australia) and reduced the perceived size of those countries closer to the equator [V5]—the latter, perhaps not coincidentally, mostly countries we today call the Third World.

Second, and more virtually significant, "maps are not purely objective" (Tyner, 2018, p. 6). In a flat map there will always be an arbitrary decision what to place in the center, thus putting other objectively equal places or things in a subjectively (virtually) inferior position [V20]. In other words, a two-dimensional map offers only the semblance of "objective truth" [V5; V6; V31]. One could just as easily place China at the center of a world map instead of the U.S. or Europe. This does not necessarily occur in a fully 3D map where the representation of the world can approach verisimilitude. However, even here the viewer must make a subjective decision what to place at the center or how to position the three-dimensional map; there is no astronomic reason to have the globe of Earth with the Northern Hemisphere at the top and the Southern at the bottom. Thus, no matter how accurate the map and how fastidious the viewer, there will always remain a degree of subjective distortion

[V5], the feeling that one is looking at a completely accurate representation, when in fact such accuracy cannot exist.

A third, and perhaps most profound, level of map virtuality emanates from a lack of information. This was certainly the case for most of mapmaking history, when full information did not exist regarding territorial outline, or the data being presented [V4]. In the contemporary world, we have managed to successfully fill in many informational blanks for most sorts of maps, but three cardinal problems remain. First, not all geographical maps have full information: social maps cannot provide complete data (not all crimes are reported, not all residents are counted in the census); neither can many scientific maps (species variety; underwater topography). Second, geographical maps can only present a picture of one level of scale. For instance, we can have an "accurate" map of the British Isles taken as a whole, but the coastline is comprised of an almost infinite regress of Mandelbrot (1977) fractals – the lower down we go on the scale, the less smooth the coastline appears, so that the picture of the British coastline we receive from a bird's-eye map of the entire country is misleading regarding its true contours. In addition, the two-dimensional map can only provide a very rough picture of the *topographical height* of the territory under scrutiny. Third and related to this: "The mapmaker who tries to tell the whole truth in a single map typically produces a confusing display, especially if the area is large and the phenomenon at least moderately complex" (Monmonier, 1996, pp. 215–216). In other words, at least when it comes to geographical maps, clarity and comprehensiveness are at loggerheads. The map has to be virtual [V4] if it is to be understood; the closer it comes to accurately representing the complexity of reality (non-virtuality), the less useful it becomes!

Of course, one can argue that the same is true of our own vision: perceptually, more is indeed less.[11] However, there is a subjective difference between human vision and a map (re)presentation. People are aware (at least subconsciously) that we cannot literally see every level of reality, but we tend to relate to maps as if they are more "objective" than our own sensory perception. Because maps are produced by professionals with specific expertise, and they look "scientific," we tend to believe that they represent reality as it "really is."

Has the situation changed in the contemporary period with computerized, digital maps becoming *more* virtual? In two obvious and banal ways, yes. First, digital maps are *physically* more virtual, given their "byte" immateriality. Second, *quantitatively* we spend more time with maps than ever before, as digital maps are far more accessible than were standard print maps in the past (and present). Today almost all of us use computers and carry smartphones containing, or connected to, massive sources of information, so that digital maps are always instantaneously accessible.

---

[11] One of the latest theories regarding autism is that it is caused by the child's inability to filter *out* stimuli. The constant onslaught on the brain causes it to shut down. In other words, autistic people might in fact be more open than normal people to external stimuli [V21]—with the overload causing them to end up with less ability to "understand," and connect to, the external world (Frith, 2003).

However, from a substantive standpoint, digital maps have also *decreased* virtuality in three ways. First, the amount of information they can hold and represent is greater than the standard two-dimensional map, because they can offer a three-dimensional perspective. In many cases, this three-dimensional picture is not merely a representation (e.g., drawing) of reality, but rather a realistic *presentation* (e.g., satellite photos), to some extent even overcoming the "fractal/layers-of-reality problem," at least on the macro (Earth, continents, countries) and meso (regions, cities, neighborhoods, streets, individual houses) levels. Second, digital maps can "layer" different types of information in infinite ways, depending on the specific need of the viewer. Thus, for any specific country, one person can combine forestry and topography on the same "surface," whereas a second person can combine precipitation and species distribution. Third, as digital maps are connected to new information sources, digital maps tend to be more updated than print maps. Indeed, they can even present dynamic, real-time information (e.g., precinct crime reports, hotline requests, police patrol distribution). The source is not necessarily "institutional" or "official"; with social media, the citizenry can add information to available maps (such as user-generated overlays of Google maps), what Farman (2010) calls "postmodern cartography." On the other hand, the very dynamism of ongoing digital map creation lends a feeling of temporal virtuality to modern maps, as they never stay the same. Digital maps are moving closer to being "simulations" [V10] rather than snapshots of a static reality.

In sum, maps have always been highly virtual and continue to be so, albeit for quite different reasons than in the past. Indeed, there is a paradox here: the more we know about our world, the more we seem to need (or want) maps to help us navigate our surrounding factual, natural, social, and/or geographical environments. The growing complexity of reality brings forth greater dependence on map virtuality.

## 9.6   Mass Communication in the Modern Era: The Printed Word

Gutenberg did not invent print. That was accomplished by the Chinese and other Far Eastern nations in incremental stages, starting more than a thousand years earlier (The history of printing…, 2021). The first stage, around the early part of the third century CE, involved ink rubbings made on paper or cloth for ornamentation on silk. Woodblock printing of letters on paper (another Chinese invention) commenced around the mid-seventh century, with the first dated book in 868 CE (the Diamond Sutra). Next came moveable type made from clay, in the eleventh century. Wood-based movable type appeared in the thirteenth century, almost simultaneously with metal movable type that first appeared in Korea. Thus, in the Far East printed books were produced for hundreds of years, on religious and secular subjects. Nevertheless, these did not become "mass" media for several reasons.

First, for most of their history, the Chinese (and Koreans) did not have a limited number of alphabet letters, but rather thousands of ideographs made up their written vocabulary—rendering movable type printing an extremely tedious affair, as the printer would have to find (like a needle in a haystack) the appropriate ideograph for placement on the tablet. Printing remained largely a carving-out, woodblock exercise. When the Koreans eventually moved to a letter alphabet, their royalty decided to maintain a monopoly on metal, moveable type printing, so that it would not (and did not) become a mass medium. Finally, even when the Koreans began using an alphabet and movable typesetting, they continued printing whole words and not individual letters, severely circumscribing the potential efficiency of the new print medium (Ong, 1982, p. 118).

In decentralized Europe, Gutenberg did not face that problem, although over time the authorities did try to censor, license and otherwise regulate the print industry. In any case, it is called the "Gutenberg Revolution" not because he was first, but because the revolutionary consequences of print as a *mass* medium occurred in Europe and not in the Far East (Eisenstein, 1983, pp. 274–275).[12]

What was the nature of the print revolution—and why did it hugely expand virtuality in the modern world? First, the print revolution reinforced and widened a trend that had started with the invention of writing. As noted, writing physically divorced the communicator from the audience that could read the text after the author wrote it and not in the author's presence. However, at least it did not divorce the communicator from the textual medium itself (stone, tablet, papyrus) that the audience would eventually use. The print medium did add this secondary "divorce" to the communication act: first the author would write on a medium e.g., vellum, and then the printer would transfer this text to another medium, the press, that produced in turn a new form of the text, both graphically (the font, page layout) and physically (bound volume with leather cover). In other words, the recipient of the communication message was now doubly divorced from the communicator: no face-to-face contact between them [V2], and also the physical object in the reader's hand (book) was not produced by the author [V3]. Yet the perception of the Gutenberg-era audience—and to some extent even today—was not necessarily one of movement to a medium with virtual effects. For instance, we still talk of the "body" of the text or a "corpus" of textual works; books still have "head"ings and "foot"notes; avid readers are "bookworms." Books are corporeal; it is their effects that are virtual.

A second virtualized effect of print has to do with readability. Pre-print writing tended to be far less legible than print. This meant that most of the time reading was done out loud to oneself, and also to an audience that might not have been able to read or to decipher the partly legible script. (Eyeglasses did not exist until the late thirteenth century and weren't used widely until much later, so that anyone over 40–45 years old, as well as younger near-sighted people from birth, would have found

---

[12] A very loose analogy to this would be Steve Jobs who was the first (along with Steve Wozniak) to invent the personal computer (Apple II), but it was Bill Gates' more popular DOS coupled with IBM's marketing power that caused the mass market for personal computers to take off, bringing on the personal computer revolution.

it difficult to read.). Thus, reading as a more palpable auditory activity gave way in the print era to a silent activity,[13] taking place not in the sensory world but rather only in the mental world within the reader [V19; V25]. Indeed, as McLuhan pointed out, "poetry could be read without being heard, musical instruments could also be played without accompanying any verses" (1965, p. 175). To which one can add: music could now be appreciated by a professional, reading the notes—without any playing or singing.

Third, and perhaps most significant from the standpoint of virtuality is the relationship of the reader to the real world, compared to the listener. As Ong explained:

> In the absence of elaborate analytic categories that depend on writing to structure knowledge at a distance from lived experience, oral cultures must conceptualize and verbalize all their knowledge with more or less close reference to the human lifeworld, assimilating the alien, objective world to the more immediate, familiar interaction of human beings. ... A typographic (print) culture can distance and in a way denature even the human, itemizing such things as names of leaders and political divisions in an abstract, neutral list entirely devoid of a human action context. (1982, p. 42)

In other words, print divorced readers not merely from the author but also, where relevant, from the entire world as they knew it firsthand. Thus, paradoxically, printed material expanded readers' horizons to include concrete things for which they had no personal experience but did so by "virtualizing" this knowledge i.e., presenting it in abstract fashion [V8, V23]. Post-Gutenberg, our horizons widened immensely, but insofar as the print revolution is concerned, the "added territory" was virtual.

Virtuality was no less affected *quantitatively* by print. First, before print was invented writers of written works were invariably solo authors. Other than the Bible (Old and New Testaments) and other religious works, there were very few important works in the ancient or medieval world that were multi-authored. After Gutenberg, the vast majority of *books* continued to be solo-authored, but: (1) printers began to serve as book editors as well, emending, revising, changing the written text; (2) print added new forms of textual communication—chief of which was the multi-authored newspaper—some not even citing the reporters' names. The reader had little idea as to the editor's input, and many times (certainly in the later modern period) also had no idea who that editor was. Thus, the multiplicity of editors (and occasionally authors), blurred and at times even significantly distorted [V5] the provenance of the written-printed word.[14]

A variation of this was the phenomenon of feedback—a harbinger of a contemporary, revolutionary development. In pre-print times it was impossible (and also

---

[13] Some books were designed for silent reading already in the twelfth century, but most continued to be read out loud until well into the print era (Vandendorpe, 1999, p. 7).

[14] Even today very few newspaper readers are aware that the journalist almost never pens the headlines of their own news items; the editor does. This is not weak virtuality [V5: distortion] but rather "strong" [V31: mistaken]. Moreover, the virtuality of journalism has recently taken on a completely new dimension: articles that are completely written by computer, with a writing style indistinguishable from human journalists (Kotenidis & Veglis, 2021). Obviously, such "journalism" is virtual in different ways [V6; V13; and, once again regarding the writer: V31].

useless) to try and contact the author of any work with suggestions for corrections, additions and the like. The chances were close to nil of a second, revised edition appearing when the writer had to rewrite by hand the entire manuscript! At best, a new work would infrequently appear, commenting on the original; more often than not such "feedback" comments would be appended into the margins of the original work, for a few future readers to see. With the advent of printed books, however, the ability to publish revised editions became economically feasible—not to mention the far larger audience that would be interested in purchasing an updated version.

In addition, a new type of co-production came into vogue: works heavily dependent on substantive, audience contributions. Atlases, encyclopedias, scientific anthologies, and especially maps (Eisenstein, 1983, pp. 76; 200) were constantly being nourished by information sent in by interested practitioners and readers. Overall, then, here too a virtual hybrid began to evolve, only this time based on the audience as "author"-contributor. None of the readers worked as a unified collectivity; indeed, most never knew or saw each other [V2]. This is even more the case with today's massive wiki-works.

Second, instead of a minute percentage of the population being exposed to the written word (mainly theologians and aristocracy/royalty) in the pre-Gutenberg era, this proportion increased significantly and ultimately dramatically post-Gutenberg. What remained for a few thousand years an extreme, "niche culture" of literacy, was transformed into the dominant culture by the eighteenth-nineteenth centuries, and highly significant well before that. Of course, it was not merely the profusion of printed books per se that made the difference. Rather, due to the increasing availability (physically and economically) of books and other reading material (newspapers, pamphlets, learning textbooks, magazines), there was growing incentive to learn how to read and write—it was the spread of *literacy* that was the prime factor in the modern revolution. The result increased cross-pollination between the narrowly defined field of intellectuals and researchers, and henceforth also between various groups and types of profession: artists and scholars; practitioners and theorists (Eisenstein, 1983, p. 45).

Third, as noted earlier, print created a new type of disembodied, de-territorialized audience [V17]. In an oral society, the audience is always found physically together—whether listening to a lecture, attending the theater, or hearing a church sermon. Writing added a new type of "audience"—the lone reader; in the age of manuscript there was generally only one copy (or a very few) of the particular text. With the mass production of each book, it was obvious to author and reader alike that many others would be reading the work at approximately the same time (and in the case of daily newspapers, certainly simultaneously). Each reader, then, acted alone but also "belonged" to a larger, virtual audience. Among other things, this necessitated a new style of writing—one that combined an intimate approach to the solitary reader with the broad-gauge style appropriate to a large assemblage (Eisenstein, 1983, p. 58).

This had a very significant, virtualizing impact in several important areas, perhaps none more than public, social, and political life. For instance, in pre-print times local citizens would get much of their news from church sermons and gossip during (or right after) the prayer service. The print era, on the other hand, saw a growing trend to reduced church attendance and greater "communion with the Sunday paper" (Eisenstein, 1983, p. 94). This tended to weaken local-communal ties. Indeed, even the coffeehouse was witness to a similar trend: from lively public discussion to the "sullen silence" of solitary newspaper reading with coffee and torte. All these led to the evolution of modern individualism and the final nail in the coffin of medieval, collective ("corporate") solidarity. To be sure, modern society did not become completely atomistic; rather, it took on the character of being "virtually communal" [V2], Anderson's "imagined community" (1991), as outlined in Chap. 8.

Fourth and finally, print not only created an imagined community but also preserved natural "communities" (in the broad sense of the word). As oral culture tends to be more fluid than print culture, it suffers from greater entropic information loss. This is especially true regarding language. Vocabulary, spelling and even pronunciation changes faster over time in spoken settings than in writing/print where "exactitude" is more highly valued. Over the course of time, exclusively oral societies will bifurcate (due to geographical distance, religious schisms, etc.) into different linguistic groups that cannot understand each other's speech. Of course, there are no purely print cultures, but societies that are heavily dependent on writing/print, conversely, will have a "lag" in such fluidity (at least on paper, if not in the street). Indeed, some will even evolve a dichotomous language: one for speaking and one for literary production e.g., Arabic.

The result of this is that here too print can lead to a larger, if somewhat virtual, community of fellow cultural members that might not be able to speak to each other but can continue to communicate through print [V2, V17]. Chinese is perhaps the best example of this, a language with several different, mutually incomprehensible dialects, and yet through a common writing system has managed to maintain cultural unity—not only between people of the same generation. As Chinese ideograms have changed so little for 3000 years, any literate Chinese person today can read texts from three millennia ago [V18] (Logan, 2004, p. 54).

Print also had a combined, qualitative-quantitative impact, in that it reinforced and hugely expanded human cognitive abstraction, systematic analysis and linear thought. As Ong explained: "an oral culture simply does not deal in such items as… abstract categorization, formally logical reasoning processes, definitions, or even… articulated self-analysis, all of which derive not simply from thought itself but from text-formed thought" (1982, p. 55). To be sure, within oral societies there were a very small number of people who were capable of such high-level thinking but they invariably were the ones who were also literate. The vast majority of illiterates constituting the bulk of society set the tone for general cultural practice and its mindset.

What precisely is the connection between writing/print and logical thinking? The connection is the same one noted earlier on the macro level regarding print being the first modern assembly line product. First, the act of writing by hand was by its very nature highly personal—no two same letters or words ever looked exactly the same. Similarly, pre-industrial production by artisans was idiosyncratic; no two products of any specific artisan were ever identical. Conversely, print produces precisely reproducible letters and words—just as modern assembly line manufacturing produces an endless number of *identical* products. Such print exactitude, however, not only influenced manufacturing technology, but more significantly it led to a sea change in science itself: "Exact observation does not begin with modern science.... What is distinctive of modern science is the conjuncture of exact observation and exact verbalization: exactly worded descriptions of carefully observed complex objects and processes" (Ong, 1982, p. 127). In other words, the *visual* exactitude of the printed letter/word carried with it the search for *substantive* semantic precision.

Second, it is almost trivial to note that reading is linear; words must follow a linear order set down by the author. Similarly, the modern, linear production process creates everything in the "correct" order repeatedly—building the product from its basic components to its final, large incarnation. However, what is far from trivial (and also not universally understood) is the fact that this linearity is exactly how science and math work inductively: building and reaching large scale conclusions or theorems from the systematic accumulation of discrete pieces of evidence or axioms. Here is another example of "transference" mentioned earlier: the act of reading print (one word follows the next) trains the brain to think linearly and scientifically. In any case, the linearity of reading as opposed to the associativity of speaking led to the more rational-logical cognition of the post-Gutenberg Age—in turn bringing on the modern scientific revolution with all the virtuality inherent in the sciences, as discussed in Chap. 5.[15]

This is not to suggest that print only brought about expanded virtuality. As just noted, print also led indirectly and directly to the Industrial Revolution that mass produced *material* goods in quantities beyond anything previously known to humanity. Indirectly—through the concept of assembly line work; directly—through universal education and training masses of workers with cheap textbooks and other printed vocational training materials. Moreover, whereas the symbolic printed text (mode) and the dissemination of abstract ideas (content) were heavily virtual, the medium of textbooks, newspapers, pamphlets, fictional novels, was very material. Like other manufactured goods, we have been inundated with quantities of concrete *media* of communication beyond anything seen before in human history.

Another area in which print reinforced *and* weakened virtuality is "historical memory." In ancient times, the authorities as well as wealthy private citizens would go to great lengths to store official documents and books of general knowledge. However, given the very limited number of copies of each work, many would invariably be

---

[15] Logan (2004, p. 69) makes the following, thought-provoking but controversial point: "The linear, sequential mode of building a system that the [Western] alphabet encouraged and Chinese characters discouraged also influenced industrial development in the East and West. Despite their technological progress, the Chinese never linked their inventions together to create assembly-line production characteristics of the Western Industrial Revolution. The lack of abstraction in the writing system reflects itself throughout Chinese thought and discourages the development of the abstract notions of codified law, monotheism, abstract science, and deductive logic." In other words, the less abstract Chinese ideograms that were also not as "linear" as letters forming words, conceptually prevented the Chinese from developing a rational-logical science as well as modern manufacturing meta-technology. If this is true (a big if), then Chinese communication (and that of other Far Eastern cultures based on ideographic writing) would tend have a lower level of content and modal (but not necessarily media) virtuality than that found in the West.

lost, or miscopied by scribes in a generational version of the modern game "broken telephone" (Eisenstein, 1983, pp. 78–79; 199). As a result, the mostly oral culture played somewhat fast and loose with history, being based on naturally imperfect human memory. Thus, in oral cultures the past was either non-existent or perpetually malleable [V5].

Print changed this immeasurably. The past could now be documented in numerous copies, held by many readers, so that each generation had a largely unified "collective memory" that did not differ too much from that held by their forebearers. From the standpoint of virtuality, this cut both ways. On the one hand, the known historical record was far less distorted and thus much less "virtual"; on the other hand, print society had far more of the past to remember i.e., virtuality increased, insofar as "living in the past" is virtual [V11], as explained in Chap. 2 (Walter Benjamin et al.).

Overall, then, print offered somewhat of a mixed bag from the standpoint of virtuality. Its impact on the material world was great. In my estimation, however, its influence on the world of abstract thought, its distancing of thinkers from their audience and concomitantly enabling that audience to respond inwardly, as well as the spread of imaginative (fictional) literature as described in Chap. 6—collectively constituted a net increase of virtuality.

## 9.7   Conclusion: The Evolution of Communication Virtuality

To sum up this chapter, let's return to its title. Without being too deterministic and inflexible, it becomes clear that *once humanity developed "civilization"* the history of human communication roughly followed the title's chronology: first, communication *to* the audience; then, *with* the audience; and finally, *by* the audience.

At first, most oral *mass* communication ran from the "top" down: political rulers, priests, military leaders, and economic elites all delivering one-way messages *to* the masses. This meant that the vast majority of communication for the average person, including dyadic/interpersonal, was not virtual at all. Only on the rare occasion when an edict was delivered from on high, would the common folk receive virtual communication. Then at a later stage—during different historical periods for each public sphere, as the socio-political unit became larger (city-state, nation)—virtual communication became more bi-directional i.e., it took place *with* the masses through social elite intermediaries capable of writing and reading. In other words, even if the vast majority could not read, they would be made aware of distant [V2] communication by those who could read (e.g., clergy sermons quoting the Bible). Finally, in the early modern period, when the scope of literacy widened significantly, virtual communication (writing, print) turned truly multi-directional—*between* the masses, as they increasingly initiated the communication (personal letters; commercial documents; novels and poems) and were not just passive receivers.

From here we now turn to the huge quantitative and qualitative boost of person-to-person as well as mass-to-mass communication in the electric era (nineteenth to mid-twentieth century) and the ensuing electronic era (mid-twentieth century to the present). In turn, the invention of the camera (still pictures and later video), telephone, typewriter, and ultimately the internet, aided by other (more top-to-bottom) electronic media such as cinema, radio, and television—all had a huge impact on, appeared parallel to, and were influenced by, the virtualization of other areas of life. All these are the subject of Chap. 10.

# Chapter 10
# Virtuality's Expansion in the Modern Era: Media and Society

> *[E]ven if it is true that there is nothing new under the sun,* **the**
> **new spirit does not refrain from discovering new profundities**
> **in all that is not new under the sun** [emphasis in the original].
> (Guillaume Apollinaire, 1971, p. 232)

## 10.1  General Overview

Until the nineteenth century, even in modern Europe, fully 75% of all residents never traveled farther than a 50 km radius from where they were born *during their entire lifetime!* In short, "real" life was extremely circumscribed—people had no direct experience with social existence other than their own, very limited purview.

Nevertheless, newspapers reported on phenomena and places that the average European had never known even existed [V29]: exotic lands, animals, flora and fauna; foreign cultures; new technologies and material goods; scientific advances; ideological treatises. The world itself had begun to "expand," not just in the literal, territorial sense of discovering and conquering the New World, but also because for the news reader this was occurring in large part only virtually. These phenomena remained mostly beyond their "real," immediate social horizon [V19]—until the transportation and especially electronic communication revolutions emerged in the nineteenth and twentieth centuries that witnessed an explosion in the number and types of new media: telegraph, camera, telephone, phonograph, cinema, radio, television, and the computer (in its myriad incarnations)—not to mention significant expansions of older media (penny press, dime novel, comic book).

Part of this overall development was directly related to, and a function of, the Industrial Revolution's breakthroughs in electricity, acoustics, physics and material sciences. Another important factor was the rapidly changing social and political

---

The *Appendix: A Taxonomy of Virtuality*—listing by numbers 1–35, e.g. [V13] or [V34], some of the various types of Virtuality mentioned throughout this chapter—is freely available to the public online at https://link.springer.com/book/10.1007/978-981-16-6526-4

milieu: expanding frontiers and the need to connect populations between distant places (telegraph in the U.S.), rapid growth of large corporations and their need for more efficient internal and external communications (telephone), filling greater amounts of leisure time for the masses (cinema, radio and TV). And finally, society's vastly transformed intellectual-perceptual mindset—more cognitively abstract, more culturally variegated, more socially and scientifically future-oriented. Whereas the first two general factors were critical in ensuring the continued invention and development of these new means of mass communication, the latter factor (along with greater leisure time) explains why the general public was attracted to these new media offering greater amounts of virtual fare.

The mode of *transmission* underwent a gradual transformation from physicality to virtuality over time: *wireless* communication instead of cable/landline. Marconi's discovery of wireless communication removed the physical connection from sight, thus transforming twentieth century new media into hybrids of the physical and the virtual. Radio, television (regular over-the-air and satellite), cellphone, WiFi, Bluetooth, GPS—the world became increasingly interconnected without any trace of what's connected to what, where and how [V25].

In addition, the "new media" offered several forms of novel *content* to be consumed passively: terse messages from far-off places (telegraph) [V2]; perfect representations of real life (photography) [V8]; disembodied voices and sounds communicating in real time (telephone; radio) [V3]; recorded, lifelike sounds emanating from an artificial disk (phonograph) [V25]; two-dimensional moving pictures reflecting three-dimensional reality (cinema and TV) [V8]. All these were virtual only from the receiver's perspective because the process of creating such content was decidedly "real": the photograph's subject was real, the recorded music on the phonograph/cassette player/CD was played by real players and instruments, the TV/movie filmed actual actors.

A third level of increased communicative virtuality involves the distancing of the audience [V2], for what are "media" if not inter*media*ries between presenter and audience, transmitting and receiving messages from a distance? The distancing trend also emerged in several new media established to present their virtual wares through some sort of audience physical presence. For example, movies were themselves virtual products [V8; V9; V22; V25] but demanded the audience's on-site theater presence.

However, as time went on newer media caused a physical decoupling of the original medium from the viewer's attendant presence: movies on TV, on DVD players, on the computer. Nor did this distancing trend affect only new media; it has occurred in almost areas of cultural consumption: "Essentially, any activity that requires us to travel to a venue, take a seat, and watch people performing in some disciplined fashion is not as popular as it used to be. We'd rather be at home, plopped in front of our 'glowing rectangles,' to quote *The Onion*" (Ross, 2010, p. 66).

Fourth and paradoxically given this "proximal distancing," virtual communication has increasingly involved the audience more deeply in the process of *creating* virtual content media, whether producing virtual content from real sources [V1] such as the camera, or creating virtual (digital) content *ex nihilo* [V28], through computers that enabled computer graphics and games, podcasts, websites, video clips, etc.

All this was accompanied by social trends from the 1830s onwards: increasing education entailed some semblance of independent (if not fully creative or original) thought; rising income levels enabled people to purchase production and consumption media. As a result, the quantity, scope, and intensity of virtual communication increased almost exponentially. By the mid-to-late twentieth century entire industries of virtual products arose, with millions of professional and semi-professional workers earning a living producing virtual products of higher quality, in addition to the hundreds of millions who did this as a side hobby. This led to a spiral of ever-increasing virtuality—creating virtual products whetted the "producer's" appetite to consume those of others and increasing consumption of virtual entertainment and information products encouraged amateurs to start producing (e.g., family photography; blogs).

As a result, people began to consider virtuality as an integral part of their reality. Our forebears mostly had to actively *imagine* a virtual life (spirits in the woods) or perform mental *creatio ex nihilo* at very specific moments and places (place of worship), whereas modern age world denizens are being surrounded, enmeshed and bombarded with virtual experiences, *visual* (photos, video) and *aural* (telephone, audio players, radio). Paradoxically, the increasing ubiquity of virtuality led it to be perceived as a "natural" part of the social landscape. The sheer *volume* of "weak" virtuality in the early Industrial Age was highly effective in paving the way for the acceptance of increasingly "stronger" virtuality in the twentieth century that combined disembodiment with fantastical representations of (un)reality—from Quantum Mechanics to 3D video games.

In turn, those virtual modes of presentation and perception paved the way for a social and conceptual revolution: by the twentieth century, virtuality was no longer a "niche" activity in human life but had become part and parcel of "real life" itself, qualitatively and quantitatively. We now turn to the main means of communication that formed the axis of this contemporary virtuality "onslaught."

## 10.2 The Virtuality of Modern New Media[1]

I will first describe each new medium and how its inherent "logic" (infrastructure, means of transmission and consumption, internal development) strengthened *and* weakened virtuality—and then show how modern/contemporary media have influenced virtuality in other central areas of life.

---

[1] I will not survey all new media, although the majority will be mentioned in various contexts throughout this chapter. Again, the book's two theses do not depend on a *complete* survey of virtuality in every area of life, but rather on the evidence that virtuality is a *central* component of human existence.

## 10.2.1   Telegraph

Despite its technological simplicity, the telegraph was one of the most important media ever to be invented as it ushered in the new age of electric communications, abolishing distance *instantaneously*. Communication at a distance [V2; V18] had been carried out for millennia: carrier pigeons, letters by courier, sequential hilltop torches, books and newspapers. However, the telegraph was a watershed in introducing *synchronous* communication [V2] between distant parties.

From one perspective, the telegraph *weakened* the virtual, communicative experience. Writing a letter to someone you cannot see, are not sure when they will receive your message, and whether/when you will receive a reply [V15; V26], is obviously a highly virtual process; much less so when the receiver gets the message almost instantaneously and replies shortly thereafter [V3].

Nevertheless, the telegraph *increased* virtuality from two other standpoints. First, given that in pre-modern history most people were illiterate, they could not engage in distance communication.[2] On the other hand, even if the telegraph was a highly circumscribed medium, it did enable even illiterates to communicate at a distance for a small fee. Thus, virtual communication at a distance expanded tremendously through the telegraph. Moreover, when using a telegraph operator, the message sender was now twice cut off from the receiver: first the operator, then the medium itself [V3]. Finally, given the cost of telegraph transmission, the contents sent and received were extremely succinct, with the receiver having to "fill in the blanks" [V4].

The telegraph had an important influence in increasing virtuality in another important medium: it provided a huge boost to the newspaper industry by enabling the audience to read really new news and not things that had occurred days, if not weeks, earlier. In the process, however, it eliminated the literary author as journalist, for the telegraph demanded (economically), and led to (stylistically), a much pithier form of reportage. While the news content was shorter, leaving out facts [V4], it also eliminated most irrelevant ("literary") writing [V9].

## 10.2.2   Photography

Photography was paradoxical: decreasing virtuality qualitatively while simultaneously increasing it quantitatively.

From the Renaissance "discovery" and refinement of perspective in Western visual representations, there was a steady drive to attain perfection in the representation of reality (Bryson, 1983). Cavell even imputes a metaphysical motive to this trend: "the human wish, intensifying in the West since the Reformation, to escape subjectivity and metaphysical isolation" (Cavell, 1971, p. 21). Photography, therefore, is the culmination of this historical process in which art moves from two-dimensional

---

[2] The non-literacy examples of jungle drums, hilltop torches, and other such distance communication techniques were "niche" media, and severely limited in message content.

representation [V8] to precise duplication [V1], and from a subjectively misperceived [V20] depiction to one of "objective" representation [V8].

On the other hand, few people could paint well, and even fewer had the resources to purchase decent artworks before the age of mass duplication. As a result, until the nineteenth century, there were very limited possibilities for works of visual representation within the home. Public visual art was somewhat more commonplace e.g., monuments, church statues, and stained-glass windows.

The invention of photography in the 1830s[3] changed this situation immeasurably, inaugurating the era of mass reproduction that included many other media invented over the ensuing two centuries. Walter Benjamin (1973) called this the Age of Mechanical Reproduction—the ability to reproduce works of art ad infinitum destroys the "aura" of such original creations. Whether this renders the reproduction more, or less, virtual depends on whether we view the artwork as primarily virtual in and of itself due to its representational character [V7; V8], or we consider it to be mainly an "original," real-world artifact. However, there is no denying that a *reproduction* of artwork is not "original" in any sense, ergo it will always have some degree of virtuality, whether as duplicate [V1], substitute [V6], simulacrum [V12] or artificial [V13].

Photography is not employed primarily to reproduce original art but rather to capture the present moment, in the process weakening and strengthening virtuality. When we open a photo album our memory is jogged into remembrance of things past [V11], but that specific memory is also less virtual because of the photo's verisimilitude (compared to a painting), reducing distortion [V5] related to that scene. Here we find a *qualitative* reduction of virtuality along with a huge, *quantitative* increase: family albums, outdoor billboards, photojournalism, and so on. In short, wherever we may be "in reality" at the present, photography also exposes us to its virtual equivalent from elsewhere and "elsewhen."

The shift to digital technology gave a further boost to photography's quantitative virtuality. Original "film" photography was expensive. Digital photography eliminates almost all costs by canceling most components and enabling the presentation of the "final" photo on different platforms. With easy erasure and very cheap digital storage we are now free to click away in abandon.

Digital editing constitutes an even more important influence on the virtuality of photography. With analog film, photographic editing was cumbersome, usually producing an awkward, "hybrid" photo [V5]. Digital technology enables easy and imperceptible "editing": color, background, shadow, blemishes, size [V19; V30; V32]. While this enormously expands our "artistic" capability in producing more aesthetically pleasing images, it also presents serious problems in journalism and law, for the adage "what you see is what there was" no longer applies [V11; V31]. Thus, the greater ease and ubiquity of digital photography is a doubly virtual phenomenon:

---

[3] The first photographic image was produced in 1827 by the Frenchman Joseph Nicephore Niepce, but it took eight hours to create and then faded quickly afterwards. Louis Daguerre formed a partnership with Niepce, and twelve years later (after Niepce passed away), successfully created the first commercially viable photograph, then called a "daguerreotype."

further expanding the quantity of two-dimensional, pictorial imagery [V8] in our world as well as introducing serious doubt as to the verisimilitude of any specific photograph [V32].

### 10.2.3   Telephone

The telegram and photograph are *visual* artifacts, successors in a long historical line of similar objects: cuneiform, sculptures, codex, paintings, book. The telephone, on the other hand, had no antecedents as it was based on *sound* "reproduction"—the first truly novel medium of virtual communication in the modern era.

However, the telephone actually *decreased* virtuality. Not only did it afford synchronous, real-time communication (no temporal delay), but it also increased the "realism" of the initial communication to some extent. As opposed to writing [V24], phone conversations remove the symbolic elements, employing natural language and voice. The leftover virtuality is rather weak: communication at a distance [V2] with partial disembodiment [V25] i.e., no body language, but still retaining intonation, cadence, and pitch.

Here too, whatever virtuality is lost qualitatively is more than made up for by a huge quantitative increase. Before Alexander Graham Bell (Grosvenor & Wesson, 1997), all vocal conversation had to be conducted face-to-face, with all its logistical difficulties for people who were not in proximity. The telephone immeasurably expanded our range of social contacts so that in the Age of Bell we converse at a distance (again, V2) with far more people than was ever considered worthwhile or possible.

As I noted above, there was another revolutionary aspect to the telephone—a step back from the mass media. For the first time in history, we had a medium for the masses that enabled the average *individual* to easily utilize a virtual means of communication. Until the telephone, all mass media "production" was beyond the technical or financial capabilities of the common person: one didn't just set up a printing house, establish a newspaper, or even run a line from home to the local telegraph office. But (almost) everyone could "talk" and learn how to dial the operator, while the cost for a local call was relatively low. This led to a significant increase in citizen communicative empowerment while also familiarizing the average person to virtual sound communication—preparation for sound recording and radio in the twentieth century.

### 10.2.4   Sound Reproduction (Music Recording)

Edison's invention of the phonograph was a return of sorts to mass media, virtualizing the entire field of sound. First, the phonograph distanced the listener from the player/producer [V2], as did the telephone. Such distancing also involved informational loss [V4]: body language of the musician(s) and the conductor, communication between the conductor and orchestra, importance of specific instruments at different points playing the musical composition—not to mention recordings of musical theater and standup comedians that remove even more critical visual stimuli. By the end of the twentieth century, the amount of time spent listening to recorded music (on cassette players, boomboxes, Walkmans, Discmans, MP3s) was vastly greater than the time spent at live performances.

Second, recorded music distanced the listener not only geographically but *temporally* as well. For the first time in human history, we could hear sounds produced in the past [V18]—beforehand found only in textual communication. Hearing music played by someone no longer alive was an astonishing "spectral" experience [V25].

Third, once again convenience trumped verisimilitude. Until the recent digital age, sound recording was not capable of reaching the quality of live music—and even when it did, physical entropic processes on the vinyl record quickly degraded the sound quality after several hearings [V5]. The mass audience was, and is still willing, to put up with this because of the far lower cost and greater convenience of listening to purchased music anywhere, anytime.

Fourth, and in the long run perhaps the most important virtualized phenomenon of all in recorded music: the creation/playing of music by non-humans, as well as the creation of non-instrumental, "unnatural" sounds [V13]. In the early 1960s, the Moog analog synthesizer ushered in an era of musical sounds without parallel in the real or instrumental world, also enabling the alteration of pitch, volume and timbre without humans replaying the music [V6]. Almost simultaneously (1957), playback of digital music by computer was developed (Grimes, 2011).

The later and more technologically advanced recording employing *digital* technology, reduced certain aspects of virtuality by removing background noise, emending playing errors, and in general producing a recording that came very close to what the player(s) viewed as "authentic" or "ideal"—a musically "cloned" expression of what the composer/musician had in mind [V1]. In a sense, this was paradoxical "authenticity," as such music was more exact than even the greatest musicians could produce in one sitting when playing live.

Electronic technology enabled a different type of virtual musical expression to flourish: "mashup" music (otherwise known as "sampling"). This was artificial, second-order, musical "composition" [V13]—different combinations of musical fragments were "borrowed" from various, previously composed, musical pieces. Thus, much contemporary music lost its internal coherence or original voice—a musical hodgepodge/mélange from various sources, put together ad hoc [V15].

Finally, voice recording involved a different kind of paradoxical reduction of virtuality. When our voice is taped, others hear the recording much as they hear our voice in person, but the individual speaker or singer does not hear the recorded voice as she hears herself live! That's because we hear our own voice through a dual aural track: 1—sound leaving our throat, reverberating in space, and then returning to our ears (as everyone else hears us); 2—sound traveling from our larynx through the bones in our skull into the middle ear's cochlea. Thus, we hear ourselves differently than everyone else hears us [V5]—but it is only when hearing ourselves "on tape" that we become aware of this discrepancy. Here virtuality is reduced for the speaker-as-listener when our "authentic" voice (in the sense of how the world hears us) is revealed to us as well.

### 10.2.5 Cinema

Whereas photography "unnaturally" froze the picture in a moment of time, without normal temporal context [V18], cinema decreased virtuality by presenting natural movement. Nevertheless, during its first thirty years cinema still lacked two critical, real-life elements: sound and color [V4] that commenced in 1927 and 1934, respectively.

However, the main virtuality element remained the *way* the technology fooled the viewer's perceptual process. With analog photography "what you saw was what you got," other than the occasional pictorial illusion. Cinema, on the other hand, is based on fooling the viewer's perception: 24 frames a second are projected onto a screen, too fast for our eyes to see the individual images flashing by. Instead, our brains "fill in the blanks" by creating a whole, continuous picture that seemingly moves seamlessly [V20]. Our eyes and ears working together also play tricks on us through "visual capture": although loudspeakers are placed off to the side, our eyes "capture" the sound as if emanating from the actor's mouth on the screen [V20] (Degault, 1994, p. 15).

The next improvement was the wide screen: Cinemascope, Cinerama (three film projection cameras on a large, curved screen covering the entire field of human vision), Ultra Panavision 70, and the IMAX system with its gigantic screens and far greater picture resolution, offering an extremely "real" sensory experience. All had the same goal: offering the moviegoer an experience as close to real-life as possible [V1].

In cinema, similar to photography, the digital revolution has affected the virtual experience, with an important difference. Whereas most digital "manipulation" of the photograph renders the picture more *realistic* or aesthetically pleasing (e.g., removing wrinkles), digital movies create fantastical or not-altogether-real images for their *entertainment* value. Thus, digital photography tends to misrepresent reality [V31] whereas digital cinema tries to present a realistic-looking unreality [V28; V35] e.g., the *Toy Story* series' realistic-looking toy characters. Newer developments in digital movies, however, are increasing cinematic virtuality e.g., "synthespians" ("virtual actors")—digitally recreated, completely photorealistic, screen characters who were not filmed but rather digitally manipulated (Tangermann, 2019). These are either deceased movie stars making a "comeback" in brand new movies,[4] or

---

[4] Food for thought: is there copyright protection on the actual image of someone deceased? For whom: the deceased's estate or the movie producer?

quasi-synthespians: (still) live actors that have been digitally "de-aged" (Golding, 2021).

Digital movies have also begun to slightly decrease virtuality through 3D cinema [V7], restoring the lost dimension. Of course, 3D movies can actually increase virtuality *qualitatively* if they strengthen our "visual belief" that what appears on the screen, not actually filmed in real life, is in fact "real" [V31].

In sum, cinema's evolution does not support continual progression towards greater virtuality. It does involve the brain's "movement reinterpretation" of sequential still pictures and visually fantastical characters (heightened virtuality), but it also added and improved sound, color, wide screen verisimilitude, and realistic picture resolution (lowering virtuality).

## 10.2.6 Radio

Wireless audio ("radio") transmission, gradually developed and improved in the late nineteenth through the twentieth centuries, was arguably the most virtual communication technology up to that point in history, for quantitative as well as qualitative reasons.

Qualitatively, wireless transmission seemed to be "magical" [V25] as there was nothing physical to connect the communication between parties (as with the telegraph and telephone) who were literally out of sight and sound of each other [V2]. True, regular human speech works "wirelessly," but this was always considered "normal"—sound waves can travel short distances without much informational loss. However, to do so over huge distances was not thought possible [V2].

For the first twenty years its central use was point-to-point communication, such as land-to-ship telegraphic transmissions and "ham radio": long-distance, person-to-person conversations. "Wireless" did penetrate the general population's consciousness as a marvelous technology of virtual communication but few saw it working up close.

Once the technology changed its emphasis to point-to-mass audience transmission in the 1920s (then called "radio"), the immediacy of such virtual communication became viscerally clear to all. At the most basic and immediate level, this medium significantly reinforced the sense of virtuality in the air—literally and figuratively—regarding other virtuality elements such as quantum mechanics, psychology, modern art etc., all of which entered the public sphere at the same, general, historical point in time, lending the world an aura of omnipresent incorporeality [V25].

Initially, radio programming was doubly virtual: phonograph music was broadcast, so that the listeners were hearing a mediated *transmission* of a mediated *recording* [two times V3]. Soon live musicians played in the studio, and later actual music hall performances were broadcast live. Nevertheless, throughout radio's history a significant portion of programming—news, lectures, or even dramatic readings—was broadcast after taping; double mediation continued to be part of the listening experience, rendered even more virtual when the audience was not told that the performance was taped, listening under the misimpression of sharing a live experience [V31].

Nevertheless, radio also weakened virtuality. Quantitatively, radio brought useful information regarding the outside world directly to *billions* of illiterate people (farmers and blue-collar workers) who were informationally cut off from the rest of humanity because of their inability to obtain minimal "professional" training and guidance. What was previously completely "unseen" and unknown [V19; V27] became potentially actualized through educational radio programs [V2].

Current events constituted a second type of virtuality dilution. Other than the occasional newspaper picture, public personalities were invisible for most of the general public. Citizens had no real sense of what these people were like [V19]— and if they did, the public "image" was in fact more likely fashioned to put the personality's best face forward [V5; V6] e.g., hiding U.S. President Wilson's stroke; FDR never shown sitting in his wheelchair. Through voice, radio lent a key human clue to the leaders' nature—important personages becoming less virtual, albeit still incomplete [V25]. Similarly, events: World War II daily radio reports brought sounds of warfare, providing a richer sensory experience than news photos—although the weekly, movie theater newsreels provided an even stronger experience [V8], albeit less frequently to a somewhat smaller (ticket-buying) audience.

Qualitative virtuality further decreased with FM radio that provided far higher audio fidelity [V1], but simultaneously increased with the huge expansion of FM broadcasting stations, enabling wider audiences to be exposed to far more of the world in mediated fashion [V3]. Contemporary digital radio further reinforces these counterbalancing trends.

Finally, radio was the first *electric* medium to introduce a previous, important virtuality-inducing phenomenon: media-use mobility. The transistor radio, not different in its virtual content from the home set—enabled all the above virtual effects to expand quantitatively from the 1950s onwards, as the amount of radio listening time increased tremendously once freed from connection to the wall outlet. This set the stage for all the ensuing recorded music players (e.g., Walkman, iPod etc.)

## 10.2.7  Television

The 1950s mass market introduction of TV continued the dual, countervailing trend to more and less virtuality. External reality now entered people's homes aurally *and* visually. The distant, outside world, previously not well sensed if at all [V19], was now merely a mildly incomplete version of the original [V4].

This trend toward demystification of the modern world was especially pronounced among population groups that previously were mostly kept out of the public loop: children and women. As Meyrowitz (1986) pointed out, television was the first major medium that displayed (albeit in somewhat distorted fashion) the world of work, politics, and other important areas of public (and to some extent even private!) life to those who in the past had little access to such social information.[5] This paralleled in news and social life what radio afforded Third World illiterates in vocational knowledge, training and education.

In several other ways, however, television heightened the average person's virtual experience. For one, television captured a major part of people's leisure time; instead of "real-world" activities, children and adults now spent more of their life indulging in virtual entertainment with the Tube. One of the ironies is that while it expanded our virtual mental life, it also led to significant "expansion" of bodily corporeality: a serious rise in "couch potato" obesity!

Second, once television set prices dropped, TV viewing turned from a social/family event into an individual experience, with each family member holed up with their own screen. This was also due to multi-channel, cable TV that provided specific television fare for each age and taste. Television viewing became more anomic, removing viewers from their natural social environment [V3]. This trend was further strengthened when the internet became the dominant medium of choice for many.

Third, early television was technically deficient—fuzzy picture, poor sound quality, lack of color, small screen; the picture was significantly different than the reality it supposedly (re)presented [V4; V5; V8]. Over time, these deficiencies were reduced and even eliminated: color, cable TV without visual distortion, High-Definition TV; stereo and "surround-sound" vastly improving sound quality, progressively larger, life-sized screens. In short, from a qualitative-technical standpoint, TV progressed from being a medium of high "noise" [V5]—to use the general communication nomenclature of Shannon (1948)—to almost perfect, two-dimensional verisimilitude [V1; V8], with 3D TV in the offing [V7].

Fourth, newspapers, cinema, AM radio and early, over-the-air television were true "mass" media. For economic and/or technical reasons the number of venues/channels were limited in any given geographical area, so that each newspaper, movie theater or radio channel on average could attract a relatively large audience. Thereafter, "narrowcast" cable television, and a bit later satellite TV, revolutionized the medium with dozens of channels available, enabling entire new types of virtual content to appear, such as music video, science fiction, romance, cooking, history, nature, shopping, religion etc.—content with differing degrees of virtuality that had little if any prior exposure on radio or early TV.

Internet streaming television (e.g., Netflix) or video (e.g., YouTube), as with internet radio and e-papers, have further sliced up the audience to the point where we can now talk of "slivercasting" (Hansell, 2006), reaching extremely small niche audiences on virtually any conceivable subject. In other words, whatever virtuality is to be found in the real world ("offline") can now be found in the hyper-virtual, "online" world as well.

---

[5] The consequences, in Meyrowitz's estimation, were revolutionary. It was not a coincidence that the Sixties' socio-political revolutions—civil rights, anti-war, feminism, and environmentalism—were produced by the first generation that grew up on television, just as they reached adulthood. TV's demystification of the world through their formative years gave them self-confidence that they knew better (or were no less intelligent) than the official policymakers.

## 10.2.8   Computer

The computer constitutes an anomaly because it is the only "medium" that was not designed to be a means of communication, but rather to aid in statistical computation. It becomes a new medium only in the late 1960s with DARPA's Arpanet, graduating a decade later to transmitting word processing, graphics, video gaming and the like.

Given its numerous sizes and functions—from the tiny music iPod, small smartphone, compact tablet, larger laptop, even larger desktop, the big microcomputer, to the supercomputer array and huge complex of servers—the "computer" is an ambiguous [V15] medium several times over: not always a medium of *communication*; constantly changing its main communication functions; metamorphosing in shape several times over.

Charles Babbage's original 1820s Difference Engine "computer" was purely mechanical, and as such completely non-virtual, transparent for all to see. A century later the computer became electric (vacuum tubes) and then electronic (transistors, microchips). At that point its "machinations" became invisible [V19], and even if the user could remove the outer metallic shell, its processing still would not have been visible or comprehensible.

Nor did the computer's intrinsic virtuality end there. As opposed to all other *analog* media that manipulated natural signals (e.g., radio's sound waves), even if in "unnatural" mechanical fashion [V13], the modern computer uses binary, "digital code"—symbolic language [V24]—to produce output. Moreover, the computer is also the basis for two other highly important, interrelated, virtual technologies to be discussed in Chap. 13: artificial intelligence (AI) and robotics. Thus, the computer also acts as a platform for corporeal virtuality (humanoid robots—V1; V6] and incorporeal virtuality (AI—V23; V25].

## 10.2.9   Internet

From the standpoint of virtuality, the "internet" could fill an entire book. Indeed, coupled with computers (= compunications) this is the realm of "hard core" virtuality, the focus of most scholars who deal with virtuality. Yet, even within this relatively focused field there are not only different ways in which the "virtual" is defined but also ambiguities in the term itself. The reason for this is not only the "amorphousness" of the term "virtuality" (addressed in earlier chapters) but also the fact that not everyone who uses the term "internet" is talking about the same thing.

The latter problem is that the "internet" can be viewed as three different things. First, a *medium*: as a channel for transmitting communication content e.g., through websites. Second, a *multimedium*: enabling and performing several quite different communicative modes of expression: email (letter writing), forums (group discussion), shopping (commercial transaction), e-books (reading), music (recorded: sound), and so on. All media in the past performed one basic function: telegraph = sending short messages; movies and TV = visual entertainment/information; telephone = audio communication. The internet, on the other hand, enables communication of many types of stimuli in almost every way possible.

Finally, it's also a *metamedium*—a platform for other media functioning outside of their respective classical mode. For example, telephony (Skype), e-news (without the "paper"), radio (without a broadcast license), video (YouTube), conferencing (Zoom), and cinema streaming (Netflix). Given its own ambiguous or multiple types of identity the internet can be considered intrinsically virtual [V15].

"Virtual space" is a good example of how the internet's virtuality is expressed on several levels. First, a "space" is digitally created in which the human sensory system is manipulated so that the person gets the simulated feeling that s/he is in a real space [V10] e.g., virtual worlds in 3D reality. This virtuality is *perceptual*.

Second, "virtual space" can also denote *affection*; we become so emotionally involved in the virtual content (e.g., downloaded movie) that we enter a "zone" in which we feel that we are "elsewhere" (Qvortrop, 2002)—"immersion" or telepresence [V3], as discussed in Chap. 6.

Third and related: a *cognitive* type of space. Here we feel "connected" to others rationally/informationally. Internet forum participants discussing a mutually interesting topic, or Facebook "friends" describing how they spent their summer vacation, are aware that they're sitting at home, but the internet enables them to feel that they share a mutual "*agora*" (public meeting) space [V6].

A fourth type of virtual space involves being "transported" to another *real* place. Here the internet complements or extends our senses: Zoom meetings [V3]; a doctor performing tele-surgery with her hands in "computerized gloves" connected to the distant operating room, clearly understanding that she is physically in place "X" but able to also "be" and "work" in place "Y" [V2]. Each of these encompass a specific aspect of the internet as a medium.

As a metamedium, the internet provides one final type of virtual space: *cyberspace* [V17]—an entire "world" that is parallel to the corpo*real* one. (Cyberspace also incorporates other sorts of virtual phenomena: virtual memory, virtual storage/cloud computing, and so on.)

To be sure, all these types of virtual space also exist among non-internet technologies and media e.g., Virtual Reality systems, cellular phones, as well as classical media: immersion in a print book; self-enclosed, video game machines. That merely reinforces the point about the internet as a multimedium, able to assimilate other media within its cybersphere platform.

Clearly, virtuality can be found in every one of the internet's many functions and types of content. Here are a few more: email is a modern form of letter writing [V9], without much tactile stimulation (and certainly no perfumed envelope!) [V23]; video clips are mini-cinema [V8]; blogs are similar to newspaper opinion columns, but with ad hoc updating [V18] and interactive audience commentary [V2]; shopping sites are humongous substitutes for brick-and-mortar stores [V6], displaying their wares in two-dimensional [V8] and increasingly three-dimensional fashion [V7], while enabling completely virtual payment [V12].

There is one additional sort of virtuality that cuts across all internet functions and most types of content: ephemerality [V18]. This is true of digital technology in general, but its greatest expression is found within the internet, as much of its content is in constant flux. News articles undergo constant updating and correcting: "a news story now represents a fluid productive process as opposed to a discrete newspaper article giv[ing] rise to uncertainty over the exact nature of news content today" (Robinson, 2011, p. 141). One can add: stock market and sports reports lasting twenty minutes until the next rolling update [V18]; items' placement on the "page" rotating periodically [V17]; wiki-based content is edited and re-edited on a constant basis [V15]; "linkrot": orphan (non-working) hyperlinks [V19]; and so on. Of course, other media "suffer" from ephemerality too, but the internet is the first significant *textual* medium not to live by the *static* word.

This leads to the final conundrum: archiving the news. Which "version" of the news from August 23, 2021, does the publisher save? The e-paper that appeared at 12:01 AM (Aug. 23) or at 11:59 PM (Aug. 22)? Internet archiving adds to the virtuality of memory [V4; V11] by forcing the editors (or other archival repositories) to arbitrarily select a specific version of what appeared—reflecting only a part of the e-news reportage in the past. And if the archiving is by specific articles, then such a digital news item becomes virtualized, decontextualized from any overall formatting framework [V17].

## 10.3 Socio-economic Trends

Media of communication do not exist in a social vacuum but rather are inextricably connected to broader socio-economic trends that influence the development of new media and are also influenced by them. The following sections look at this nexus more closely, focusing on virtuality as a running theme in all these worlds over the past two centuries.

The Industrial Revolution led to a huge migration from the countryside to the city. In rural areas, the locals came into contact with a limited number of people, almost all of whom belonged to the same socio-economic-cultural-religious background (*Gemeinschaft*). Moreover, the agricultural mentality was tied to the unchanging land and natural seasons. The farmer was almost by definition (at least for most of human history) a deeply conservative person by dint of the cyclical and concrete nature of the profession. In the burgeoning cities, these (im)migrants lived and interacted in a far more dynamic, heterogeneous socio-cultural milieu (*Gesellschaft*). Such contact was not virtual at all—it was palpable, concrete, and ongoing—but it did predispose the new city resident to think in terms of "alternative life," whether national or international, cultural, or social. This was a perceptual and conceptual revolution. Perceptually, city residents could see variety in all walks of life; conceptually, they would start thinking less in terms of inevitable stasis and more in terms of mutable change. In short, the idea of life's *potentialities* (a la Bergson and Benjamin) turned the here and now into much more than an inexorable "given"; the present was rife with the "virtual" of what could be [V26].

Such a potentiality came to fruition through rising levels of "economic surplus," thereby freeing an increasing proportion of workers to produce "culture"—mostly virtual in one form or another—for a population also benefitting from more purchasing power and leisure time. Still, in the nineteenth century most people labored in manufacturing. Ironically, even this reinforced virtuality; as opposed to farming i.e., dealing with Nature and its natural products (plant and animal; food and clothing), the Industrial Revolution involved a man-made, "artificial" work environment. True, it was all very concrete (drearily so for the most part), but such artificiality and its non-natural products led to a mindset pre-disposed to accepting and later demanding other forms of artificiality, increasingly of the virtual variety: radio, cinema, television and so on.

The inexorable march of greater work productivity led to the next massive economic shift: the service economy. This too had an impact on the changing mindset related to virtuality. Agriculture and manufacturing are involved in very concrete activities involving "things." On the other hand, service does not deal with things but rather (in large part) with *people*. Of course, people are not at all "virtual," but in servicing them the worker must use a level of thought—intellectual, creative, intuitive—far beyond anything found in the world of agriculture or manufacturing. (It is important to note that the true creative workers in manufacturing—inventors, tinkerers, scientists, technologists—were in fact "service workers" by profession, even if they were employed by a manufacturing company in the Industrial Age.) In short, this type of work involves human imagination, thinking about new ways of coping and increasing economic efficiency, if only because the "object"—human customer—is not passive (as the world of Nature was largely to the farmer, or the machine to the proletarian worker), but rather an active "subject."

To enable such a budding service (and proto-information) economy to flourish and expand, society also expanded higher education to the American "masses" in the mid-to-late nineteenth century (College Land Grants) and in twentieth century Europe. Moreover, the nature of higher education changed—from almost exclusively handing down accumulated knowledge from generation to generation, to *creating new* knowledge i.e., the research university, started by Humboldt in early nineteenth century Europe.

Here again we see the Bergsonian idea of the "virtual as potential" [V26]—a mindset tied not only to the "what is" but also the "what can be." The result: for the growing number of researchers, the future was seen in increasingly "virtual" terms: new potentialities. Similarly, the common person seeing the growing number of amazing inventions also started to consider the idea of "future as different from the present." Among other things, the increasing popularity of late nineteenth and twentieth century World's Fairs is testament to the growing appetite of common folk to virtualize the present or think about the virtual future. It is not coincidental that precisely at this historical moment a new literary (and later cinematic, radio and TV) genre became wildly popular: science fiction.[6]

Ultimately, all this led to the Information Economy that continued the historical trend to greater abstraction in the workplace. This reinforced virtual cognitive and behavioral activity on the individual micro level, as well as constituting a very powerful macro-intellectual infrastructure for virtuality in many fields.

## 10.3.1   Science: The New Religion

From the standpoint of overall virtuality, the early modern period was marked by a deep irony. It was witness to an increasingly accelerated decline in religious institutional power and also religion's hold on the individual mind. Superstition did not disappear completely and neither did religiosity, but in general the religious ethos had less of a hold quantitatively (proportion of the population) and qualitatively (the extent of its influence on various aspects of individual life).[7]

What took religion's place? Two broad and somewhat interrelated perspectives on life: humanism and rational-empirical science. The first placed its emphasis on the person and worldly benefits to attain the "good life" in a material, not necessarily metaphysical sense; the second looked to a more reasoned, *physical* understanding of the world, instead of the metaphysical faith underlying Biblical and Papal "authoritative" declarations of how the world was organized (around God). To a great extent, these two "modern" approaches complemented each other, for the material good life could really only be attained by developing a scientific and technological mindset and social infrastructure that ultimately provided the resources for such humanistic advances: public hygiene, better nutrition, advanced medicine—all leading to human life span extension; material surplus enabling more leisure time and flourishing of the arts and entertainment; better information transmission technologies (media) leading to an exponential explosion in knowledge and higher education among the general public and intellectuals (each on their own plane); etc.

From the standpoint of virtuality, it might seem that the modern era was ineluctably moving in a reverse direction—towards greater attention to the "here and now." In fact, however, it resulted in the humanistic and scientific approaches replacing the central metaphysical/religious approach of the pre-modern era with a few different

---

[6] The term science fiction has no set definition. There are those who argue for its lineage going back millennia, even to the *Epic of Gilgamesh*. However, these works were not informed by any scientific ethos but were rather fantastical projections or fabulations [V35]. Even Thomas More's novel, *Utopia* (1516), that some consider to be the first proto-science fiction work of the modern era, was more of a *social-moral* tale than one imbued with any significant science.

[7] A minor, if not insignificant factor in the decline of superstition, was the literal "enlightening" of the night commencing in the seventeenth century when the major European cities gradually introduced street lighting: Paris in 1667, Amsterdam 1669, Hamburg, 1673, and so on. Among other things, this reduced the ability of people to "see ghosts" and to believe in nefarious creatures that supposedly emerge only at night (Koslofsky, 2011).

forms of virtuality. Indeed, paradoxically, modern science that begat modern technology—all based on a greater understanding of the "real world"—in turn gave birth to types of virtuality heretofore undreamed of.

"Paradoxical," though, is not "contradictory." If "virtualizing" is a profound human tendency, then it should not be surprising that mankind will utilize more powerful technologies to expand and deepen the experience of virtuality. This suggests that the move from metaphysical religion to physical science did not produce technologies of virtuality "by accident." While we did not exactly develop virtualizing technologies "on purpose" through some thought-out "plan," our need for experiencing virtuality in the wake of the demise of religious sensibility ineluctably led us to *replace* our metaphysical loss with things secular that could take its place.

In this we succeeded beyond our wildest dreams (or to some: nightmares), as life from the nineteenth century onwards took on an ever-increasing hue of virtuality in almost all fields of endeavor touched by science and technology. This constituted the underlying "socio-psychological" factor in the virtuality explosion over the past 200 years—the desire to experience something beyond the here-and-now, beneath the in-your-face, in a world that could no longer find the ineffable in traditional religion and other metaphysical outlets.

### 10.3.2 Science: Discovering the Objective Virtual World

One would expect to see a major decline in metaphysical thinking once the modern world began to develop and appreciate science in its newer, empirical mode. For sure, this was and continues to be the goal of modern science: real-world understanding. However, that contained some very unexpected surprises for the modern scientist, in several important fields—leading us back to the imaginary, the incorporeal, the "magical" and the virtual. We might ultimately succeed in making our world completely comprehensible in a material and objective sense, but over the past two centuries the reverse has occurred in a few central fields of scientific endeavor. The result: virtuality has not only *not* disappeared, but it now has an empirical veneer and possibly an outright scientific imprimatur!

This occurred on two opposite planes: the macro-world of the cosmos and the micro-world of nature incorporating the very, very small. In both cases, in the early modern period we began to discover that the "real" world was full of phenomena beyond the unaided perception of human beings, but no less objectively real for that. Thus began an understanding of the virtual not as something "unreal" but rather altogether real, however imperceptible to common human perception [V19]. In that sense, scientific discoveries during the early modern period based on the microscope and telescope (continuing apace to the present) opened up worlds of "objective virtuality."

The impact was contradictory. First was the sheer beauty and strangeness of an unseen world, literally under our noses forever. From one perspective, the virtual (unseen: V19) had become real to a certain extent (now perceived)—constriction of the virtual, and expansion of the real.

In a deeper psychological sense, however, it served as a double form of expanded virtuality. Even if we know knew that many more stars existed than we thought, and that life on earth was infinitely more variegated than we imagined, in a practical sense both these worlds remained out of our perceptual field—except for the very few wielding the telescope and microscope. The average person understood *intellectually* that there was far more to the world than met the eye, but *perceptually* the world remained the same for all intents and purposes, indeed it had even *shrunk* from a psychological point of view because we now were aware of how much of the real world we couldn't normally see! On the other hand, and psychologically no less "problematic," if these objects had existed all along without our realizing it, who was to say what objective reality still lay out there to be discovered? Until we managed to uncover such hidden phenomena they would continue to exist virtually (as potential discoveries: V26; V27), as opposed to the situation in the pre-modern era when we could not even imagine the possibility of their existing.

Moreover, there were also "things" in the universe that we could *never* "see" but that existed nonetheless e.g., gravity—an unseen force [V2; V19; V25] that underlies the entire foundation of universal existence, cosmologically and locally. This was a watershed discovery on the road to modern virtuality; what else might be "out there" that is inextricably tied to our (corporeal) life that we are not cognizant of? Two hundred years after Newton's epiphany regarding gravity, Maxwell discovered electromagnetic fields, Marie and Pierre Curie revealed radiation, Marconi employed radio waves, and so on. And as we saw in Chap. 5, this was but the start of future speculations and revelations regarding many forms of radiation, a fourth dimension (time), still more dimensions (string theory), not to mention black holes, dark matter, and dark energy later in the twentieth century.

In short, if billions of stars exist and then we discovered other types of heavenly bodies such as galaxies, quasars, black holes that we had never seen; and if in our body and under our feet there are untold numbers of microscopic species of plant and animal life that the human race had no knowledge of; and if everything is held together by a gravitational force that we had no inkling of for millennia—then theoretically *anything* could exist! The central irony here was that precisely those intellectual processes and tools designed to uncover *reality*—science and technology—had the ultimate effect of reinforcing and even expanding our belief in things and processes *virtual*, not just scientific but "meta-physical" as well (Blum, 2006).

Indeed, it is no coincidence that the first works of science fiction emerged in the early modern period e.g., Francis Bacon's *The New Atlantis* (1626), parallel to the development of the telescope and microscope. As newer worlds were discovered, this literary genre expanded by suggesting other "worlds" yet to be uncovered or created: imaginary journeys to the moon, as in Kepler's *Somnium* (The Dream, 1634), and Godwin's *The Man in the Moone* (1638).

Thereafter, scientific discoveries of virtual phenomena never ceased, further reinforcing the idea that most of reality was hidden from our ordinary, human perceptual apparatus: electricity (or more precisely, electro-magnetism); atoms, molecules and chemical reactions; gravity; gases such as carbon dioxide, hydrogen, oxygen—even the very air we breathed was full of "hidden" elements! The list could continue but the point is clear: uncovering (newly discovered) reality paradoxically actually strengthened the feeling that (the unknown) virtual world was vast. As neuroscientist Eagleman put it: "…at the end of the pier of knowledge are uncharted waters; what we know is vastly outstripped by what we don't know" (Richards, 2010).

The nineteenth and twentieth centuries continued this trend but constituted a qualitative difference as they upended "scientific paradigms," entire systems of thought. Whereas previous discoveries added discrete knowledge about the world, several now upended the very way we thought about our world. From the mid-nineteenth century onwards, a veritable tide of Eternal Truths Overturned washed over modern society—moving from the scientific mainstream into popular imagination. The overall effect was to place a giant question mark on everything that we thought we knew. Reality as we had understood it from time immemorial was not necessarily "real"; accepted natural processes were not what they seemed to be—and if potentially everything that we had considered to be real was not, if the world actually worked in a way different from what we always understood (and even occasionally counter-intuitively), then we were living in a virtual world, at least in the sense of constant intellectual uncertainty and physical camouflage [V20]. Plato's Cave had returned in modern guise, with the important difference that as the cave dwellers we were now *aware* that the shadows on the wall were not necessarily real.

## 10.4 The Expansion of Virtuality in the Late Modern Era: Natural Sciences

The purpose of the following highly schematic survey of these revolutions is to display them in holistic fashion i.e., what the modern mind incorporates wherever it turns to in science. It is the *totality* of these intellectual revolutions that is important, some in the social sciences as well. Indeed, because the common person has difficulty understanding these abstruse or complex "natural science" theories, the greater influence on the public is probably felt in the social scientific revolutions—to be analyzed in Sect. 10.4.2.

### 10.4.1 Science: Darwin and Evolutionary Potentialities

Darwin's revolutionary theory was not virtual but corpo*real*; biological evolution is very concrete. There is nothing innately virtual about the physical interplay within and between species, not to mention the interaction of the genome with its immediate environment, organic or topographical. Indeed, in one important sense Darwin's Theory of Natural Selection can be understood as a "regression" from mystical/metaphysical explanations of life to one solely based on very corporeal chemical, biological and environmental processes.

What, then, is "virtual" about Darwinian Natural Selection? First, biological evolution is all about "potentiality"; nothing is pre-determined or pre-ordained [V26]. What exists in the present is indeed "real." However, species arrive in the present not because that was their ontological "destiny" but rather because of many internal and external factors interacting in myriad ways. Indeed, once we acquired a general understanding of this process's huge complexity, we had to accept the fact that the relative force of each factor behind any specific evolutionary story becomes essentially unknowable [V15] because evolution depends heavily on "chance": random mutations, freakish meteorological phenomena (e.g., the meteorite that caused the dinosaurs' extinction), unexpected geological events (tectonic plates separating continental land masses), and so on. The general process, then, is comprehensible, but the practical nuts and bolts of how future evolution will play out is beyond our ken, ergo it is virtual [V16].

Second, Darwin's theory posits flux and indefiniteness. Flux: each species is constantly changing—even if the time span is too long for any human generation to see such change. Indefiniteness: as species evolve in extremely slow, incremental fashion,[8] they traverse a spectrum in which it is very hard to discern where and when they have become a new species [V15]. At what point did the Great Apes (*hominids*) cease to belong to that family and started becoming early humans? At what developmental stage did somewhat domesticated wolves (*canis lupus*) evolve into dogs (*canis lupus familiaris*)? Thus, all species have an "approximate" identity i.e., we are virtually human, dogs are virtually canine—if only because we are continually in the process of further evolutionary development.

## 10.4.2   Science: Quantum Physics, Relativity and the Underlying Virtuality of Reality

Quantum Mechanics and the theories of Relativity (Special and General) were already discussed in Chap. 5. Here I shall only briefly repeat the main aspects as they relate to our understanding of reality and virtuality, given their centrality to contemporary thought. First, quantum physics posits (and experimentally has proven) that the sub-atomic world is based on an inherent measure of unpredictability, based on probabilities alone [V15] (each atom holds within it "potentialities").[9] In other words,

---

[8] Evolutionary theory also has a sub-theory called "punctuated equilibrium" (Gould & Eldredge, 1977), more colloquially known as "catastrophe theory," whereby evolutionary change speeds up (by geological time standards) as a result of a radical change in the environment, but these tend to be rare occurrences in the vast scheme of evolutionary history.

[9] It bears repeating that because of its incredible strangeness, many of the key physicists in the early twentieth century and even today had/have a hard time themselves believing that quantum mechanics is indeed what reality is all about. Nevertheless, while the last word has not been said and there are some corroborating experiments that still need to be carried out, at present all the

there is no bedrock constancy to anything but rather an everlasting throw of the dice (to paraphrase Einstein) through which reality "comes into being." This is not to say that when reality does finally appear it is not "real"; it is to say that at the most basic level we cannot be sure what the precise nature of that reality actually "is."

Second, the most constant element of reality, *light*, is inherently contradictory in nature. Light can be both a wave and photon at one and the same time [V15], an obviously absurd and counterintuitive—but nevertheless true!—state of affairs. Third and related, reality (at least at the atomic level) must "make up its mind" as to which of its states it will be *once an intelligent observer focuses on the specific situation.* Although humans do not decide for the atom what it will be, the mere act of human observation forces the atom to "make a decision"—but until then, this result is never actualized [V27]. We are therefore enmeshed in the world around us in a way that is beyond pure Newtonian determinism.

Fourth, the enmeshing—or "entanglement" as it is called in modern physics—goes well beyond the relationship between intelligent observation and atomic activity. Even more strangely—one is almost tempted to say *impossibly*—two atoms that are a huge distance apart seem to be capable of "communicating" information one to the other [V2] much faster than the speed of light (actually, instantaneously). This is called by physicists themselves (note the irony of scientists using language heretofore relegated to mystics and clairvoyants): "spooky action at a distance"—a situation where one major part of modern physics, quantum mechanics, directly contradicts another: general relativity!

Fifth, the theory of relativity posits that there is no "objective" vantage point from which one can immutably measure our world. Everything is relative to everything else [V15]. Sixth, Einstein's most famous equation e = mc$^2$, among other things postulates that reality is not necessarily "material"; it can just as easily be viewed as being "energy." In short, the "substance" we are made of has a non-material nature to it as well [V25].

Beyond these six physical law verities, there are a few other aspects of modern physics that are still in the realm of unproven speculation. The first is "String Theory," a highly complex mathematical construct (without any empirical evidence for the time being[10]), positing that we live in 11 dimensions, 7 of which we cannot sense [V19], a position that reverts to the discussion in Chap. 3 regarding the lacunae in our sensory apparatus. A second theory talks of "multiverses" i.e., universes outside

---

empirical evidence we have garnered and all the predictions emanating from quantum physics have "proved" it to be true. Of course, we seem to be still quite far from a satisfactory "conjoining" of "micro" quantum physics with "macro" cosmological physics (e.g., gravity), but the difficulty in combining the two does not in any way undercut the veracity of the quantum description of how things "are" on the atomic level. Indeed, many modern technologies (e.g., lasers) are based on this "weird" scientific theoretical structure.

[10] There is also no empirical evidence that string theory is wrong. The problem is that given the state of our scientific, experimental, physics technology, we have no way of empirically testing the theory. As a result, many physicists do not consider string theory to be a bona fide theory at all as it does not meet the scientific requirement of proof testing. However, that may only be a problem regarding the current state of our technology and not an obstacle to empirical verification in principle.

of our own universe (our language has not quite caught up with scientific theories). If this turns out to be true, we are living in a situation whereby the vast majority of the "world" (multiverse) is totally beyond our ability to observe [V34]—most of "what is," is literally "virtual" to us! Moreover, other universes might not even act based on known physical laws but rather could in principle harbor completely different sets of physical laws [V29].

Third, because of how the cosmos works, most physicists and cosmologists today believe that the known universe ("ours"—the one we can observe) is comprised mostly of something called "dark matter" and "dark energy"—"dark" because they are invisible not only to the naked eye but to any form of observational technology we can devise [V19]. So far, we have been unable to find empirical proof that these two "phenomena" exist. In any case, in terms of our subject the existence of dark matter and dark energy comprising the overwhelming part of the university's "mass" means that corporeal reality is but a tiny part of the universe, ergo for us mostly virtual!

Fourth, and more speculatively, several serious physicists believe that the universe is nothing more than one gigantic "computer" and that we are all basically "information"—reality has no real "concrete" component to it [V25] (Lloyd, 2006). A brief personal anecdote is highly illustrative of this concept. While writing my PhD dissertation at Harvard in the early 1970s, I had the great fortune of being present at one of Heisenberg's last public lectures (he died in 1976). At the end of his fascinating review of the early, heady days of Quantum Theory, he took questions from the audience. There was one question and answer that has stayed with me:

*Question*: Prof. Heisenberg, what you would consider to be the most fundamental basis, the underlying foundation, of all reality?

*Answer* (after pondering the question for a few seconds, Heisenberg looked at the audience and responded in a somewhat defiant tone): "Mathematical equations!"

Heisenberg was clearly not referring to any specific equation—Einstein's or Dirac's—but rather to his belief that at base the universe was held together by mathematical laws and not by physical matter. I view this as hoisting Einstein with his own petard: it is not the gravitational "attraction" of mass[11] that constitutes the universe but rather pure mathematical "correspondences." Thus, contemporary physics has returned to a modern form of abstract Platonic Idealism [V17; V25; V30; V33].

The sum total of all this contemporary physics is to place virtuality front and center within our world—both on the atomic level and on the cosmological plane. Several theories are considered as "rock-solid" as science can get, verified several times

---

[11] In fact, Einstein did not call gravity an "attraction" between two bodies, but rather explained that the heavier body "indents/depresses" space–time, and thus the second body "rolls down" the bent "surface" towards the heavier body. As the inordinate use of quote marks here indicates, our language has not fully caught up with these concepts, so that we still must use terminology that does not quite describe reality in truly accurate fashion. In short, even our language uses virtual (roughly corresponding) terms [V4; V6] and not "real" (fully accurate) ones when describing what we think of as "reality."

over in various experiments. Others offer a convincing explanation for perceived phenomena, but these theories have not yet been empirically verified. And others remain in the (highly?) speculative stage, awaiting proof or a more convincing theory.

Whether they ultimately pan out is only tangentially important for our purposes. Even if none of this scientific theorizing were found to be true (something quite possible in the far-off future), the "psycho-cultural" influence on the contemporary world remains the same: we are told by "those-in-the-know" that what we see is *not* what we get, nor is much of it even comprehensible! The fact that these theories are bandied about so widely in the general public realm means that they have an enormous influence on the general *zeitgeist*, with the effect that in many respects we feel as if we live in a highly virtual world.

Before moving on, we should note a countertrend in contemporary society: scientific "denialism."[12] Paradoxically, this anti-scientific approach parallels the same scientific ethos of "what we see is *not* what we get": just because science claims to show A, this does not mean that B (or in the case of many denialists—Z) is not the real story. As Specter (2009) notes, a significant minority of people in the West are not willing to take scientific theory and fact at face value. Conversely, denialists do believe things for which there is no scientific evidence: homeopathic, "natural" cures; UFOs; and so on. While some of this can be traced to vestigial religious beliefs (e.g., Intelligent Design), in most cases it seems to be a function of the need for some sort of mystical or metaphysical explanation of our world. Thus, the anti-science "school" is a throwback to a pre-scientific mindset that attributes much of the world's processes to hidden, virtual forces beyond human ken. Which brings us to the next area of scientific scrutiny...

## 10.5  Late Modern Virtuality Expansion: Psychology and Ideational Philosophy

Humans live in two "worlds": physical environment and social system. Parallel to the greater perceptual virtualization of our physical world in the modern age, we find a concomitant virtualization in many areas of social life, whether "sociological" or "artistic." This is not a historical coincidence. Expanding media and the broadening of learning (books, higher education) in the nineteenth and twentieth centuries led to significant intellectual cross-pollination, even if the purveyors and consumers of sundry virtualized products were not always completely aware of the mutually reinforcing influence of other fields on their own.

---

[12] Scientific *denial* is the refusal to accept a specific theory or empirical fact; *denialism* entails a wholesale mindset that refuses to accept science as the ultimate arbiter explaining our world. Thus, a person who does not yet accept that humans are the main cause of global warming, can be in scientific denial—but if that person also does not accept Darwinian Evolution along with other generally accepted scientific theories, that would be scientific *denialism*.

The following schematic survey will touch briefly on a few of the more important, even "revolutionary," social scientific and intellectual developments in the modern age that were directly related to, and further influenced, the general expansion of virtuality.

## 10.5.1  Social Science: Freud and the Mind's Hidden Inner Life

Before discussing the revolution of Freudian psychology, the same point regarding physics needs to be reiterated, especially for the reader in the early twenty-first century with some knowledge of the latest advances in psychology, psychiatry and neurobiology. For the purposes of this virtuality survey, it is not that important whether Freud's theories are largely correct or mostly bunk. What is germane is the public effect of his theories. In any case, as seen in Chap. 3, his underlying insight of an unconscious at work has actually been strengthened by growing neuro-biological evidence.

With Freudian psychology we move from the external world to the internal. He pointed out how our internal world is also "virtual" in that it is hidden from us—right underneath our eyes, as it were [V19]. The idea of the unconscious entailed the virtualization of human thought—our conscious thoughts might be "real" but they are influenced and manipulated by other strata of the brain that we have no control over and are hardly aware that they even exist.

Freud's theory of the unconscious received a warm public reception, possibly because by the early twentieth century people had already seen virtual phenomena discovered in many other areas of life. Indeed, regarding the "unconscious," Freud himself acknowledged his debt to prior thinkers: "The poets and philosophers before me discovered the unconscious. What I discovered was the scientific method by which the unconscious can be studied" (MacIntyre, 1958, p. 6).

Freud's insight (pun intended) was therefore not only a reflection of what happened in the external social world throughout the nineteenth century, but also constituted a prototype of what would occur over the ensuing twentieth century: the "virtual" increasingly became a significant component within the "real," influencing our concrete life to an increasingly greater extent. Thus, Freud's theory of the unconscious can be seen as a metaphor for virtuality in the modern era: the widening of the "virtual sphere" had and has a self-fulfilling dynamic to it. The more virtuality we deal with, in more of areas of life, the more "natural" it seems—thereby enabling people to "transfer" it to still other fields where it may not have been felt in the past.

## 10.5.2  Ideational Systems

The loss of religious-metaphysical sensibility, mentioned earlier, was not only replaced by concrete technologies and arts that enabled us to have virtual experiences in secular fashion (e.g., camera, telephone, radio, TV, spectrometer), but also by *ideational systems* that emphasized the virtuality of "life" (in its social and physical senses). I already mentioned one example of this in Chap. 7: Adam Smith's "invisible hand," the basis of capitalism. Here is a sampling of four others from the world of politics: Liberal social contract theory; Marxism[13]; Romantic Nationalism (Fascism and Nazism); and Conservatism.

The idea of an abstract "social contract" [V23] holding society and the polity together, became a central pillar of the emerging nation-state. However, the three main proponents of the idea offered differing elements of virtuality. Hobbes offered *realpolitik*: we do not have to ask the citizens if they are willing to subsume themselves within the social contract tendering absolute power to the monarch because their self-interest (physical survival in the jungle of the state of nature) implicitly assumes agreement with such a system of political authority. Here the basis of political obligation is highly "realistic," but also virtual in that the contract might never be written down and certainly isn't signed by each ensuing generation [V27].

Locke's philosophy was also based on each citizen's needs—the preservation of life, liberty and property. Citizens of a country fashion a real social contract in which they freely give up certain liberties in return for a government that guarantees their remaining liberties [V7]. For Locke the social contract is real, but he injected a strong element of virtuality in the "rights" that all citizens retain [V26]. Property rights also constituted the early basis of modern capitalism, so that Locke's influence on virtualization in the economic realm was also significant.

Rousseau's "system" is probably the greatest mixture of virtuality and reality imaginable. His social contract is an ongoing, tacit agreement between the people and the rulers/legislators, with the basis of decision-making the "general will"—an extremely amorphous, highly "virtual" concept that represents the actual good of society and not necessarily what the majority declares to be the good [V15]. This is an almost Platonic conception of the "Ideal." Where, then, is reality in Rousseau's system? He argued that the people must actually take a real vote in order to express the General Will: "Every law the people has not ratified in person is null and void—is, in fact, not a law" (Book III, Chap. 15).

Overall, the revolutionary and highly influential social contract philosophies and systems of government injected a modicum of virtuality into the political system—either through an historically fictitious written "constitution" [V32] or by creating civil rights where none existed before [V33]. Indeed, this last aspect became a central pillar of Liberalism from the nineteenth century onwards, a philosophy that continued adding virtual human rights (privacy, education, speech) to the political basket.

---

[13] Chapter 7 also dealt with Marxism—as the antithesis of virtuality. Here I will offer a somewhat more nuanced examination—not contradicting but rather complementing that analysis.

Although Marxism is generally viewed as the philosophical, polar opposite of Liberalism, it does share one central common denominator: *rationality*, through its Scientific Materialism. However, Marxism attempted to cleanse politics (and political ideology) of all semblances of virtuality, what he called "false consciousness," by ostensibly laying bare the real forces that prevent true understanding of social reality. Marxist Socialism (a/k/a Communism) was based on a deterministic *materialism*, whereby economic forces work their way through history, influencing human consciousness in the way we perceive our social condition. However, despite placing "matter" at the center of his system, and notwithstanding his attempts to undermine the workers' false consciousness, Marxism did not eliminate virtual elements.

First, Marxism was heavily influenced by Hegel's dialectic theory of Spirit in history [V33]—ideational thesis, antithesis, synthesis. True, Marx "stood Hegel on his head" by substituting matter for spirit, yet the dialectic itself incorporates virtuality in that whatever our present situation, it harbors the potential to be transformed into its opposite [V26], thereby neither stable nor static but rather indeterminate [V15] from a temporal perspective. Second, despite Marxism's materialistic bent, it was secularly "eschatological" [V34], looking forward to the "end of days" when completely collective ownership reigned supreme, and history i.e., human struggle for material survival and social status, ends. As Rothbard (2009) noted: "For Marx and other schools of communists, mankind, led by a vanguard of secular saints, will establish a secularized Kingdom of Heaven on earth" [V30]. Marx provided no details as to what this communist society would look like; the final outcome remains murky [V15].

The more emotion-laden political ideologies also added elements of virtuality. Romanticism, represented powerfully in the political sphere by "Nationalism," was a general cultural reaction to the hyper-rationalism of the Enlightenment and Liberalism. The "nation" was deemed by nineteenth century Romantic ideologues to be a collectivity with a common history and culture—obviously a mythical fabrication [V32], especially in Europe that had seen mass migrations, invasions, ethnic intermarriage and intermingling of cultures for many hundreds and even thousands of years. The nation-state was the final political expression of that ethnic-cultural collectivity, brought into being to protect the national group's solidarity and "cultural purity." Nationalism reified the state, placing it on a pedestal as the citizens' required main focus of loyalty. This metaphysical entity—the "Nation" [V33]—was no longer one element among many (real or imagined), but rather public life's *central* element.

Ultimately, this was taken to its extreme in Fascism, with its mediated glorification of the supreme leader [V3; V32] and Nazism with its idealization of the state itself [V30] as the supreme value (*"Deutschland über alles"*). Nazism further extended the national myth into a construct of racial purity, thereby turning the population against other groups: the Jews were "proven" to be a race apart by such outright fabrications [V32] as *The Protocols of the Elders of Zion.*

Not all modern philosophical systems are laden with virtuality. At least one is non-virtual in most important respects—Burkean Conservatism[14]—and (to speculate) that could be one reason it is not usually taught as part of the modern *philosophical* canon.[15] Burke's Conservative approach should not be confused with classical traditionalism based on blind acquiescence to received authority. Rather, as a leading eighteenth-century British parliamentarian, Burke was profoundly realistic, accepting the possibility of many different systems of government and types of social order. He argued that society—*any* type of society—is an organic whole, evolving over long periods and as such should not be too readily or radically tampered with lest the woof and warp of the finely bound social structure come apart at the seams.

Burkean Conservatism, therefore, is not against reform; it simply opposed radical reform.[16] Burke recognized that all societies are imperfect, and that human misery is a fact of social life. He was not impervious to such misery; indeed, he pointed out the most egregious examples of such in British society (the miners). He was also not averse to political and social reform but argued that this had to be carried out with great caution. One step forward and then we wait and see whether that leads to two steps backwards or a somewhat better situation.[17] In the Conservative scheme of things, therefore, social and political change was always attached to the real world and not to some overarching, monistic, idealistic theory as to how society should look and behave.

Clearly, the Conservative approach is intrinsically non-virtual: it posits no "ideal" society or state, but rather supports all societies that "work"—focusing on the "is" while profoundly suspicious of any new "ought."

## 10.6 Holistic Virtuality

The survey of various disciplines and areas of life (Chaps. 4–9), coupled with the various scientific, philosophical, intellectual, and ideological systems discussed in this chapter, all provide evidence of a significant expansion of virtual activity, ideas, and content in the modern age. Virtuality was never, and is not, a unidimensional phenomenon; rather, it can be found in most fields of human endeavor. However, what

---

[14] I spell Conservatism with a capital "C" to denote a distinct, socio-political, philosophical approach. When spelled with a lower case "c," conservatism denotes an individual's psychological attitude.

[15] "It is not surprising that Burke has been quietly ignored by many recent thinkers, or dismissed from consideration by being labelled as a 'conservative'—but it is of great interest that he has found many admirers amongst those who succeed in the conduct of practical politics" (Harris, 2011).

[16] Burke supported the American Revolution because he felt it was not "revolutionary"—the American colonists were simply demanding that their rights as British citizens be restored.

[17] Interestingly, this seems to presage Darwin's theory of evolution: small, incremental changes over time, each either succeeding and thus forming the basis for the species' next step "forward" (survivability), or the specific change "failing" (undermining the individual creature's survival), in which case the "mutation" is lost and another one comes into play at a future date.

separates the modern age from previous eras is not merely the depth and breadth of virtuality but also its interrelatedness within and between many of these fields. True, in the ancient and medieval worlds some interrelationship existed between several areas: religion and art were closely linked; physics and philosophy were considered almost the same "area" of intellectual endeavor; political empires could not have functioned without the means of distant communication.

However, these existed in parallel, working in tandem when expedient. The ancients did not understand that these activities were part of the same underlying phenomenon. In the modern age, however, various new, virtual concepts and processes *directly influenced* the development of other concepts and processes, most often in somewhat *conscious* fashion. I have occasionally hinted at such interaction in surveying several of these fields; I shall now devote greater attention to this type of trans-disciplinary interaction and mutual influence. Why did this happen?

The ancients were certainly not any less perspicacious than us moderns. Indeed, if anything we would expect them to be more aware of the underlying, common denominator of virtuality in various spheres, for they possessed a holistic view of life (e.g., science and philosophy were viewed as one: "natural philosophy"). The late modern world, on the other hand, has turned to greater disciplinary specialization. The answer lies precisely in the fact that the ancients did not view the real and the virtual as being fundamentally different. For them, God was as real as the wheat in the field. Thus, without a clear demarcation between anything and everything, there was little attempt to find commonalities between different areas of life because they were considered a priori cut from the same cloth. The post-Kantian modern mind,[18] on the other hand, following Kant's irrevocable splitting off metaphysics from natural science, compartmentalized each area of life (at least professionally and intellectually), a form of disciplinary specialization. In this modern situation, when underlying commonalities between seemingly disparate phenomena were discovered, it occurred due to conscious adoption and adaptation from one field to another, not necessarily immediately or even quickly, but rather osmotically, slowly filtering from one area into another. The leading thinkers/doers/inventors in one area become aware of virtualizing trends in another, and slowly adapted them to their own field of endeavor. Herewith several examples of this cross-pollination.

### 10.6.1   Dimensions: Math, Religion, Literature, Art, Psychology

The idea of multi-dimensionality (more than three dimensions) was in the intellectual air from at least the eighteenth century—Henry More's "*Spissitude*": a fourth spatial dimension of the spiritual realm (1712). Kant (1910) mentioned the possibility of

---

[18] The young Kant was one of the last of the great thinkers to try and combine metaphysical philosophy with what we today would call empirical science. His later, three monumental "critical" tomes rejected this classical approach (Schonfeld, 2000, pp. 4–6).

multiple universes (p. 22) and extra dimensions (p. 25) that would not be accessible to our perception. The mathematicians d'Alembert in 1754 (time might be the fourth dimension) and Möbius in 1827 (the fourth dimension could flip left and right), also speculated about the extra-dimensionality [V14, V19] of the world. The demise of religious sentiment specifically and metaphysics in general, possibly motivated thinkers to subconsciously search for a virtual-"empirical" substitute, and the idea of a fourth dimension proved enticing.

Nevertheless, only in the later nineteenth century does the issue become a major element of the *zeitgeist*. Carrol's *Through the Looking Glass* (1872) suggested that other three-dimensional worlds might exist right near our own, with some "portal" to traverse from one to the other that must be four dimensional! Thus, even though these "neighboring," three-dimensional, virtual worlds are similar to ours—they are conditional on a four-dimensional situation that we can't see.

Abbott's *Flatland* (1884) sought to show how unsensed [V19] higher dimensions [V14] are possible by postulating a world of only two dimensions [V8]—and how we three-dimensional creatures would (not) be perceived by them. This concept of four dimensions was adopted by religious thinkers as well: "We conclude, therefore, that a higher world… may be considered as a world of four dimensions; …that the spiritual world…by analogy would be the laws, language, and claims of a fourth dimension" (Schofield, 1888, p. 44).

Finally, two important, late nineteenth century thinkers pushed the boundaries of "four-dimensionality," each in a different direction. The British mathematician and physicist Hinton produced numerous articles and books regarding the fourth dimension (Hinton, 1884). Some of his ideas anticipated Einstein's theory of relativity ("world lines") and even seemed to have influenced Borges, as Hinton is mentioned in one of his stories ("The Secret Miracle").

All of this was but the overture to Einstein's special theory of relativity (1908) and the inextricable relationship between three-dimensional space and the fourth dimension, time. Most germane to the survey here is his influence on other fields of endeavor—the parallel interplay between physics, media technology, and art.

It is surely more than a coincidence that at the precise historical time that Einstein and colleagues were promulgating a radically new vision of the physical world, and when the world was absorbing new, "virtual," electronic media technologies (photography, telephone, cinema, radio, phonograph), a similar revolution was occurring in the world of painting, sculpture, music, literature, and theater. As many of these were already noted in Chap. 6, I will focus here on the main trends and draw some important parallels with the emerging, new perspectives of the scientific and technological world.

Incorporeal virtuality expanded as a result of the emergence of new media that brought the "real but foreign" world to the consumer from afar, in addition to producing visually imaginary content. This entire phenomenon reinforced (and perhaps also influenced) trends in the general culture that added another component to the overall virtual ambience: the drastic reduction of "representational art" and the collapse of classical form and structure in all the arts [V23]. Painting and literature no longer attempted to "copy" or "perfectly/ideally represent" reality and the natural world, but rather to undermine it or at least to present a parallel, "subjectively (un)real" world [V5].

With realistic representation [V1; V8] now in the victorious hands of the camera (static, and later moving, pictures), painting began to take on a more "impressionistic" air. The goal henceforth was not to portray reality, but rather to enable the expression of a subjective "impression" of that reality [V5], as in Van Gogh's painted flowers that were far more colorful than in real life. However, Impressionist art was not merely reacting to the camera's technological competition; it also found legitimacy in certain aspects of Jamesian psychology, especially relating to psychopathology.

The more radical artistic break, however, occurred at precisely the same time as Einstein's relativity (and a bit later, quantum mechanics) revolution. With Picasso and Cubism, the world of art began a journey where classical rules break down completely [V22; V28]. *Man with Violin* (1911–12)[19] shattered the traditional plane into a fractured picture with multiple perspectives [V14]—a perceptual relativity that captures the essence of Einstein's relativity in space. Subsequent artists—Duchamp, Braque, Gris—turned Cubism into the major avant-garde art movement of the second decade.

Moreover, the connection between their work and that period's scientific theories/speculations was explicit. The Cubists did not just paint; they also spent nights discussing the major scientific currents of their day. For example, Duchamp explained the conceptual *raison d'etre* of his major work: "What we were interested in at the time was the fourth dimension. Simply, I thought of the idea of a projection, of an invisible fourth dimension, something you couldn't see with your eyes. The Bride in The Large Glass was based on this, as if it were the projection of a four-dimensional object" (Art in the 'fourth dimension', 2004).

At approximately the same time, but running well into the 1920s, another art style took hold. "Expressionism" was less interested in spatial deformation and more in psychological distortion [V5] as it portrayed the world from a completely subjective perspective. The underlying goal was to vividly present personal ideas and moods by distorting the subject matter to cause a sharp emotional viewing effect [V28; V29].

This directly paralleled Freudian psychology. The roiling emotional undertow that Freud described and analyzed in his pathbreaking works was put to canvas by the Expressionists. Among other iconic works, Munch's *The Scream* (1893–1910) exemplified the emotionalism of Expressionism—and perhaps became very popular as the quintessential expression of the public's angst that all classical order was breaking down: politically, socially, economically, intellectually, and scientifically. In short, it was a public shriek of Freudian proportions.

Expressionism incorporated almost all fields of artistic endeavor: (1) *literature,* portraying a distorted, nightmarish vision of the world that reflected the protagonists'

---

[19] Art historians normally point to Picasso's classic *Les Demoiselles d'Avignon* (1907) as his breakthrough work. This is true from Picasso's personal standpoint, but while one can discern the beginnings of a new art style in the distorted faces of the women, this iconic painting still maintained the normal plane perspective found in traditional paintings i.e., "Protocubist" (Cooper, 1970). In any case, for those who prefer to see *this* painting as the watershed, its greater proximity to Einstein's 1905 Special Relativity publication only strengthens the nexus between physics and art in this case.

(and society's) psychological conflicts (Kafka); (2) highly stylized *dance* replete with emotion (Jooss; Bausch); (3) twelve-tone *music*, without a tonal center (Schoenberg). All these combined two themes: lack of a grounded, core "reality" [V6] and intense subjective emotion—objective and subjective relativity.

Other disciplinary crosscurrents manifested themselves well into the twentieth century. A new music idiom arose—*jazz*—a subjective, improvisational genre [V15]. Literature began to turn further inwards with "*stream-of-consciousness,*" free association novels and writings (Woolf; Joyce). Dreamlike *Surrealism* tended to present distorted landscapes, an artistic metaphor (and perhaps attempted representation?) of non-Euclidean, Einsteinian space–time curvature. Indeed, Dali's famous painting *The Persistence of Memory* (1931) clearly reflects the relativistic amorphousness (V5; V22] of Einsteinian "time," painted a decade after Einstein became world famous because of his gravitational light-bending prediction being empirically proven correct.

Nihilistic, "anti-art" *Dadaism* followed [V23]. After World War II, *Abstract Expressionism* with its "action painting" (splotch paint on canvas) offered more than a hint of the anarchy of Quantum Mechanics, its art suggesting nothing less than the indeterminacy of position and momentum of the atom [V15].

Elsewhere, sculpture began to lose its representational form as well, breaking up into several major styles: *Cubism, hybrid "constructions,"* Surrealism, geometric abstraction, and more. These had the common denominator of reinterpreting corporeal reality, some in grotesque but recognizable ways [V22], others in unrecognizable, completely abstract fashion [V23].

Virtuality interpenetrated other fields of endeavor after WWII, with the main nexus occurring between "post-quantum" scientific theory (vast cosmological and tiny dimensional space), and other forms of advanced engineering technology along with the cybernetic revolution on the one hand, and popular and high culture on the other.

Space travel reopened the entire space–time issue, as people began thinking what it would take to colonize the very distant stars [V17]. Others imagined "wormholes" through which we could take a shortcut from one part of the curved universe to another (not a completely impossible concept [V28], but certainly beyond any conceivable technology in the foreseeable future). The iconic TV show *Star Trek* (1966) used the wormhole concept to great effect in order to overcome the space distance problem, expanding it to include time travel as well [V18].[20]

During this period, virtuality reached an apex of sorts (but not the only one, as I will shortly show) with the naming and identification of black holes. The whole concept of a part of our universe that can never be seen [V19] proved to be very fascinating to the public as witnessed by the incredible popularity of Hawking's *A Brief History of Time* (1988).[21]

---

[20] For a wide-ranging survey of the nexus between physics and popular culture, see Krauss's *The Physics of Star Trek* (1995). All sorts of virtual phenomena gradually filtered into public consciousness through this television show and its subsequent popular series of movies.

[21] The book was on the British *Sunday Times* bestsellers list for a record 237 weeks!

1969 proved to be a watershed year for virtuality, coincidentally in two other areas. First, the moon landing reduced virtuality—outer space was no longer beyond the grasp of Mankind. Second, virtuality received a tremendous boost—unbeknownst at the time—with the creation of the internet and the start of "cyberspace" [V17], a term coined later by Gibson in his pioneering novel *Neuromancer* (1984). Thus, the world of "formal virtuality" (what most people today understand by the word "virtual") shifted from outer space (cosmological world) and inner space (sub-atomic world) to the human artifact called the computer and all its accoutrements. The public's attention regarding virtuality had begun to shift to a new "world" in front of their face and under their fingers.

Physicists became involved in three major areas of virtuality, two already discussed in Chap. 5: string theory and the multiverse—neither with much resonance in the popular imagination. The third area, though, struck a deep scientific, technological, sociological, and cultural chord: aliens [V29], a general, virtuality phenomenon on several planes—from the most serious scientific speculations and theorizing to the more populistic and sensational superstitions and pseudo-panics.

Pre-World War II, there were a few notable examples of alien "penetration," the most famous being Welles' fake radio news broadcast (Halloween 1938) of Martians landing in New Jersey. The issue of aliens only began to appear as a "serious" public issue in 1948 with the supposed sighting of a UFO in New Mexico—not coincidentally a short distance from Ground Zero where the first atomic bomb test was held. Thereafter, there was a tremendous increase in the number of UFO sightings, personal claims of alien abductions, morning-after large circle rings in British fields, along with accompanying books and other forms of popular literature on the general subject e.g., von Däniken's hugely popular *Chariots of the Gods* (62 million copies sold around the world). The issue of aliens continued to capture the popular imagination e.g., the huge success of the movie *E.T.: The Extra-Terrestrial* (1982), and several others such as the *Alien* series from 1979 onwards.

Why the modern interest in aliens? Kaplan (2013) suggested that age-old human insecurity, projected onto mythical creatures in the distant past, has been resurrected in the modern age in the form of outer space aliens because of growing human *alien*ation in modern society.

Serious scientists also began to speculate about the chances of intelligent life in the universe and the ways of discovering/communicating with them [V2]. The first to seriously analyze the issue was the physicist Fermi whose "Fermi Paradox" has egged on all subsequent efforts. Fermi asked a simple question: given the humongous size and age of the universe and number of stellar objects within it, technologically-advanced extraterrestrial civilizations should exist. If so, why is there a total lack of observational evidence for that?

In 1961, Drake strengthened the paradox but also tried to turn it on its head. The "Drake Equation" was based on gross probabilities of the existence of galaxies, with suns, with planets, with atmospheres, with the "right" chemical matter etc.—similar to ours. Given the almost incommensurable [V16] number of stars (70 *sextillion*: $7 \times 10^{22}$) and accompanying planetary systems in the universe, and so on, the chances *must* be quite high for intelligent life to exist many times over in the cosmos. Based on this premise, the astronomical-technological project dubbed SETI (*S*earch for *E*xtra-*T*errestrial *I*ntelligence) went through several systems, each increasingly more sophisticated and advanced, to scan the skies for non-random radio frequency signals that would suggest intelligent life [V2]. This culminated in the SETI@home project started in 1999 in which (by 2020) close to two million participants around the world volunteered their computers to work on deciphering such signals on an ongoing basis [V3]. Meanwhile, we were not only trying to pick up alien communications; in 1977, NASA sent to outer space on Voyager 1 a few gold-constructed records with different, embedded messages from the human race.

Let's consider the virtual aspects of these projects. First, we are searching for alien life with whom we almost certainly will not be able to communicate directly for a very, very long time [V18] given the vast expanse of space and the relatively "slow" speed of light. Second, we are doing so to communicate with something that we are not at all sure exists [V15; V28]—even if logic (or math) dictates that it must exist. Third, even if we did discern intelligent life in the cosmos, it is not clear that communication is possible; it could be (some would say quite likely) that their form of life and modes of communication are totally different from ours [V29]. Fourth, based on known optical physics, it will be at least several centuries until we could produce telescopes sharp enough to be able to actually *see* (and not just indirectly infer) the aliens themselves on their planets (Choi, 2010) [V19]. Finally, Hawking suggested that perhaps we should not wish to find and communicate with aliens because they could be decidedly antagonistic (think Spanish Conquistadors and Native American "Indians") (Leake, 2010).

## 10.6.2   Post-modernism (and Its Heirs)

While all this was going within the world of physics and popular culture, another seemingly unrelated trend was evolving in the intellectual world: post-modernism (P-M). This doesn't mean that there was a *direct* influence of physics theory and our understanding of reality on P-M but rather that there were several underlying similarities between these two parallel ideational schools of thought. Nevertheless, given the osmotic effect of intellectual currents in one area of life permeating others—despite the putative split between science and the humanities (Snow, 1959)[22]—it is more than likely that such intellectual cross-pollination did occur. Indeed, it could hardly be coincidental that the term "post-modernism" began to appear in various cultural contexts in the first few decades of the twentieth century, precisely when classical physics was being decimated by relativity and the "mystical" strangeness of quantum physics.

---

[22] A few years after his famous lecture in 1959 and subsequent book publication, Snow somewhat backtracked from his strong assertion of a huge divide between the world of literary intellectuals and scientists, suggesting the development of a "third culture": an educated, non-scientific audience.

P-M is a wide-ranging, trans-disciplinary, philosophical, and cultural "movement" that became increasingly dominant from the 1960s onwards. It influenced, and in some cases, turned into the reigning approach within such disparate fields as literary criticism (Derrida), literature (Borges), religion (Caputo), epistemology (Baudrillard), sociology and history (Foucault), political science (Said), linguistics (Lyotard), anthropology (Geertz), music (Cage), and even some "less soft" disciplines such as architecture (Graves) and science (Kuhn). Thus, P-M provides a fitting coda to the present sub-chapter, for it's in P-M that we see the full expression of such a cross-disciplinary and all-encompassing perspective on virtuality.

P-M is opposed to "modernism" and the latter's belief in objective truth. Rather, it upholds the *subjective* relativity of truth [V32] in the fields of culture and ideation in their widest sense, accepting each local culture as being of equal value because all social reality is at base a social construct [V13] whose expression is a function of objective factors connected to specific environments and periods of time (among other things, this was the intellectual basis for Anderson's nation-state as "imagined community," discussed in Chap. 8).

A survey of how this plays out in each discipline would take us too far afield, not least because of the wide range of P-M thinking and conceptualizations. Suffice it to say that such a general approach is in tune with the contemporary world's virtuality *zeitgeist*, in which traditional institutions, longstanding value systems, former scientific truths, conventional social norms, and the like, have come under attack and/or have undergone irrevocable transformation. We live in an era of great social ambiguity [V15] and intellectual uncertainty [V31]—and in no small measure P-M has contributed to this. Virtuality, then, as expressed through P-M theories and analyses in their sundry manifestations, reflects the kind of world we live in.

However, notwithstanding the universal presence of various virtual, ideational elements in our world, virtuality (and P-M, for that matter) would not have had the same impact on our consciousness without what has become the axis around which all other fields revolve: modern, electric/electronic media of communication—the subject with which this chapter began.

These two pillars of contemporary society—P-M and the media—reach their "apex" (for now) in one of the less positive manifestations of virtuality: disinformation (purposeful lying; as opposed to "misinformation," a result of mistaken error or ignorance), "truthiness," fake news, deepfakes, and other forms of informational fakery [V32], all leading to a potential "infocalypse" (Warzel, 2018). There is little need here to expand on the general phenomenon, as it has received more research and media attention[23] in the recent past than almost any other social phenomenon,

---

[23] Based on Google's book n-gram (https://books.google.com/ngrams) and using their first significant appearances as the starting point, the term "disinformation" has increased 500-fold since 1960 (when P-M began to hit its stride), "fake news" 450-fold since 1990, "truthiness" 300-fold since 2000, and "deepfake" 2500-fold since 2012!

given its outsized influence on politics (domestic and international) and society (anti-vaxxers; QAnon; etc.). The main underlying "culprits" in disseminating such disinformation are social media that enable all citizens[24] to "publish" and transmit with ease, and digital technologies that facilitate textual, visual, and aural manipulation with increasing ease.

Nevertheless, the centrality of technology and especially the media for the huge expansion of many types of virtuality does not mean that such influence is unidirectional—only from media *to* society. Without the rich, non-media menu of virtual experiences and intellectual trends as outlined above, it is questionable whether media technologies and contents that produced so much virtuality in the last 150 years would have become so popular so quickly among the broad, general public. In the final analysis, media virtuality rests on a bedrock of virtual aspects in many other fields of human endeavor as detailed throughout Part II of this book.

---

[24] A thriving business has evolved through which individuals can hire others to create "realistic" disinformation on social media (Fisher, 2021)—a case of *triple virtuality*: in cyberspace [V17], with false information [V32], not written by the putative communicator [V12].

# Part III
# Virtuality and Reality: Functions and Connections

# Chapter 11
# Why Do We Virtualize?

## 11.1 The 6 W's of Virtuality

Why is virtuality such a central pillar of human existence, as illustrated and analyzed in Part II? The answers can be found on different levels, although admittedly somewhat more speculative than the evidence regarding the when, where, and what of virtuality.

The most basic level is biological: pre-human and human. As already noted, there are several types of "virtuality" in Nature: on the micro-level, all unconscious; on the macro-level, without prior intention. In evolutionary terms, most accidental (mutated) traits that confer survival and procreation advantages to the creatures involved will tend over time to increase across the specific population. One such trait would be camouflage [V19], the ability of the living thing to "hide" in its environment, so that predators have a hard(er) time finding it. From a predator's perspective (literally), a twig grasshopper is a distortion of the grasshopper [V5] and a misperceived incarnation of the twig [V20].

The panoply of tricks in the plant and animal world is impressive. Whether structurally or behaviorally, there's a lot of "prevarication" in Nature. Even here, a world unaffected by humans—where we might expect, a priori, the greatest amount of verisimilitude—virtuality is an integral part.

Moving up the evolutionary ladder, other types of virtuality appear. Among mammals we find much playacting—especially, but not only, among the young. The cubs might "fight" with each other, but neither they nor their parents are fooled or worried by such activity—they instinctively understand that such virtual play is an important component of their progeny's social (and perhaps hunting) skills development.

---

The *Appendix: A Taxonomy of Virtuality*—listing by numbers 1-35, e.g. [V13] or [V34], some of the various types of Virtuality mentioned throughout this chapter—is freely available to the public online at https://link.springer.com/book/10.1007/978-981-16-6526-4

Through the eons, misrepresentation has been rewarded; the most effective types of deception [V31] and species of deceivers win out. Without stretching the point too far, Nature rewards virtuality no less than it does reality.

Given such evolutionary "preparation," humans too have a "built-in" psychology of virtuality. As *Homo Erectus* "ascended" from the Great Apes, they *naturally* would know how to paste leaves on their body or paint it in order to blend into the environment. Later on, they developed additional types of virtual behavior not seen in the animal kingdom. For instance, with the advent of agriculture, humans transplanted their aptitude for dissimilitude from their own body to an external object, e.g. building a scarecrow.

At this point in humanity's development (actually, quite before the advent of agriculture, e.g., cave wall paintings), it was a small leap to cultural artifice, this same virtuality mindset now employed to purposes not connected with survival—"culture": art, literature, religion, philosophy, communication—even economics and warfare.

Which brings us back to virtuality as an artifact of human brain functioning, somewhat different than that found in the animal kingdom. Human brains have far greater powers of analysis, abstraction and self-reflexivity—all the mental powers that underlie conscious virtualization. Thus, although human brain psychology is partly based on what we emerged from, it also constitutes a *novum* in Nature, evolving to the extent that virtuality is no longer merely a mental survival skill but rather an integral part of the way we absorb, comprehend, and create our world.

In other words, we not only *perceive* the world partly in virtual fashion, but we also often *mold* the world along virtual lines, through mimicry and/or of the need to "fill in the gaps"—to supplement elements that are not perceivable with our ordinary sensory apparatus. As the neuropsychologist Gregory pointed out: "Our brains create much of what we see by adding what ought to be there" (Berger, 2019).

The second level of the "why" question relates to motivation: what is the *purpose* of virtualizing? One answer involves reversing cause and effect: instead of looking at how we used our "virtualizing" brain for specific human purposes, we should look at how early humanity's needs evolved our brain to develop virtualizing capabilities. To survive in a hostile environment early *homo sapiens* developed symbolic language for more efficient communication, especially to transmit and retain knowledge ("where's the nearest watering hole?"). In a never-ending Darwinian cycle, our developing and expanding brain virtualized more areas of life, further increasing survival chances of the better "virtualizers," and so on.

Nevertheless, the early evolution of virtualizing brain capabilities does not necessarily explain why we continued to expand the areas of virtuality and deepen those already virtualized from the start. At some historical point, and certainly regarding certain fields of endeavor (e.g., philosophy), virtualizing ceased being only a matter of increasing our chances of survival. Something quite profound was involved, clearly universal to human nature. Probably, this third level of "why" isn't a completely unconscious reaction to the environment, but to some extent can be consciously volitional.[1] Herewith, six different functions of virtuality for humanity.

## 11.2 Why Virtuality? Escaping Boredom and Dreariness

*...human kind*

*Cannot stand very much reality.* (Eliot, 1943, No. 1)

Paradoxically, the development of the human brain enabling greater understanding of the world also gave humans a greater appreciation of the different possibilities inherent in their life. Not only did our brain become more complex but as a direct result, so did human society. In the hunting and gathering stage *homo sapiens* lived a very circumscribed life—forage/hunt for food, cook meat for eating and occasionally prepare skin for clothing, tell stories around the fireplace, go to sleep, and the next morning repeat.... Once we established civilization the situation changed drastically, for civilization meant division of labor and expansion of goods and services. For each local clan of earlier *homo sapiens* there was no visible example of "living life differently"; for human beings in early civilized society, increasingly there were many such examples.

The uneducated and poor preferred the life of those in a higher station, but those in better stations probably also imagined what living as "another" might be like. As inter-cultural contact increased, the possibilities multiplied. Social diversity, of course, was not merely a matter of education, profession, status or language; it involved myriad forms and elements of culture, broadly defined.

Unfortunately, the real world—from the perspective of any individual—is severely circumscribed. Humans cannot choose to do anything they want. General social mores ("thou shalt not..."), the limitations of our personal skills and resources ("my vocabulary is poor to be a poet..."), demands of family and professional needs ("eight mouths to feed—I have to work twelve hours a day"), and even physical restrictions (gravity, weather, terrain etc.), all render the human condition exceedingly "frustrating." Making matters worse, as we approached the modern age and certainly within it, people began to live longer, so that the boredom of sameness became ever more difficult to bear. Add our ability to preserve and remember the past (oral storytelling, written histories, electronic media)—with their somewhat different lives—and one can see how any person's life would seem to be "limited" compared to all the other types of living that existed past and present. We became aware that the "virtual" (as potential: V26) far outstripped the "actual."

---

[1] This level echoes Popper: "I hold that there are three (or perhaps more) interacting levels or regions or worlds: the world 1 of *physical* things, or events, or, states, or processes, including animal bodies and brains; the world 2 of *mental* states; and the world 3 that consists of the *products of the human mind*, especially of works of art and of scientific theories" (1977).

This was half the problem—perhaps the easier half. For the vast majority of humans throughout history, life was a Hobbesian "nasty, brutish and short": back-breaking work, 14 hours a day, 7 days a week; illness, pain and debilitating conditions (e.g., loss of teeth, declining eyesight); death of loved ones (one in four children dead by the age of 5; very high rate of breech births killing the mother); and other assorted tragedies and depredations e.g., plague, drought, incessant warfare, natural disasters, tyrannical rulers. There was little chance of evading one's condition—other than suicide (a cardinal sin) or... mental escape.

To relieve that boredom and occasional dire distress, people started using the same mental facility of virtualizing that had served them so well for sheer physical survival. They used imagination, symbolizing, and artistic creation for "mental survival" i.e., preserving sanity through virtuality. Physically and mentally "creating" virtual artifacts, ideas, and phenomena, became the central (and occasionally the *only*) way to temporarily overcome and circumvent our world's rigid physical constraints and social restrictions. As Maugham put it: "To acquire the habit of reading is to construct for yourself a refuge from almost all of the miseries of life" (1977, p. 29)—not an abrogation of personal responsibility to the real world, but rather a necessary activity to enable proper functioning within our world: "Rather than a pathological disavowal of duties, escapism can be an important means to regulating our emotional states and enable us to sustain an otherwise unbearable situation or to turn a tedious stretch of time into a cognitively and emotionally engaging activity" (Calleja, 2010, p. 347). This is not only a matter of consumption but production as well. "The virtual...is a fecund and powerful mode of being that expands the process of creation, opens up the future..." (Levy, 1998, p. 16).

In short, "escape" is a core aspect of our humanity: "Culture is more closely linked to the human tendency not to face facts, our ability to escape by one means or another, than we are accustomed to believe.... A human being is an animal who is congenitally indisposed to accept reality as it is" (Tuan, 1998, p. 5).

Contemporary culture has grasped the importance of virtual escape in psychologically surviving our constricting reality. Perhaps the most extreme expression of this was the seemingly oxymoronic "Holocaust comedy" film *A Beautiful Life* (1997): to psychologically protect his son, the father overlays a virtual world on the deathly reality in which they exist. The eventual liberation of the camp through the appearance of an American tank entails "real reality" reinforcing the reality of virtuality. Indeed, only through such a fabricated virtuality was the child able to survive the horrors around him![2]

---

[2] This is not to say that all cultural expressions are uniformly and completely positive about virtual escape. For instance, in the film *Pleasantville* (1998), the male teenage protagonist escapes his humdrum reality by watching the 1950's soap opera "Pleasantville" where everything is pleasant. A strange TV remote control (virtuality at a distance!) puts them inside the town, in black and white—suggesting that virtual escape is not the panacea we might think. However, as his wilder sister does her thing (1990s sexual mores in the 1950s), the town begins to sprout color—another suggestion that virtual life is not always as exciting as real life. In any case, the movie highlights the fact that virtuality is an escape from the humdrum of reality—and that the two experiences have become intermingled to the point that each can contribute to the other.

Highly imaginative, virtual consumption can also alleviate severe physical pain and overcome phobias. Hoffman (2004) describes how patients undergoing wound care reported their pain dropping dramatically when actively engaged (immersed) in virtual reality scenes [V10], and similarly in confronting their terrors through realistic simulations of the feared thing (spider) or frightening act (flying). This is paradoxical, essentially employing technologically induced external virtuality to overcome mentally generated, internally virtual feelings.

The need and means to escape reality were on full display when Random House asked readers to vote for the "best" 100 novels of the twentieth century (100 Best Novels, 2000). Fully one-fifth of these chosen books were "fantasy" of one sort or another (science fiction, dystopic futures). Clearly, many people want to escape not only from their reality to some other imagined reality, but also much "further out" to an "unreality" or "surreality." As Cohen & Taylor (2007) suggest, imaginative explorations—whether *daydreaming* or *fantasizing* (Campbell, 1987)—enable us to deal with the routine by imagining our being somewhere else, even if we're aware it's only temporary.

In sum, to alleviate the humdrum and occasional misery of reality we turn to imaginative virtualizing because: 1—fictional people tend to be more entertaining than people with whom we have daily contact; 2— life has long stretches of banality, so that we seek narratives that have more "action"; 3—humans are able to mentally create an escape through their own imagination; 4—technologies that employ our imagination provide stimuli impossible to experience in the real world (Bloom, 2010, p. 175).

## 11.3  Why Virtuality? Efficiency

The second general purpose of virtuality is to render our material life easier, better, more fruitful. Here too the main factor is the world-as-is. At any point in historical time, people live their lives within the existing cultural, social and technological framework, faced with problems therein. Virtual mentalizing (originality, creativity, recombination of existing elements) enables us to find solutions to these problematic areas of life.

To be capable of such solution-oriented virtualizing we first have to develop our brain's cognitive ability. To that end, virtuality also plays a role. As Gopnick asserts: "For human beings the really important evolutionary advantage is our ability to create new worlds... that's just what our human minds do best—take the imaginary and make it real" (2008, p. 14). Tooby and Cosmides (2001) suggest that the arts and imaginative activity in general are *brain organization* adaptations enabling us to continue learning as new life challenges arise: "We think that the task of organizing the brain both physically and informationally over the course of the lifespan is the most demanding adaptive problem posed by human development..." (2001, p. 14).

Once the individual reaches the requisite level of brain organization and cognitive "maturity," the process of creative thought and original invention/solution occurs by the slow accretion of information, many times seemingly unrelated, and then combining those pieces of knowledge—while sleeping (Johnson, 2010, p. 101) or more often when *not* thinking about the problem but rather letting the brain wander.[3] Thus, before something is invented it is merely a "non-actualized potentiality" [V27]; the inventor strives to turn that potentiality into some form of actuality. Put simply, a future idea's components are usually already found in the brain but need a process to combine them into something "new." Literally and figuratively, invention is the mental act of developing to *real*ization the potential [V26] inherent in unrealized components.

As noted throughout Part II, striving for greater "efficiency" through virtuality is not merely a matter of scientific conceptualization or technological invention. Many of humanity's greatest advances involve "social reorganization": democracy, feminism, abolition of slavery, rights of minors (and minorities), money, bureaucracy, organized religion, academies of learning, and so on. Each of these demanded their creators/developers imagining a different way to organize human affairs to leverage the (mostly human) resources at their disposal.

Here we have moved from virtual input to the virtuality of output: one cannot touch "civil rights"; one can't "feel" democracy; the "nation" is a figment of our collective imagination ("image-in-nation"?); the corporation is a legal fiction designed to carry out highly efficient commerce and trade through economies of scale. Of course, such virtual ideals and institutions have a "real" component. In most cases, the social invention could only be effectuated through concrete venues: school buildings, government offices, offices and stores, bank branches, etc.—but the underlying concepts emerged from humanity's virtualizing ability in the pursuit of a better life materially and psychologically.

## 11.4  Why Virtuality? Curiosity

The third general motivation for virtual behavior is "curiosity." Although many animals also evince curiosity, this mental trait seems to be insatiable, deeply embedded in the human psyche—at least among the young. Children will spend hours taking things apart or driving their parent crazy with an incessant "why?" Many people never lose their curiosity about the world; if they did, we would not have myriad newspapers, magazines, books, current events programs, and social media produced in never-ending numbers.[4]

---

[3] Research (Thatcher et al., 2008) indicates that the human brain has a built-in "chaos mode" when neurons go out of sync; then "new" connections are made between disparate "facts/knowledge" that lead to "original insights." The longer this "chaos mode," the higher the person's IQ!

[4] Not all forms of "curiosity" are the same. One type is *social* ("What are my friends doing? Which celebrity is having an affair with whom?"); the other *informational* ("What are the latest scientific findings? What legislation has been passed recently?"). I refer here mainly to the latter; the former is not unique to humans, as many animal species also elicit a keen need to know what their fellows are doing, and with whom. None (as far as we can tell) want to know why leaves are green....

Why this curiosity drive? The human mind works like a perpetual motion machine. Once we started on the road to thinking in abstract, symbolic, representational (i.e., virtual) terms for survival's sake, an exaptation occurred: this mental process continued apace even when survival was no longer an overwhelming problem. In certain fields, the search for knowledge can be its own motivation; there is no immediate utilitarian purpose in studying cosmology, philosophy, musicology, history, and a few other intellectual pursuits. Yet few would argue that they are not, or should not be, intrinsic parts of our intellectual life.

In many other areas of life, the curiosity motivation is connected to the efficiency function and to the sixth one (existential angst). "Survival" continues to motivate our curiosity through virtual thinking to find ways to improve our material quality of life, the basis for much of science. Moreover, as we haven't conquered Death, we try to extend life through the "scientific imagination." This has certainly brought some success as our understanding of Nature grew less imaginative [V28] and metaphysical [V33], but still virtual e.g., gravity (unseen) [V19]. Quantum mechanics—with its wave-photon duality [V15] and other "mystical" phenomena (e.g., entanglement) [V2; V25]—provides an even better example of how modern science not only doesn't repudiate external-world virtuality but rather reinforces it.

In any case, until such time when our knowledge about the world is complete (if ever), curiosity will continue to drive our need to discover what lies "over the horizon" i.e., the virtual world just beyond what we currently perceive and understand as "real"—whether for functional reasons or just because "we want to know."

## 11.5  Why Virtuality? Theory of Mind and Social Cohesion/Conflict[5]

Think about this (literally): why are you (sometimes) sure that you know what your spouse or best friend is thinking? Unlike any other non-primate, humans have developed a "theory of mind"[6]: understanding that others think like us so that we can figure out what they are thinking ("what would I do/think in this situation?") and our past experience with them ("what did they do/think in previous, similar situations?"). As

---

[5] Section 11.5 focuses on the *social* consequences of virtualizing. However, the theory of mind can also be turned inward to beneficial purposes. For instance, envisioning how one would react to a strange phenomenon or in an unusual situation—thereby learning about one's own mind and preparing "it" for such potential eventualities in the future. Indeed, it is likely that "self-awareness" is merely another form of employing mind-reading inwards to our own mind (Rosenberg, 2016).

[6] One of the ways in which we develop this is by creating imaginary "companions" at a very early age, carrying on a complex (and somewhat long-lasting) relationship with them (Taylor, 1999). Approximately half of all children do this—and it turns out that they have a greater ability to distinguish between reality and fantasy than children who do not, and in addition understanding other people's emotional and general mental state.

a famous *New Yorker* cartoon put it (a husband talking to his upset wife): "Of course I care about how you imagined I thought you perceived I wanted you to feel."

Such virtual projection of—and into—others' minds has enabled human society to expand and function in highly complex fashion, well beyond the animal kingdom's instinctual social interaction. As previous chapters elucidated, such "mind communion" has enabled humanity to develop many social quasi-virtual fields of endeavor (religion, the arts, government) in addition to several important human professions e.g., psychology. Zunshine (2006) asserted that we enjoy narrative stories of all types because they help us understand other people's minds. In certain circumstances with the proper mental attitude our theory of mind can even lead to mutual toleration: "mental simulation… enables us… to gain more tolerance for the thinking processes of people we fundamentally disagree with" (Ryan, 2001, p. 111).

Indeed, our innate theory of mind might constitute the basis of universal values and patterns of behavior such as social justice, altruism, charity and the like. By putting ourselves in others' shoes, we are moved to help them through the idea of "there but for the grace of God go I." A strict view of "survival of the fittest" (that Darwin did *not* ascribe to), would have us helping genetically close relatives exclusively. The fact that we go well beyond this, attests to our ability to transcend pure biological self-interest, understanding that all human minds work similarly.

Why, then is there warfare, savagery? Humans are the only known animals to kill members of their own species for purposes other than food and procreation. What accounts for this? The same phenomenon: theory of mind. As we are well aware of our own aggressive tendencies, we easily imagine the same of other people: "I am vulnerable at the moment, so that if I were him I would attack me right now; I better attack him first!"

How do we live with such contradictory outcomes of the same ability to virtually picture what's going on in others' heads? By dichotomizing humanity: "my group" (here too a somewhat virtual construct [V31; V32][7]) vs. "them," the "other" (similarly, virtually defined). In the former, we see many more positive elements; in the latter, the baser aspects of our being. Of course, the "other" can be in some distant land or exist right here at home. As human societies became increasingly stratified socially, the higher status sub-group began to view the lower status groups as having a different "mix" of mental elements e.g., rational vs. emotional. Thus, *within* the larger society the theory of mind tends to separate one group from another; *between* larger societies it leads to social cohesion within the larger internal group and greater antagonism to external macro-groups.

From this I offer a pure speculation as food for thought: world peace will only arrive (if ever) when we are threatened by an alien species [V29] from outer space. At that point, each person's theory of mind will tend to see all other humans in terms of their common, benign modes of thought, in contradistinction to the very

---

[7] Of course, the distinction between "us" and "them" can be logically understood as reflecting real ideational divides: Christian vs Moslem (in the twenty-first century), Capitalist and Communist (nineteenth and twentieth centuries), Catholic and Protestant (sixteenth/seventeenth century). However, these are "virtual" in that the primary basis of these differences is metaphysical [V33].

alien extra-terrestrials.[8] However, we might well not need wait for this unlikely event to happen. As Pinker shows (2011), there are several macro-historical trends and factors underlying a steep decline in wars and social violence, in part due to our expanding "theory of mind," as well as some other phenomena I have noted in previous chapters book e.g., the enlargement of, and interconnection between, political units (Nation-State, Regionalism, nascent Globalism).

To sum up, for better and for worse, when we view other humans, we are essentially looking in the mirror—and projecting behavior towards the "other" based on our own deepest feelings of empathy and/or fear [V6].

## 11.6 Why Virtuality? Thinking Beyond Present Reality

Animals are usually very good at discerning threats to their existence or sources of sustenance—here and now. With some very few, narrow instinctual exceptions, animals are not capable of envisioning, and planning for, the future (although, see Nicholls, 2008). This is probably one of the most important traits of humans, individually and collectively.

Our ability to imagine a reality different from the present can be found in all important areas of life. First, this is what most artistic expression is all about—certainly "non-representational" art. In the case of the narrative arts—literature, theater, cinema—a central goal is not to present a picture of "what is" but rather "what can be" [V26, V27, V28], to show society the path towards a different way of structuring and ordering itself. Among *homo sapiens*, social structure differences are numerous. Some of these are due to environmental pressure (e.g., polyandry where there is a dearth of present males), but in large part different social systems stem from the many ways we can envision human society functioning.

On an individual level, we all have the ability to think ahead beyond our past or present experience. These may seem to be straightforward types of decisions, but they involve "theories of the world" that have two main sources. First, our ability as young children to "make believe" enables us to "try out" the world's possibilities without directly experiencing them. Second, we acquire knowledge from virtual sources

---

[8] This "aliens" comment is relevant to intra-human interaction as well. Another reason for human aggression is that occasionally our theory of mind goes to the extreme of perceiving the "other" as not having the same "mind" at all—whether demeaning and discounting their level of intelligence (e.g., slaves, racism), or demonizing their way of thinking (e.g., "yellow peril"). Here it is not the perceived *similarity* of (malign) minds that leads to our aggressive behavior vis-a-vis the "other" but rather the perceived *alien*-ness of the other's mind. One can speculate that this is the reason for the popularity of science fiction and horror narratives starring aliens; they are thinly disguised substitutes for the "foreigners" in our midst. In short, not so much V29 ("alien") as V22 ("quasi-representational"). Of course, an extra-terrestrial alien might not be "aggressive" i.e., aggressiveness might be a uniquely Life-on-Earth phenomenon. There are exobiologists and science fiction writers who suggest (only partly tongue-in-cheek) that the reason we have not been visited by aliens is that they want nothing to do with a planet riven by so much (or any) inter-species, and even intra-species, strife.

of information about the wider world (e.g., books). In both cases, we utilize virtual means to expand our understanding of the larger world for planning our future actions. This, as opposed to animals who "test" surrounding reality in direct fashion (eating the tasty-looking mushroom), and more often than not suffer dire consequences. As Popper (1997) put it regarding the way humans deal with our potentially dangerous world: "...we may formulate our hypotheses and criticize them. Let our conjectures, our theories, die in our stead!"[9]

Such "theories" are not always a reaction to our real environment. Humans are (most probably) unique in that we employ imagination to mentally create any scenario, possible and impossible. Often, we do this consciously (e.g., brainstorming with others), sometimes unconsciously (dreaming), and very often semi-consciously (as noted in Chap. 3, "mind-wandering" daydreaming occurs close to half our waking hours: Klinger, 2009; Killingsworth & Gilbert, 2010).

Modern science works similarly. We do not merely observe the world; rather, from the limited facts at hand we derive hypotheses that can best be described as educated guesses as to why the world is (or at least seems to be) as it is. Thus, scientists develop a virtual concept (hypothesis [V26]) that takes factors and/or variables into account beyond those we know at the moment to be relevant, and then test the hypothesis [V10] on new data, situations, phenomena, by seeing whether (and how) the broader picture fits the hypothetical construct.

Of course, in a few fields such as mathematics and physics, the hypothesis need not be immediately tested experimentally but rather mathematically i.e., whether it is self-consistent and/or follows accepted "rules." Moreover, in order to try and understand whether an initial theory (hypothesis) or even a proven theory (system, principle, rule)[10] makes sense—or how it makes sense—we will occasionally invent a "thought experiment" [V9] that has no parallel in the "real" world (e.g., the quantum indeterminacy of Schrödinger's cat—alive and dead at the same time), but forces us to think about the implications of the hypothesis/theory. Real-life applications of

---

[9] Taleb calls direct experience of reality—the only type that animals know—"first-order natural selection." Humans indulge in second-order natural selection, using mental imaginings in place of direct experiential contact: "...projecting allows us to cheat evolution: it now takes place in our head, as a series of projections and counterfactual scenarios" (2007, p. 189).

[10] There are two general meanings for the word "theory," and this can cause great confusion. The commonly held definition is something that we think might be true but there's no real proof, as in "I have a theory why the Democrats won the election..." In truth, this is really a "hypothesis" or speculation. The scientific meaning of "theory" is something that has been tested empirically or mathematically and found to be in accord with the available evidence. In this sense, Darwinian Evolution is a "Theory"—one that has stood the test of many experiments and empirical investigations. Of course, this doesn't mean that any such Theory is the Absolute Truth, for we can never know if further evidence in the future won't undercut the Theory and lead to a newer one. For example: Ptolemy's Geocentric Theory held for 1300 years but eventually gave way to Copernicus's Heliocentric Theory that itself gave way after about 400 years to the present Theory of a universe with no "center" and expanding uniformly in every direction. This famous series of examples, as with Newton's Theory of Strict Determinism that gave way to Einstein's Relativity and Heisenberg's Uncertainty Principle, reinforce the human drive to continue hypothesizing about how "reality" works, notwithstanding any present Theory that explains the present situation more than adequately.

hypotheses or abstruse mathematical proofs can emerge much later on, but this too shows the power of mathematical virtualization.

Two examples: (1) Riemann's geometry of "curved space" posited that the shortest distance between two points is *not* a straight line, thought to be at the time (nineteenth century) unreal with no way of showing its applicability/reality in the world—and then later Einstein's curved space–time showed exactly that! (2) Regarding the epidemiology of the Covid-19 pandemic: "Because of pure mathematics, people didn't suddenly have [to] develop a whole theory of how diseases spread.... The mathematical side was ready because pure mathematicians had done the work" (Cunliffe, 2021).

Finally, our ability to project into the future and imagine very different, virtual (i.e., alternate) realities [V28] extends even as far as envisioning a world without humans (Weisman, 2007), due to ecological disaster, and/or the ultimate replacement of humans by our cybernetic "progeny"—artificially intelligent constructs. It can even involve physicists investigating the universe's eventual end trillions of years hence, the result of its never-ending expansion leading to entropic "heat death" (single atoms everywhere), or the reverse—the entire universe's collapse into an unimaginably dense singularity. This brings us to our final category explaining the need to virtualize.

## 11.7 Why Virtuality? Resolving Existential Angst

*As far as we can discern, the sole purpose of human existence is to kindle a light in the darkness of mere being.* (Jung, 1961, p. 326)

From time immemorial, humans have looked at the stars and wondered "what's it all about?" Of course, it was enough to see their loved ones buried in the ground to wonder whether "what we see is what we get" from this world, and no more than that. The idea that "this is all there is," was—and for most people continues to be—too horrible or depressing to contemplate.

Here began the search for greater meaning. Obviously, though, it could not be found in the purely material world so replete with death and decay. So, humans turned to inventing, creating, seeing, and believing elements beyond the concrete world. These took two main forms.

First, most people sought "ultimate answers" in religion broadly defined, as discussed at length in Chap. 4: gods/God, devil, angels, immortality, soul, Oneness, and myriad other virtual constructs [V25, V28, V33, V34]. This general search was clearly produced by fear of the unknown in general and our personal unknown future specifically—or as the theologian Niebuhr put it: "Religion is always a citadel of hope, which is built on the edge of despair" (1932).

Others of a less metaphysical bent, or perhaps less fearful of the unknown, sought solace not in religious belief but in personal creativity. If they could not fully understand the world and our personal place within it, at least they could strive to alter it or add something of value. These efforts produced works of imagination, of artistic expression. Here too the purpose was to offer non-economic sustenance for the mind; not to better understand how the world works but rather to provide it with some worth. As Krauss (a hardcore scientist) put it: "All these creative human activities reflect... the spark that raises our existence from the mundane to the extraordinary" (2005, p. 6). In fact, some argue that this even serves an evolutionary function (survival) for humankind: "Playing...helps us face our existential dread. The individual most likely to prevail is the one who believes in possibilities—an optimist, a creative thinker.... Imaginative play, even when it involves mucking around in the phantasmagoria, creates such a person" (Henig, 2008).

The use of virtuality to overcome "what's-it-all-about" questions is not only a matter of *active* creativity; *passive* consumption of creative works also serves the same function. As Oliver and Raney found in their empirical study of motivations underlying entertainment consumption: "eudaimonic [truth-seeking] motivations may be particularly well suited to assist individuals in their search for and understanding of 'larger' questions surrounding the human condition, including existential question surrounding issues of mortality" (2011, p. 1001).

While religion, artistic expression, and scientific re/searching constitute different approaches, one should not negate the underlying common need to resolve the same human problem. As Einstein put it: "All religions, arts and sciences are branches of the same tree. All these aspirations are directed toward ennobling man's life, lifting it from the sphere of mere physical existence..." (Einstein, 1956, p. 9).[11]

## 11.8  Virtuality's Multiple Benefits

These six general motivations for virtualizing—*alleviating boredom, improving efficiency, satisfying curiosity, engendering social cohesion, projecting the future, and allaying existential angst*—are not islands that stand alone. For instance, whereas music and storytelling may serve the primary purpose of relieving boredom, they can also serve an efficiency purpose by teaching us how others do things, as well as strengthening the cultural glue holding society together and making members of a specific society more willing to help each other. Another example: much theoretical science (curiosity) ultimately leads to practical science and technology (efficiency), as well as the ability to think beyond the here and now (future planning). Honing the mind (curiosity) ineluctably also leads to benefits of a more practical kind (efficiency, social cohesion, planning).

Even the search for existential answers can have efficiency effects. Religion can bind together humans within a specific society (but also create conflict between them); the creative arts can highlight societal flaws that if fixed can strengthen the

---

[11] Krauss goes even further, suggesting that scientific speculation can help religion: "the fact that the possibility of the existence of extra dimensions was at least allowed by science... also meant it allowed a place for a world of purpose, the spiritual world, to exist" (2005, p. 78).

collectivity; religious ritual can relieve boredom and defuse potential social protest, thereby aiding social cohesion (Marx's "religion is the opium of the masses").

As we are complex beings influenced by sundry external and internal forces, so too our virtual functions (emphasis on the plural) should be viewed as integral parts of an interrelated whole. Certainly, the motivating factor underlying all these is not unitary. Just as there are numerous *ways* in which we virtualize, there are no fewer *reasons* for doing just that.

# Chapter 12
# Virtuality and Reality: Two Sides of the Same Coin

## 12.1 Virtu(re)ality

It is a truth that since the time of Pythagoras – an especially since that of Plato – society has recognized the existence of a realm of pure ideas, worthy of respect or even worship in its own terms.... . It is too much, however, to expect the human race to undergo a wholesale switch to Platonic idealism just because technology has given more plentiful access to information. (Jonscher, 1999, p. 227).

One might get the impression from an uncritical reading of this book that virtuality is "taking over the world." True, the emphasis on virtual experiences is designed to illustrate the universality of virtuality over time and among sundry fields of human endeavor. However, this does not mean that virtuality is everything, or even that it is more important or more prevalent than reality. We do not live in a world of "*virtuality über alles.*" Nor can we, given the largely corporeal condition of our social world and our embodied biology.

Instead, virtuality and reality constitute two sides of human life: "virtu-reality" or "virtu(re)ality," to coin a neologism. Whereas each has its discrete and unique characteristics, they cannot be separated—almost like conjoined twins who develop their own personality despite their "inextricable attachment." Some scholars argue that virtuality can have no autonomy: "the virtual does not constitute an autonomous, independent, or 'closed' system, but is instead always dependent, in a variety of ways, on the everyday world with which it is embedded" (Malpas, 2009, p. 135). This is too narrow a definition of "virtual," and certainly doesn't include the human mind's imaginative, creative virtuality; nevertheless, it does describe most types of virtuality.

It is also possible to posit the inextricability of reality and virtuality for the opposite reason. As we have seen in several chapters, one can argue that virtuality and reality are intrinsically the same thing because *reality* as we generally conceive it *is at base*

The *Appendix: A Taxonomy of Virtuality*—listing by numbers 1-35, e.g. [V13] or [V34], some of the various types of Virtuality mentioned throughout this chapter—is freely available to the public online at https://link.springer.com/book/10.1007/978-981-16-6526-4.

*a virtual phenomenon*, or at the least emanates from a type of existence that can only be defined as virtual. Here are a few phenomenological ways by which we see this at work.

The first resides at the lowest level of our "physical" world: the dualism inherent in quantum mechanics that simultaneously exhibits reality and virtuality. As explained in the Chap. 5, light can be either a photon or a wave [V15]; when we "force" it to "decide," it then becomes either a "real" thing (photon) or a "virtual" thing (wave). Here the "virtual" is not the potential that has yet to be realized but rather the potential that has not been (and no longer can be) realized [V27]. However, given that the next time we run this photon/wave light experiment the result might be the exact opposite, one cannot argue that the non-occurring wave (in the first experiment) is simply "non-real"; rather, it is the virtual shadow of potentiality that still can become real the next time around [V26]. And again, Heisenberg's Uncertainty Principle posits that all atoms are "unclear" as to their final state until we actively observe them — and then we can only determine with absolute accuracy either their velocity or position [V16]. This means that beforehand they are both "here" and "there"—pluripotent, or if one wishes: multi-virtual, collapsing into the "real" upon human observation.

Intrinsic virtuality can also be found above the atomic level. In numerous unrelated disciplines on the meso-level of human existence, researchers have recently come to realize that the biological and social worlds are imbued with the phenomenon called "emergence." We are all familiar with the aphorism "the whole is greater than the sum of its parts." The narrow (common) understanding of this: the larger entity does things that each of its "member units" cannot accomplish by themselves. However, "emergence" has a broader—and philosophically more important—meaning: the end product cannot be strictly deduced from its constituent parts. In other words, even a complete understanding of what each part is comprised of, and what it can do, would still *not* enable us to predict the way the final entity functions: "They [complex systems] are *emergent* in that we did not anticipate the properties exhibited by the whole system given what we knew of the parts" (Bechtel & Richardson, 1992, pp. 266–67).

This turns out to be the case in almost every important sphere of existence (Morowitz, 2002): psychiatry—chemical molecules leading to ephemeral qualia we call "consciousness"; biology—one can't simply extrapolate from DNA coding to the whole biological organism; society—the organizational functioning of social organisms (each ant is very stupid; thousands together display highly "intelligent" social behavior); physics—the translation of probabilistic micro-quanta into "deterministic" macro-matter; economics—understanding the quasi-"selfish" economic behavior of many individuals does not enable us to predict overall market development; meteorology—chaos theory par excellence; and so on.

Where is virtuality here? Whereas each sub-component is itself "real" (e.g., a brain neuron), it is also virtual in that it contains the potential [V26] to join with other components in order to create something "above and beyond" itself (e.g., a mind). In other words, every small-scale entity has a basic "reality" but also a potentiality—in most cases many possibilities—that it cannot express individually, but only in connection or conjunction with others.

And even the above statement "every small-scale entity has a basic reality" might not be altogether correct, for once it is embedded in the emergent larger entity its "reality" changes as a function of the profound influence the larger whole has over the constituent part. By itself it is "incomplete" [V4] relative to what it can/will be within the larger whole. This too holds true in various areas of life. For instance, quiet, law-abiding citizens can do nasty things when caught in the frenzy of the mob. Are they "really" law-abiding citizen or rioters? There is no absolutely correct answer: standing alone, these citizens really are the former (and virtually/potentially the latter), and as part of a larger whole they are really the latter (later, virtually/potentially reverting to the former). The same holds true for almost any other element in the social or biological world e.g., attached to a living body and its sensory inputs the brain acts with "mind"; attached to a dead body (heart not pumping) the same lump of organic brain matter has no mind.

Even more confounding is virtuality as a result of combined identities: a chimera. These used to be the stuff of mythology: "beasts" combining parts from different animals [V15; V35]. However, with today's technology these can be created through genetic engineering i.e., organisms with a mixture of genetically different tissues. Although, they are wholly corpo*real*, their identity is ambiguous [V15]. And in fact, a normal human being can also be called a chimera, given that close to half our body's cells are the cells of bacteria, viruses, fungi, and archaea living within (and on) us and not "our" cells at all (Gallagher, 2018). Even "worse," while a human being has approximately 20,000 genes, our microbiome contains 2–20 *million* different genes [V19]! People are virtually not who they think they are; indeed, more "we" than "me."

The third aspect in which we can see virtuality and reality as almost the same thing depending on the context is not a "phenomenon" but rather something much deeper: what the world is made of. We have seen that modern physicists realize that energy and matter ($e = mc^2$) are basically the universe's identical constituent. However, even this might not be the last word on (the) matter, as "an alternative view is gaining in popularity: ...*information* is regarded as the primary entity from which physical reality is built" (Davies, 2010b, p. 75) by being transferred from one "material" entity to another. (This is similar to the story related in Chap. 5: the Heisenberg lecture that I attended at Harvard University back in the 1970s where mathematics starred as the universe's underlying foundation.) If so, we will have come full circle from early Platonism in which the Ideal Form was considered to be the ultimate "reality" and our material world a rough reflection or representation of that reality. In any case, "information" is virtual in its ineffability and its constant mutability [V17; V25].

Can one discern any connection between these virtual phenomena and principles on the micro, meso, and macro levels? Do these constitute three *different* phenomena with some analogous similarities, or do they both stem from some underlying, fundamental "law"? I would like to suggest that perhaps the latter is the case, based on another phenomenon, "entanglement," also discussed previously: "Recent hypotheses in quantum physics suggest that the whole physical universe is 'entangled' in such a way that the parts of a system—even the behavior of elementary particles—cannot be fully understood without seeing their role within a greater whole: ultimately the whole of space–time" (Ward, 2010, p. 284).

Entanglement does not merely occur between "atom" and "atom" but between everything and everything else. This could be the "logical" basis, among other things, for Heisenberg's Uncertainty Principle—without a "connection" to the observer the atom is stuck in a virtual state of pure potentiality, but once "connected" to human consciousness through observation it becomes "real." How is such a connection effectuated? In other words, what is the trigger for a particle becoming what it is? *Information transfer.*

It is not altogether clear in which direction this information transfer occurs; perhaps the observer absorbs information from the particle or conversely the particle is "informed" that it is being observed. In any case, the "reality" of "matter"[1] turns out to be profoundly virtual: it is not material but rather ineffable [V25]; it is not strictly deterministic but rather quantum-probabilistic [V15]; it does not have an unchangeable identity but rather its "core" is deeply connected to (and influenced by) the world around it.[2] All this does not render what we perceive to be reality an illusion. Rather, it changes our understanding of reality from something deterministically Aristotelian/Newtonian to a phenomenon more Idealistic/Relativistic/Probabilistic as well as Universally Networked. In short, the underlying structure of our world—as it emerges from bottom to top, and also as it behaves at the bottom—is a profound mixture of real and virtual.

Returning to the connection between the micro and the meso levels of the world, research (Brooks, 2011) provides tentative evidence of a deep connection between quantum behavior in the atomic world and our sensory capabilities, through "vibrational sensing" that can explain such conundrums as how humans can distinguish 100,000 different odors with only 400 olfactory receptors [V21]. This same connection might explain otherwise impossible phenomena related to energy production in mitochondria and photosynthesis.[3]

In short, the "material" world in which we live and interact with concrete objects, might work on the same (quantum) principles as the sub-atomic world that we now know to be replete with "magical," that is virtual, processes.

---

[1] My apology for all the quote marks, but they signify meanings that differ from common usage.

[2] One can substitute "person" for "matter" and reach the same conclusion: humans are not merely "meat-matter" but have an ineffable consciousness; they are not puppets on a string, mindlessly acting deterministically because of their physical environment; and they do not have a monistic identity but rather one that is flexible, plastic and plural—willfully changing their identity depending on the social context and interaction with others.

[3] If all this is true, what remains to be explained is how such meso-level quantum processes do not degrade immediately under environmental influence. Quantum theory asserts "decoherence" of quanta the millisecond anything from the outside impinges on their state. Smell receptors, plant cells, etc., incorporate a jumble of such factors, so that immediate decoherence should (in theory) negate the quantum connection between micro and meso.

## 12.2 Virtuality Impinging on Reality

We now turn to several examples of virtuality impinging on reality. This involves human activity individually and collectively-socially. What contemporary human society is doing—and to a lesser extent did in the past—might be a rough reflection of the deeper reality-virtuality nexus just presented. The word "reflection" is key: one cannot state that what happens on the level of human society is *directly caused*, or emanates from, the fundamental "virtual reality" of the universe. Nevertheless, some of the parallels are striking.

Let's start at the micro-individual level. Assuming that contemporary neuro-science is correct, there is no Cartesian mind/brain duality.[4] Our subjective virtualization both from within our brain (e.g., dreaming) and from external sources (distorted perception) has a physical component. As suggested in Chap. 4, even the most metaphysical experience of all—religious/spiritual ecstasy and transcendence—seems to have its source in certain parts of the brain. However:

> We're beginning to realize how much human and animal cognition runs via the body. Instead of our brain being like a computer that orders the body around, the body-brain relation is a two-way street. The body produces internal sensations and communicates with other bodies, out of which we construct social connections and an appreciation of the surrounding reality. Bodies insert themselves into everything we perceive or think.... The same hill is assessed as steeper, just from looking at it, by a tired person than by a well-rested one (de Waal, 2009).

Moreover:

> A rapidly growing corpus of research indicates that metaphors joining body and mind reflect a central fact about the way we think: the mind uses the body to make sense of abstract concepts. Thus, seemingly trivial sensations and actions—mimicking a smile or a frown, holding smooth or rough objects, nodding or giving a thumbs-up—can influence high-level psychological processes such as social judgment, language comprehension, visual perception and even reasoning about insubstantial notions such as time (Carpenter, 2010).

For instance, drinking something hot will make us feel more "warmly" towards other people in the vicinity at that moment!

In other words, not only our immediate, organic sensory organs (eyes, ears, nose, tongue, skin) can distort perception for physical reasons, thereby rendering external reality for us somewhat virtual [V3], but also physiological factors (our entire body) can do so as well. Moreover, parts of the body can even "produce" abstract thought: "I act, therefore I think" to inversely paraphrase Descartes (Ananthaswamy, 2010b). The mind is no less a slave to the body than the body is to the mind.

Additionally, the mind is also heavily influenced by the mind. In other words, the embodiment need not be physical—it too can be virtual. Experiments have shown that the personal avatar a person creates for her online activity can influence the way she perceives herself (her body) and also can influence its consequent action. For

---

[4] For an excellent precis of this issue, see: Mind–body dualism (2021). Dualism does not have satisfactory answers to some of the central critiques against it, but the monist school is not much more successful explaining how physical processes turn into "qualia" that we feel.

example, adding height to a personal avatar increases the chances to succeed in real-life negotiations (due to greater self-confidence); adding age to an avatar increases the real person's propensity to save money as opposed to spending it (as people age, they perceive the need for saving for their children or for their own retirement); and so on (O'Brien & Kellan, 2011).

How to ameliorate the perceptual problem engendered by the embodiment of mind? It is not clear that we want to do anything, for if our perceptual and cognitive apparatus is inextricably tied to physical (or even virtual) embodiment, then no matter what we do to "upgrade" the physical sensory organs and/or our body, they will always "distort" objective reality (e.g., it is almost impossible to conceive of an eyeball—organic or even inorganic—that will be able to pick up *everything* on the full light wave spectrum). Moreover, given that the human brain (indeed all life-form brains on Earth) evolved in conjunction with, and "adapted" to, the body, changing the body too much—or eliminating it altogether—might "crash" the brain.[5]

Nevertheless, we continue to invent external tools (generally called "media") to help us better perceive our world, and they are made of corporeal material. Therefore, whether through our own body parts or artificial extensions of our sensory organs, virtual perception is directly based on, and filtered through, corporeal media (many of which have elements of virtuality, as seen in Chaps. 9 and 10). In addition, our entire perceptual apparatus is also in constant need of physical nourishment (food) and protection (housing, clothing), and when these are satisfied, we move on to other concrete goods that will constitute the basis for non-material social status and other psychological-hierarchical needs (Maslow, 1943).

However, this life process from materiality to virtuality does not follow a strict gradation or is necessarily dichotomous; we are not faced with either virtuality *or* reality. Rather, the issue is one of "equilibrium" for the person and for society at large—individuals and society ultimately finding the "right" *balance* between the two poles.

There is no specific answer because most phenomena in life are intrinsically not purely "real" nor are they completely "virtual." Rather, each phenomenon falls somewhere between these two poles, and this holds true over three or four different dimensional aspects of the term "virtuality."

"Economic value" is a good example at the macro-social level—the way we tend to measure the worth of most everything in the modern world (not necessarily the best or most appropriate way to assess something's value in every case). Over the past few decades, the world's advanced economies have produced a greater percentage of their GNP in work that involves virtuality than materiality. However, the numbers are exceedingly difficult to precisely quantify for several reasons. A typical corporation (Toyota, Walmart, General Electric) produces and sells material goods (cars, stores, jet engines) but also is involved in much virtual work (design & marketing, logistical

---

[5] One example of this is the phantom limb syndrome (mentioned in chapter 3). The mind refuses to accept that the body part is gone! One can imagine the disastrous result if the *entire* body was eliminated (a la mind uploading, to be discussed in chapter 13). However, there is also contrary evidence: quadriplegics who have lost all feeling in their body are still able to perceive the world in normal fashion (sight, hearing etc.—other than touch) and also do not suffer cognitive decline.

planning, finance). Moreover, the same employee can be doing virtual and real work simultaneously—designing a new product and creating the tools to manufacture it, or copywriting advertisements that will appear on a billboard. The same holds true for government and public service workers: teachers transmit virtual information but also use material equipment (projector, textbooks). The academic researcher creates new knowledge by handling test tubes, working in physical labs, and manipulating materials.

Another virtuality/reality nexus involves "information," by definition virtual [V9; V24]. However, virtuality is also a matter of process—not only *what*, but also *how* information reaches us. When we stand before a horse, that's a real (corporeal) experience; when we gaze on a representation of the same horse [V8] through a camera photo or artist's canvas, we are internalizing the same "information" but the way we do so, mediated and distanced [V3], lends an air of virtuality to the process.

Thus, if we return for a moment to economic valuation, even "hard-core" manu-facturing is partly virtual if the human does it from afar [V2]—whereas even math-ematical invention is partly material when undertaken on a computer screen, with subsequent paper printouts. Moreover, as noted in Chap. 7, increasing amounts of virtual work is done for "free" (collective open code collaboration, joint "wiki" production), so that increasingly large swaths of the modern economy do not even enter the financial ledgers of the official statistics [V16]. The bottom line: we cannot easily use the common system of "market value" (whether product price or worker salary) to measure the amount of *macro*-virtuality/reality found in contemporary society.

That leaves us to analyze virtuality without quantitative specificity, and even in somewhat impressionistic terms. Nevertheless, this doesn't render virtuality less "real"; it is not merely an alternate, "secondary sphere" of life but rather inextricably linked to reality. As Hansen put it: "all reality is mixed reality" (2006, p. 5). Moreover, as illustrated in many of this book's chapters, the intermingling of virtuality and reality has become increasingly pronounced in the past two centuries to the point that in the information age, "when our symbolic environment is, by and large, structured in this inclusive, flexible, diversified hypertext in which we navigate every day, *the virtuality of this text is in fact our reality*, the symbols from which we live and communicate" (Castells, 1996, p. 403; my added emphasis).

## 12.3 Examples of Virtu(re)ality

The following survey is meant to fill in some holes regarding areas of life that were not covered earlier, and/or to highlight some (obviously, not all) of the most recent examples in contemporary fields of life in which this nexus can be clearly seen.[6] A

---

[6] The most obvious "omission" here is the technology called Virtual Reality, combining reality and virtuality, or at least offering a complete reproduction of reality [V1; V6] through virtual, sensory illusion [V6; V12].

few moments of thought as well as social observation should suffice for any reader to come up with other examples beyond those listed here.

### 12.3.1   Sex

It would be hard to think of a more corporeal, indeed carnal, activity than sex. Body meets body, body part penetrates body part (or at the least, stimulates body part). And yet, sexual activity is full of virtuality as well, as "the brain is the sexiest organ." Arousal is only partly a function of physical stimulation; without some exciting thoughts the body will have difficulty becoming aroused. Indeed, usually we don't need too much physical stimulation if one's sexual imagination is in high gear.

Sexual activity can involve several different types of virtuality. First, fantasies of sex with another person—sometimes during the sex act itself—is not that unusual. This almost always occurs during masturbation (the physical sensation might not be sufficient for orgasm) [V10]. It also occurs some of the time when we engage in a semi-virtual *menage-a-trois*: one-on-one with a physically present partner while simultaneously one-on-another in one's own mind with a third partner, whether real without any personal familiarity (a celebrity [V6]), or ex-partner from one's own past [V11], or even totally imaginary [V28]. Another variation is virtual sex within virtual worlds—my avatar coupling with your avatar [V6, V12, V13, V17]—with occasional real-life consequences for the person involved: divorce due to virtual adultery (Morris 2008). Finally, there is the classic "phone sex" with a real person on the other end, playacting a hyper-sexual role for the customer seeking sexual relief. This is double virtuality: sex with someone from afar [V2] who is only verbally acting out the required role [V32].

Non-distanced, physical sex can also be virtual—with sex robots. This has become a burgeoning industry in different areas of life beyond individuals' own homes or brothels: nursing and retirement homes, prisons, monasteries/seminaries, the military (Bendel, 2021).[7] Here the virtuality lies in the artificiality of the partner [V5; V13]—and some psychologists would add the emotional lacuna such a "relationship" entails [V4; V10; V32].

A second example is pornography—something that has become far more widespread for purely logistical (and also perhaps, changing moral) reasons in the internet age. Here we have sexual arousal through still photos or video depictions [V6; V8] of *others* "performing" sex (usually merely simulated [V10]), thereby stimulating the viewer's mind to bodily arousal. The same is accomplished by the aforementioned interactive phone sex or computer cam-sex (the "other" sexually acting out what would normally be done face-to-face). There is ample scientific evidence that pornography does not increase rape and other sex crimes, actually reducing physical pedophilia (Diamond et al., 2011).

---

[7] Bendel notes that the idea of "artificial love servants" has been around for millennia e.g., Pygmalion's Galatea (2021, p. 3).

Somewhere in the middle of these two categories lie two other "phenomena." The first is the "lovebot" (also called "slutbot"): artificially intelligent, completely human-like avatars[8] that function as sexbots, but on one's screen (Ellis, 2019). A second (bizarre, even by contemporary standards): human-with-avatar (permanent) marriages through virtual reality technology (Sharma, 2017).

Finally, the most "realistic" type of virtual sex would occur through virtual reality technologies. Despite the lack of any physical contact between the partners, here they experience real, physical, and emotional fulfillment through the virtualized sex act. The exact technicalities need not be detailed here (see: Zhai, 1998, pp. 44–46), but suffice to say that such virtual reality sex would be (almost?) indistinguishable from the classic way of doing it, proximately face-to-face.

In short, one of the most fundamental biological needs and functions—sex— has undergone a mini-revolution in enabling humans to fulfill it in various ways, either quasi-virtual (with all the sensory and emotional experience of the real thing, through virtual reality or sex dolls [V1]), or semi-virtual (pornography [V7, V8, V10], or through distant stimulation [V2]) as we usually do seek physical climax; or even completely virtual (through avatars or by people suffering from impotence [V6, V25]).

## 12.3.2   Live Event Video [V8]

One of the more interesting quasi-virtuality trends in contemporary culture is the combined use of video during live performances: huge screens in rock concerts and other large hall venues where distant and partial view seats reduce or obstruct viewing the stage performance. Moreover, given the huge size of sports stadiums—along with the attending audience's habitual need to see their sports "on the screen"—many simultaneously display the field action and video replays on humongous screens (Hairopoulos, 2009), and even live video from competing cities' games, plus updated statistics of the present game (Schmidt, 2010). Indeed, the use of video playback in sports events at times will determine the outcome of the game itself e.g., the Hawkeye line calling system in tennis; the VAR goal scoring determinant in international football ("soccer"). Such close-call determinations (aids for referees) are also found in baseball (home runs), U.S. football (il/legal catch) and other sports as well.

Theater and dance performances now also incorporate video that accompanies the stage performance (Kelly, 2007). In this case, the virtual does not merely "replay" or "enlarge" the action [V6] but rather complements it as an integral part of the performance [V12].

---

[8] As noted in chapter 10, we now also have "synthespians": visually "resurrected" and animated *deceased* actors. These too could be exploited as sexbots with even greater verisimilitude.

### 12.3.3  Virtual/Physical Activity

Whereas video games are an example of "strong virtuality" (the player does move fingers, but most of the "action" occurs on the screen and in the player's head), another trend combines physical action with virtual stimuli [V10]. Some examples: Nintendo's Wii Game Console in which the participant performs physical activity in a virtual environment (e.g., playing tennis with a "racket" in hand, swinging at a virtual ball, against a virtual player—or a real player, presented virtually on screen); Microsoft's "Kinect" system for the Xbox, that "reads" human body language, enabling the individual to "communicate" with the computer system through body language alone (hand gestures etc.).

These hybrid activity systems are significant because they obviously address a cogent criticism of the world's growing penchant for media virtuality. Children are becoming heavier (in part) due to greater sedentary compunications use and decreased outdoor play; adults suffer no less from lack of physical exertion at work (and life in general). Yet in our hyper-mediated world most of us do not seem capable of abandoning virtuality even for a short while in order to "get back to (physical) basics"— even our treadmill workout has to have a TV screen, not to mention streaming music we listen to while jogging, and so on. Thus, if we cannot completely leave virtuality for our own physical benefit, at least virtuality can tempt us back to "real" body exertion by incorporating virtual experiences into such activities.

A totally different form of duality is found in location-finding systems such as GPS. Here the individual "contributes" to the virtual information noösphere by simply existing and moving around. By aggregating such location data, society becomes more efficient: avoiding traffic jams (Waze); dealing quickly with a nascent flu outbreak (based on a high level of school absences); and so on. The converse is also true as anyone can take personally relevant, location information from the noösphere. Apps enable us to find out where the best pizza is sold in our neighborhood; or to locate strangers nearby with similar hobbies or profession, whom we might be interested in meeting. In short, twenty-first century humans have become so much a part of the information environment that we need only live our physical life in normal fashion (not necessarily interacting proactively with our media), and still influence—or be influenced—by our surrounding virtual noösphere (de Souza e Silva & Sutko, 2011).

### 12.3.4  Gaming the Real World

A book title sums up the present chapter's point succinctly—*Online Worlds: Convergence of the Real and the Virtual* (Bainbridge, 2010). This convergence is a function of the online virtual world's replication [V1; V12] of many real-world elements and activities (virtual newspapers, buildings, money transactions, schools, embassies and so on) to further real-world activities, and not just to escape the real world.

The converse is also becoming significant in the contemporary world: real-life activity that is rendered more enjoyable and even efficient by incorporating some element of game or virtual play:

> Welcome to your entire life as a game, where everything you do can be played. The point-scoring, treasure-collecting and fantastical elements of video games are no longer restricted to the virtual world of consoles and computers: they are leaping out of the screen and into the physical world, thanks to smart phones, cameras, sensors and the increasingly ubiquitous internet (Campbell, 2011, p. 37).

To further real-world goals, we increasingly turn to virtual games for motivation strengthening and improving our skill set for regular activities. Educators include games in lesson plans (Klopfer et al., 2009); dietitians devise games to keep dieters on track; ad infinitum.

### 12.3.5 Artificial Intelligence Events

Chapter 13 will survey the future of artificial intelligence (AI) [V6; V12; V22]. For now, we can note how AI programs and robots are becoming part of the news and entertainment field. Its first major boost came in the famous Kasparov vs. Deep Blue "world championship" chess matches back in 1996 (world champion Kasparov won) and in 1997 (the IBM computer, Deep Blue, emerged victorious). Since then, numerous other such human vs. computer (or robot) competitions have taken place. Two additional examples: in 2011 IBM's "Watson" computer beat humans in the popular TV show "Jeopardy"; the annual Loebner Prize competition pits AI programs against humans to see which of the former can best fool the latter into thinking it is human (the Turing Test). Other times such contests pit computers or robots against each other: the Pentagon's DARPA Grand Challenge between driverless cars; an annual Computer Poker Competition; the World Computer Chess Championship.

### 12.3.6 Education

Teaching others has always been a highly labor-intensive enterprise. Most people learned (for those who underwent any kind of education) by listening aurally to a teacher, while a fortunate, literate few were able to supplement such face-to-face education with reading books, a quasi-virtual, educational vehicle. In the modern age, this did not change much, except that textual media became more widespread; the formal, oral teacher could demand the student "self-educate" (homework) through textbooks and other types of educational literature. The brunt of teaching, however, continued to be face-to-face.

This has begun to change significantly, and some might argue radically, especially in higher education. Distance learning [V3], sometimes entire degrees based on distance education, has become widespread, accelerating during the Covid-19 pandemic.[9] These courses do have a flesh-and-blood educator with whom the students have some virtual contact—Zoom lectures and discussions, email, or chat Q&A's, etc.—but most online course pedagogy is based on virtual materials: text, audio, video.

Moreover, given the need to keep up professionally and personally in our hyper-dynamic age, there is a growing need for continuing education throughout one's lifetime. Thus, non-degree online courses and other ad hoc educational offerings are proliferating as well.

Finally, virtual technology is being put to use in novel ways within the "classic" classroom setting. One example: teaching a class within a virtual world (e.g., Second Life)—teacher avatar to student avatars [V12]. With current pupils and students growing up in a saturated environment of virtual media, it is only "natural" that educators use many of the same technologies to keep the students alert and motivated to learn.

## 12.3.7   Training Simulation

Whereas "simulation" in general is a strong form of virtuality [V10] in which the entire process takes place within the computer and the human is a passive bystander reading the results, there is another form: "interactive simulation" in which the human takes an active part (Simulation 2021).

Interactive simulation technology is increasingly used for all sorts of training, educational and psychological purposes. If in the past, vocational education was taught face-to-face by a schoolteacher, university lecturer, or professional trainer, today we are increasingly using virtual simulations—at times accompanied by a human teacher, at times not—through which the student learns the material. However, beyond vocational training (e.g., flying and driving lessons), social science education uses simulations as well, such as role playing for international relations policymaking. Even more interesting is the growing use of simulations in psychology e.g., to reduce phobias the patient will undergo in stages increasingly longer or more intense doses of realistic simulations of their fears: sitting in the terminal, getting on the plane, relaxing in the plane's seat, going down the runway for takeoff, etc.

---

[9] Beyond the need for social distancing during a pandemic, routine reasons for increasing widespread use include: lower costs (no campus), easier logistics for the student (no drive to campus and easy adjustment of work and family), and perhaps even better educational outcomes.

### 12.3.8 Online Shopping

Buying goods and services is obviously a central phenomenon in modern, capitalistic life. In the past, purchasing and trading was done almost exclusively by the consumer or a physically present agent. This changed for the first time in the twentieth century with the introduction of several novel sales approaches, each with a different virtual element. The first was the sales catalogue—a printed "book" enabling purchase at a distance [V2]. Telemarketing by phone added another element to the "shopping-at-a-distance" trend, although this at least maintained a (disembodied) human voice [V3]. TV home shopping came next in the trend to virtualizing buying, with a human salesperson on the screen [V8].

Virtual shopping has reached a pinnacle of sorts through the internet. Online shopping incorporates a few virtual elements: carried out at a distance from the seller [V3]; usually no human is at the sales end with whom the shopper can interact [V27]; the products can be compared by price, quality, service, etc., without having to run from store to store [V2]; in the case of clothing, customers can "try it on" their virtual/avatar "body" [V12]; in some cases (clothing, appliances), the potential customer can vary several elements to see which of them look best (color of a shirt, height of a fridge) [V6]; the purchaser can read aggregated personal recommendations of other, prior customers [V9]; purchase, of course, is finalized virtually through some form of electronic payment [V17; V25]; and in the case of a digital product (music, video, book/magazine) the product itself can be delivered instantaneously straight to the buyer's internet-connected gadget [V2]. In short, while the goods in most cases are completely, concretely "real" (except for the digital ones), the ways in which we obtain them have become increasingly virtual.

All of this does not mean that virtuality is merely a *means* to a material end. It can be—and many times is—an *end* in itself e.g., the pure enjoyment of intellectual and/or creative virtualization (creating an original story, the "rush" of succeeding to "win" in a virtual, multi-player game world). In short, in order to understand the full impact of virtuality on contemporary life, it is not enough to look at patterns of consumption and/or transmission/transfer, but increasingly also how our own "production" (as a leisure and not only work activity) has become virtual.

## 12.4 The Real Becoming Virtual

Reality and virtuality as "two sides of the same coin" does not only mean that the two can co-exist at the same time. It can also mean that the exact same thing can be real at one time and become virtual another time. I have already discussed at length the most prevalent phenomenon of this: a life event is real when it occurs, but later as recalled memory it becomes virtual [V11].

The following list refers to something else altogether: the materially real can be real at first and then later become virtual *without changing its material form*. A few examples will suffice to illustrate the point.

### 12.4.1   Ghost Towns

Towns with real inhabitants at one time can lose all their population and fall into desuetude. If they were simply "out of sight, out of mind," they would not much interest us. However, they are major tourist attractions and come in several forms: fragments of cities [V4]; rundown towns in their "pristine" state [V5]; towns restored with "authentic" materials [V6]; hybrid places in which various elements of different towns are (re)placed in the former urban complex [V7]; and so on (Eco, 1986, p. 40).

### 12.4.2   Historical Sites

These are specific, longstanding sites of important historical interest: Parthenon, Pyramids, Great Wall of China. Their physical form is as real as the day they were constructed (albeit frayed and sometimes partly destroyed). Their virtuality lies in the lack of spatial context, for they are divorced from their socially "natural" environment [V4]—akin to a long-married spouse who lives long enough to lose her husband and children; at that stage her "wifeliness" is purely virtual.

### 12.4.3   Archeological Sites

These are long-lost sites rediscovered through painstaking archeological digging. They are almost invariably highly incomplete [V4] (except for rarities like Pompeii, buried whole beneath the volcanic ash) but maintain original materiality—after archeological cleaning. Unlike ghost towns and major historical sites, archeological sites are interesting from another virtuality standpoint, as they invert the normal process of memory dissipation [V11]. Here the site has been lost to history (or remained as some distant mythology) and then "reborn." In this sense, they decrease virtuality from something that is not sensed [V19] to merely incomplete [V4], although depending on how much is missing they also involve a fair amount of ambiguity or indeterminacy [V15] regarding who, what and when lived there.

### 12.4.4  Museums

Much the same is true of authentic objects displayed in museums; of necessity they are not in their original environment. True, the best museums do yeoman's work in reproducing the earlier social environment, whether by textual description [V9], 3D dioramas [V7], or, increasingly of late, computer animated, reconstructed representations [V17; V25]. Nevertheless, given space limitations most museum artifacts are shown as discrete objects one after the other, perhaps conceptually or chronologically connected with each other but without the wider socio-historical context.

In sum, given the law of entropy, today's "real" will inevitably become tomorrow's virtual, even when it retains much/some of its original materiality. How we maintain, reconstruct, combine, and present the decayed "objects," determine their eventual type and strength of virtuality.

## 12.5  Blended Life: Indistinguishable Virtuality/Reality

I have used several terms to describe the interconnection of reality and virtuality: dual, hybrid, combined, double-sided coin. In each case, we are still able to distinguish what is virtual and what is real. However, as time and technology advance, this is becoming more difficult to sustain, as Baudrillard's concept of the simulacrum suggests. With the virtual merging into the real, it is possible to lose track of what is real and what not; reality and virtuality become blended. This can occur in fields that start as mainly virtual enterprises with reality impinging—or where things begin as purely "real" and then have virtual elements thoroughly integrate themselves. Here are three examples, in ascending order of gravity.

The first type is something especially pronounced in today's media/entertainment world. The following blog post excerpt illustrates this "problem"[10]:

> We'll soon be seeing video games based on reality shows—you know, the shows that employ script writers to make sure the reality sounds real. These are different writers from those who were protesting that they weren't paid enough and had to falsify their time sheets to show that they worked fewer hours.

> Then there are people making real money running Second Life businesses that sell virtual goods to others inside their virtual world. There are others who auction virtual items on eBay that enable game players to advance to higher levels; these items are assembled by real, low-wage workers who spend their days playing the games to accumulate them......
> There are also real Web pages for fake people, some of which were created by advertising and PR people who want to push a particular brand or agenda....

> Then there is a pseudo-documentary that ABC-TV aired recently about 9/11, which interwove fictional dialogue spoken by actors playing real people.... (Strom, 2006).

---

[10] The word is in quote marks because it is not clear that this is indeed a "problem"—at least for future generations.

And this is but the start of future (positive) concatenation, (neutral) confusion or (negative) consternation—each of us can choose our normative perspective—as we become more enmeshed in the smartphone world (as a generic metaphor) of mashups, recombinations, and "borrowings"; as avatars and robots become completely lifelike; as automated synthetic voices begin to sound perfectly human; as augmented reality technology (and biotechnology too) starts to distort "for the better" our environment and own body; as our virtual social network expands geographically and the physical one contracts proximately; and so on. Even more significant for the future is the inability, unwillingness, or simple apathy of the younger "digerati" generation to care about such virtuality/reality blending—and for some, even an eagerness to hyper-blend the two worlds into total indistinguishability.

The second type of blend occurs when reality is "invaded" by virtuality and "taken over." Biology has always been almost the exclusive province of the real. Life forms might be extremely diverse and even weird, but all are based on the same metabolic principles and identical organic material at the molecular level. However, this has begun to change very recently through the emerging fields of synthetic and artificial biology.

Synthetic biology is the utilization of existing organic compounds to create biological molecules or higher-level organic units that either exist in nature or are alien to earth nature as we know it (Jayanti, 2020). This is a direct (albeit far more complex) offshoot of synthetic chemistry that had its start a century ago. Genetic engineering (the manipulation and/or recombination of genes) is one example of this. The virtuality here is in the "unnaturalness" of the resulting product [V13], but not in its constituent components. Artificial biology is far more radical: the creation of organic compounds—and ultimately "life forms"—based on *non*-organic materials i.e., a creation-from-scratch approach.

Today's synthetic biology is leading us towards tomorrow's completely artificial biology—virtuality as *creatio ex nihilo* of artificial biological life as substitute for natural, biological life [V6]. If and when this occurs, it would obliterate the age-old distinction between nature and artifact, as organisms have always been a product of historical, evolutionary processes. Artificially created organisms would sunder that connection for the first time, thereby blurring to indistinguishability what we generally call "organic" and "inorganic" chemistry.

A final type of blend is the ultimate in virtu(re)ality: the post-human (also "trans-human"). Chapter 13 will expand on bio-technologies—in addition to cybernetics, nanotechnology, and artificial intelligence—as central components of future forms of virtuality. Here I merely note that humanity might be moving to a complete blend of the artificial and real in body and "soul" [V5; V6; V13; V15]: "The posthuman view configures human beings so that it can be seamlessly articulated with intelligent machines. In the posthuman, there are no essential differences or absolute demarcations between bodily existence and computer simulation, cybernetic mechanism and biological organism, robot teleology and human goals" (Hayles, 1999b, p. 3).

This view is not the ranting of a solitary voice in the scientific wilderness. Reputable social science researchers, technologists, and scientists in physics, chemistry, electronics, computers, as well as the human sciences of biology and psychology, have discussed in all sincerity the various possibilities (and dangers) of post-humanism. Many post-humanist technologies are already in place: brain and body implants of computer chips and nano-machines, bio-molecular and genome engineering,[11] human-like robots, artificial or bio-printed organs (Mironov, 2011), and future technologies only now in their initial stages promise even more radical changes (e.g., life extension, mind reading). Thus, the demarcation between subjective and objective i.e., internal and external, virtuality—a leitmotif running throughout this book—might not hold much longer as the external and internal worlds merge materially and existentially. If and when we reach that point (there is no guarantee that this will happen), the distinction between virtuality and reality will be almost completely obliterated.

## 12.6  Conclusion: The Expansion of Both Reality and Virtuality

Looking at the macro-picture, in the past and increasingly in the contemporary present and towards the future, we can discern the outline of two complementary worlds: the physical *biosphere* (everything in our world that supports life) and the virtual *noösphere* (or the increasingly more colloquial term: *infosphere*).[12] The former is found exclusively in the corporeal, "real" world; the latter "online" (cyberspace; digital media) and "offline" (libraries; analog media) with no clear demarcation between the two spheres, nor within the infosphere itself. The biosphere holds vast amounts of information (DNA) whereas the infosphere also produces huge amounts of data,[13] produced by physical technologies and mostly embedded in some form of artificial physical media. Moreover, the lines are blurred within the infosphere itself: offline data are increasingly being copied/stored online; much online data are incorporated in offline activities by way of paper-print, slide presentations, etc.

Another way of thinking about this is to return to modern physics:

...life is to be seen as a layered system in which each layer has, and indeed demands, its own vocabulary. On one side...there is the firing of neurons. On the other, there are symbols,

---

[11] Not all these technologies will necessarily heighten artificiality as virtuality [V13]. For instance, bio-regeneration of a person's own body parts will enable the replacement of faulty organs and joints with cells genetically compatible to the original. The only "artificial" aspect of this is that the new body part will be a close-to-perfect duplicate of the original [V1]. For an example of this, see: Biological joints..., 2011.

[12] "Logosphere" connotes human-initiated rational thought; "infosphere" connotes all types of data—not necessarily originating with humans (our intellectual machines now absorb, transform, and produce huge amounts of information) and not only for "rational-logical" purposes. The Web carries far more digital bytes related to entertainment (YouTube; pornography) and social association (Facebook; TikTok) than to news and information (NY Times; HuffPost).

[13] For example, the CERN Hadron Collider produces approximately 10 petabytes of data annually—1000 times more than is held by the Library of Congress's print collection, itself a product almost exclusively of human and not machine output.

representations of the physical that also have a physical reality.... Here we have Bohr's complementarity on a larger scale, two modes of mutually exclusive descriptions making up a single system.... There are two realities to this thing. Just like a light is a wave and a particle at the same time. Information and construction, structure and function, are irreducible properties... that exist in different layers with different protocols (Gazzaniga, 2017).

Nor is this a zero-sum game. The biosphere and the infosphere have both increased tremendously over time. The biosphere has seen a huge growth in human beings, and in many (not all) cases, a concomitant rise in animal life (e.g., ruminants for meat). Human population increase brings with it a parallel expansion of the infosphere, if only for numerical reasons (assuming that each person produces only the same amount of data). However, the last two centuries have brought a huge *per capita increase* in information, with the infosphere expanding faster than the biosphere, but certainly not taking its place.

They also do not work on parallel tracks, as blended reality/virtuality illustrates. Even without futuristic "post-humanism," we have steadily merged into our virtual environment: "[t]he threshold between *here* (*analogue, carbon-based, off-line*) and *there* (*digital, silicon-based, online*) is fast becoming blurred" (Floridi, 2007, p. 60). Beyond such technological systems as augmented reality, ubiquitous computing, the internet of things, and so on, from the perspective of the infosphere user there is no longer a clear demarcation between the biosphere and infosphere as there was in the days when we received information by opening up a printed encyclopedia or walked to the nearby library. Once we moved into digital technology (computers, CDs, hard drives) (Allen-Robertson, 2017), the two worlds began to mesh. Today, we are all attached to the infosphere through the mobile media that we carry with us (G4 or G5 cellphone) or advanced technologies that move with us (RFID tags on consumer products; GPS location finding devices on cars; Nike running shoes wirelessly connected to our Fitbit)—even to previously "pure biosphere" places such as vacation spots or our own home bathroom.

Floridi (2007) calls such objects "ITentities," seamlessly connected "a2a" (anything to anything), that works "a4a" (anywhere for anytime). The more human beings become an integral part of this system, the more we turn into "inforgs" (information organisms)—not much different (informationally) from the artificial, digital inforgs/agents (AI search engines) that we have created to aid us navigate this new bio-infosphere world. Should we be sanguine about this? Not necessarily. This chapter (and certainly the next) can overwhelm us by the sheer force of past, present, and future trends towards greater virtuality and the merging of humanity into a blended world.

In order not to give the impression that all this is necessarily positive, I conclude this chapter with somewhat of an admonitory warning by Heim, one of the earliest analysts of virtual reality, using Gibson's path-breaking, sci-fi novel *Neuromancer* as his springboard: "As we suit up for the exciting future in cyberspace, we must not lose touch with the Zionites, the body people who remain rooted in the energies of the earth. They will nudge us out of our heady reverie in this new layer of reality. They will remind us of the living genesis of cyberspace, of the heartbeat behind the laboratory, of the love that still sprouts..." (Heim, 1993, p. 107).

# Part IV
# Present and Future Perfect: Where Do, and Should, We Go from Here?

# Chapter 13
# The Ultimate Virtual Frontier: Neuroscience, Biotech and Compunications

## 13.1 The Virtuality of the Future

It is a banality to note that the future (as a generic concept) is virtual—a truism, for the future holds potentiality [V26] but by its very nature it has not yet been actualized [V27].

However, this truism's very banality is strong evidence that virtuality is an inherent aspect of humanity; as already noted, *homo sapiens* is the only species on Earth that thinks about the longer-term future in "systematic" fashion, regarding many areas of life and functions. Therefore, the present chapter's gazing into the future is itself a virtual experience.

Future prognostication has been a common, multi-faceted thread running through human history. Focused future-thinking is not an exclusively modern enterprise: social "prediction" institutions have existed for thousands of years: the Jewish High Priest's *urim ve'thumim* (breastplate jewels) and of course the biblical Prophets; the Greek Oracle of Delphi (reputedly in direct connection with the gods); astrologers (heavenly bodies influencing life on Earth); diviners (interpreting animal secretions); and so on. The modern era included Nostradamus' (in)famous predictions hundreds of years forward in time, followed by fortune tellers at every corner, and a variety of future-prediction paraphernalia: tarot cards, crystal ball, tea leaves, Ouija boards, palm readers, etc.

Yet all these were (and still are) "niche" areas of life not impacting most people. Nevertheless, something changed in the past two centuries with the penetration of futurist thinking into general culture as well as its incorporation within central social, economic, and political institutions. Novelistic science fiction caught the public's imagination: Mary Wollstonecraft (Godwin) Shelley's *Frankenstein* (1993), Jules Verne's adventures in space and beneath ground, H. G. Wells' alien attacks; and in

---

The *Appendix: A Taxonomy of Virtuality*—listing by numbers 1-35, e.g. [V13] or [V34], some of the various types of Virtuality mentioned throughout this chapter—is freely available to the public online at https://link.springer.com/book/10.1007/978-981-16-6526-4

S. N. Lehman-Wilzig, *Virtuality and Humanity*,
https://doi.org/10.1007/978-981-16-6526-4_13

the twentieth century, Heinlein, Asimov and Clarke, along with sci-fi magazines, comic books, and wildly popular films (Lucas, Spielberg et al.). Non-fiction "futures studies" regularly make the best-sellers' lists, led by Toffler's *Future Shock* (1970) and *The Third* Wave (1980). Even academic-professional associations have been established to "study" the future through periodicals and annual conferences (e.g., World Future Society). In short, contemporary popular culture is saturated with the near and far-off, the likely and improbable, future.

Futurist thinking has even been incorporated by the "establishment." Perhaps as a response to World War II's devastation, myriad types of institutions arose whose sole purpose was predicting the future: think tanks (e.g., Rand Corporation); academic centers for future-oriented research; types of Office of Technology Assessment units in parliaments[1]; and even the media have become more future-oriented in the "news" than what it had published in the past (Neiger, 2007).

In order to carry out such prognostications, many simple as well as highly sophisticated "prediction methodologies" have been developed: trend extrapolation [V26], scenario building [V9; V26], Delphi Technique [V2] (reiterated surveying from afar of experts), simulation [V10] (mechanical, computer, metaphorical, game, and computer-agent, crowdsourcing), decision tree [V15; V26], social physics [V19] (S-curve, envelope trend curve), among others.

What are they trying to predict? The weather; inflation and unemployment; election winners; new technologies to invest in and their impact on the economy; the future of war; and the huge world of "futures" exchanges where almost everything and anything can be bought and sold at a "future price."

As a result of the accumulation of all this institutional, methodological, and substantive futurism, public sphere debates are being increasingly taken up by future-oriented issues: global warming; fishery depletion; population explosion (and implosion!); food production and famine; species extinction; deforestation; alternative energy; ecological sustainability; the promises and perils of nanotechnology, genetic engineering, and artificial intelligence; and so on.

In sum, we can*not* accurately predict the future, but we have always *tried* to. It is not prognostication success or failure that defines our virtualizing here but rather (and merely) our constant *attempts* to do so. There are those (Taleb, 2007) who deny the possibility of any significant sort of prediction (other than "the sun will rise tomorrow at 6:43 AM"); if indeed they are right, it would even further reinforce the "virtuality" of future prediction as having somewhat of an illusory nature [V31].

---

[1] The U.S. Congress's Office of Technology Assessment existed from 1972–1995. Israel's Knesset established a unique Commissioner for Future Generations (2001–2006) whose purpose was to analyze all significant legislation and consider its impact on the future. Other parliaments maintain research units that do similar things, if not within a single, discrete, mono-functional body.

## 13.2  The Future Fallacy

As several chapters have shown, there has been no letup in the increasing breadth and depth of virtuality in modern times. The trend is clear: virtuality will most probably continue to expand in the future—not necessarily because virtuality is *ipso facto* beneficial, but rather because present technological developments augur new, and in some cases quite revolutionary, directions for virtuality.

If revolutionary, then these technologies also raise serious questions: ethical, moral, philosophical, and social. The main purpose of this chapter is not to address the emerging dilemmas, questions, and problems, but rather to note how "futuristic" technologies are connected to, and perhaps will also possibly expand, virtuality.

However, ethics and virtuality are connected because of what can be called "the future fallacy": *what each generation considers to be unethical, immoral, dangerous, harmful, unnatural, or at least problematic, may not be similarly perceived and evaluated by ensuing generations.* Therefore, before we arrive at any restrictive conclusion regarding the possible harmful influences of a new technology, we have to "think out of the box"—in this case, to try and view the matter from a perspective outside of our present generational, social milieu, asking the following: "if our grand/children will be living in circumstances different from ours today, might not new technologies be viewed in a different light?"[2]

How is this related to "virtuality"? It is quite possible that the expansion of virtuality through new technologies induces a process of socio-psychological reinforcement i.e., the more that people live in a world surrounded and imbued with virtuality, the less they (will) view it as something "alien(ating)," "divorced from reality," "unnatural," "nonhuman" (as opposed to inhuman), or "harmful." One sees this today with parents lamenting their children spending far more time in *virtual* communication and play than they did in their own youth. However, this new generation brought up on multi-player gaming, social networking, cellphone texting, WhatsApp chatting, does not seem to view their behavior as anti-social and "unreal." Indeed, how quickly today's adult generation seems to have forgotten our parents' plaintive cries from the 1950s through the 1980s: "get off the phone/close that TV—and go play outside with your friends!"

Thus, it is likely that the next generation(s), having grown up with virtuality all around, will see no problem (or at least no greater problem than the pre-computer parents did with their TV-obsessed kids' generation) in the development of ever newer and more radical technologies that promise to deepen the virtual experience in ways that an older, mildly virtualized generation might find appalling. This doesn't mean that we should simply let these technologies "evolve naturally" and not ask

---

[2] The 2010 Nobel Prize for Physiology/Medicine was awarded to Dr. Edwards for being the first to successfully carry out In-Vitro Fertilization (IVF). The awards committee pointed out that "several religious leaders, ethicists, and scientists demanded that the project be stopped" (Press release, 2010). As the press release then noted, IVF ultimately became widely (almost universally) accepted. An even thornier issue is now emerging in this field: growing babies outside the womb, from conception to birth (Ball, 2021). This is triply virtual: using a duplicate womb [V1] that is a substitute for the original [V6], artificially created [V13].

probing questions along the way. However, it does suggest that any assessment of the benefits and/or harmful effects of such technologies must consider various possibilities, especially the different socio-cultural milieu in which future generations will be living.

Envisioning such a milieu and then seriously analyzing possible accompanying ethical dilemmas would take us here too far afield. Therefore, this chapter's purpose is modest: outlining several emerging technologies that could well constitute the next stage of humanity's seemingly unending quest for richer virtual experiences.

A final introductory point: the technologies surveyed below are mostly in their early stages of development, some even at a very primitive stage and therefore either highly cumbersome, expensive and/or not very user-friendly at this point. This, of course, will change with time—just as computers have evolved from the gigantic and very expensive machines in the 1950s (with punch cards, memory storage on large magnetic tapes, and having no screen, no pictorial icons, no mouse, no keyboard!). What we see today in the newest compunication and neuroscientific technologies is certainly *not* what we will eventually get. On the other hand, some of today's incipient technologies will reach maturity with fantastic capabilities whereas others may take a lot longer to emerge full blown and/or with less-than-predicted usefulness—and a few might never reach fruition. Overall, however, these contemporary breakthrough technologies are far enough along that we can be relatively sure that most, if not all, will change the world in the medium-term future, and particularly add immeasurably to our internal and external virtual worlds.

## 13.3  3D Holographic TV

Although three-dimensional television (on 2D screens) failed in the marketplace for several technical and marketing reasons (Silva, 2021), the next generation technology—holography—holds greater promise, as it is geared not only for the home entertainment market but for professional ones as well: medical diagnostics, education, etc. (Holographic TV Market, 2021).

Holography eliminates the need for "screens" altogether. Instead, visuals are transmitted as spectral 3D holograms of pure light [V25] in the immediate space facing, within the circle of, or surrounding the audience, with 360° circularity—precisely what one would see were the figure physically present: walk around a holographic projection of a person and we see their back. Moreover, holographic *haptic* technology already exists, enabling holograms to be even "touchable" (Holograms you can touch and feel, 2021).

It is not unlikely that within a few decades the screen will disappear altogether. If 2D "Zooming" is popular today, one can imagine the future scope and extent of holographic videoconferencing [V2; V3; V7]—for business meetings, educational classes, and everyday conversation. This would add another nail in the growing coffin of reduced face-to-face, physical interaction. Moreover, communication becomes even more virtual from a corporeal standpoint, sans a material medium [V25]. True, we might still require some sort of "projector" (or an array) but these will probably be hidden away within the walls of the room. Our communication and entertainment will have turned into a light and sound show without any noticeable material platform. On the other hand, compared to today's 2D screens, subjectively this will seem less virtual to the audience given the hyper-reality appearance of such "light" figures and objects [V1; V7].

## 13.4    Artificial Intelligence (Micro)

### 13.4.1    Virtualities of AI

Human communication, unmediated or mediated, has always had one common denominator: the communication took place between humans (notwithstanding the highly truncated, incomplete, and distorted communication [V4; V5] between humans and animals). We are now entering a completely new stage in human communication with the development of Artificially Intelligent (AI) programs, machines, and objects.[3]

How does AI exhibit virtuality? Physically, AI can be found in computer programs located in either physical objects like a laptop, cellphone, tablet—or hidden away within the surrounding world in everyday objects such as traffic lights, refrigerators, air conditioners and the like, through "ubiquitous computing." In the case of a computer, the virtuality is not corporeal but rather in being "artificially human" or "duplicating/simulating" human thought [V1; V10]. As microprocessors have become an integral (but unseen) foundation of the general world, and as electricity is mostly out of our sight, so too AI also involves a lack of materiality—at least from the individual's perceptual standpoint [V19].

At a deeper level, though, all AI is virtual given that the essence of such intelligence is to be found in software algorithms that are purely symbolic in nature. Of course, just as human "consciousness" and "intelligence" are based on some physical substrate (networked axions, synapses), so too AI's algorithm's need hardware to run on. Nevertheless, the *essence* of AI is the virtual (symbolic) software [V24] and not the hardware.

Finally, AI not only seeks to duplicate (or surpass) human thinking, it also is moving in the direction of mimicking [V6]—and some would even claim, replicating [V1]—human beings in most respects by taking on some sort of human form and human behavioral patterns. AI, then, constitutes quasi-human, virtual *thought*,

---

[3] For our purposes here, I shall use the terms "entity," "automata," and "robot" interchangeably – the latter being an AI entity with the power of locomotion. Other terms are "homunculus," "android," and AI "being," signifying an AI entity that moves and has recognizable *human* characteristics, whether behaviorally or in physical appearance.

and a virtualized human *body*—potentially rendering our relationship with AI some-what virtual as well, a sort of simulation [V10] or simulacrum [V12] of real human relationships.

## 13.4.2   The Pre-history of AI

The concept of an artificially intelligent being is not new. As Geraci noted: "From mythology to mechanics, robots have antecedents from the ancient world and the early modern period" (2010, p. 148): ancient Egyptian and Babylonian temples with their oracular talking heads and moving statues; Hero of Alexandria (first century CE) built automata that moved around the stage; da Vinci in 1495 created an automaton in knight's armor with the ability to sit and move its neck and arms; Japanese artisans built tea-serving dolls and other automata in the late eighteenth century; in addition to famous literary creations of homunculi (quasi-humans) such as Pygmalion, the Golem of Prague (2010, pp. 149–159), and Frankenstein. Thus, the concept of a virtual being [V4; V6; V13] has a long history. Nevertheless, it was not until the mid-twentieth century's cybernetic revolution that a "true" self-directing automaton became theoretically possible: AI.

## 13.4.3   Types of AI

AI comes in three essential forms: weak, strong, and full. The latter two are usually combined into "Strong AI," but they really require separate categorization because Strong AI is more about *appearance* whereas Full AI is about *capability* and *essence*.

In weak AI, a cybernetic entity is capable of analysis within a specific field or related fields and provide answers and/or acts. These are generally called "expert systems," and they perform for us (some at a very high level) in dozens of areas of life: medical diagnostics, traffic control, oil exploration, financial planning, stock market trading, machine translation, and so on [V10].

Strong AI, as suggested by Alan Turing (1950), involves a computer that can carry on a "blind" discussion with a human to the point where the latter cannot tell whether the former is human or computer. This does not mean that the computer has *become* human, but rather that it "understands" humans and the external world well enough to think and communicate like a human [V6]. Among other things, this entails knowing a lot about "trivial" matters (dark gray skies lead to rain), but not so much that obviously the "entity" was an artificial (clearly non-human) "walking encyclopedia."

Finally, "full AI" is where the AI entity claims to have "sentience" (the ability to *feel*), "consciousness" (subjective thought), and "self-awareness" (perception of self with a separate, somewhat unique identity). If such a "being" demanded full civil rights as the moral equal of humans because of its intelligence, social empathy, and so on [V1], in short because of its "human-ness,"[4] we would have a difficult time finding a good reason to deny granting such rights.

From the standpoint of virtuality, this progression from the first level to the third—we are presently only at the very start of developing "strong AI" (the second level)[5]—exhibits a *regression* in the intensity of virtuality in two different ways: objective and subjective. First, whereas expert systems are largely "hidden" from view [V19], as they are deeply embedded in computer systems behind the scenes of our economy and society, Strong AI robots and certainly full AI androids will be in full view and in constant physical contact with human beings as they move into what today is called the "civilian robot sector" (child caretakers, gerontology aides, classroom teachers) [V7]. Second, as they become more like us in physical form and behavior, we will view them as more human-like and at some point, perhaps as "human" [V6]—even to the point where we might have serious emotional and sexual relationships with them (Levy, 2007).

### 13.4.4  Future Computers

For technical reasons having to do with the laws of physics,[6] computer scientists are fast approaching absolute limits to the rapid continuation of Moore's Law (doubling of computer speed and power every 18–24 months)—within the classic form of silicon-based microchips.

Heretofore, computer chips have not been virtual—they are made of silicon and other inorganic materials. However, with the impassible nano-obstacle approaching, scientists have begun to develop three replacements. First, the *optical* computer uses light (the fastest "thing" in the universe) instead of electricity. Light has no mass, so that such a computer's basis would be materially virtual [V25]. Second, the *quantum* computer, potentially offering *unlimited computer speed and processing power*. Its basis is sub-atomic particles, with extremely small mass but with actions that are decidedly "strange" as explained in Chap. 5 [V17; V18; V19]. Third, the *biological* computer, made from organic material, or some combination of inorganic and biological matter.[7]

---

[4] There are all sorts of arguments as to what sorts of human *psychological* characteristics it would need (or at least present) to qualify for such legal and moral "equality." See my early essay on this subject: Lehman-Wilzig (1981). Coeckelbergh (2010a) argued against the possibility of a truly "moral robot," contending that without real consciousness and feelings robots cannot think morally. However, he agrees that robots can create the *appearance* of having emotions and of being moral – *precisely as do humans!* [V10]. For a recent, extended argument in favor of granting robot rights (under certain future conditions), see Coeckelbergh (2010b).

[5] Not everyone agrees that Strong AI will ever be achieved. Searle's (1980) Chinese Room scenario is the most famous argument against it.

[6] A chip with myriad transistors compressed in a space only a very few nanometers apart would start behaving in quantum probabilistic fashion and not in Newtonian deterministic fashion.

[7] There are two types of biological computers. The first manipulates pure DNA for very specific computational purposes. It is already in use (Cohen, 2020). The second grows biological material on

The last two new types of computers are particularly relevant to the issue of AI. In the case of a quantum computer, the unlimited power and speed would mean that "Full AI" androids are a definite possibility, even a probable eventuality. Again, such a robot would constitute a very weakly virtual being, as its intelligence level and social capabilities would be indistinguishable from humans except that they could theoretically be much *higher*. Once we add a body that looks like a perfect replica of a human—whether made of biological material or not—this quasi-human would be externally indistinguishable from a real person [V1].

With biological computers the situation becomes even more complex. Once we create AI from organic material then at least in principle (in practice, there are extremely complex technical challenges) we could build a human "from scratch." Instead of traditional, sexually initiated gestation, this AI entity would be "conceived" and "grown" in the lab, based on "synthetic cell creation" technology that already exists (Artificial life forged in a lab? 2021): *creatio ex [biological] nihilo*. From a purely *organic* standpoint it would not be virtual at all. However, if we define "human" by the *process* in which people come into the world, then such beings would be *procedurally* virtual due to the artificial way they are "conceived" and "born" [V13]—somewhat similar (but not entirely) to the increasingly numerous ways that humans are conceived (e.g., in vitro) and gestated (e.g., surrogates).

Beyond the philosophical import of all this, the real-life consequences are more relevant to virtuality: the way we will perceive, interact, and relate to such AI objects, especially those with elements of materiality to them. Contemporary social science research clearly shows that for us to have a "real" relationship with any sort of "seemingly sentient" artifact, it is not at all necessary that they highly resemble a human being in form or even behavior.

It might be that to distinguish the latter from real humans, future societies will insist on a clear differentiation in appearance. However, this would not significantly change our "relationship" to such beings because of the amazing human penchant for anthropomorphizing even material artifacts. We can *literally* fall in love with security blankets, dolls, cars, shoes (Lastovicka & Sirianni, 2011). And when something even *suggests* the "appearance" of being alive (e.g., the robotic dog "Aibo") or human-like as in computer synthesized faces (Reeves & Nass, 1996) and voices (Nass & Brave, 2005), our innate response is to relate to them "naturally." This explains the cinematic popularity of robots/androids, as in the movies *Blade Runner*, *Star Wars*, *AI*, *Wall-e*, *Ex Machina*, and many others, where we easily become emotionally involved in what we know are not "humans." Thus, the future holds within it a full array of "interpersonal" relationships with virtual humanity, and the more they also look like us—not just think and feel like humans—the less virtual they will become from our subjective perspective, and perhaps from an objective (philosophical) standpoint as well.

---

a silicon "lattice," or combines "a neural network in the brain of a living animal and a silicon neural network of spiking neurons on a chip... by neuromorphic synapses, thus enabling co-evolution of connectivity and co-processing of information of the two networks" (Kramer et al., 2008; Hybrid network, 2020). Ethicists have begun discussing the implications of robots with biological brains (Warwick, 2010).

Even purely digital "objects" [V25] can have a powerful emotional hold over humans. Paradoxically, at least for today's teenagers, virtual possessions can have enhanced value compared to material, private property—for three related reasons. First, because of easy access from anywhere compared to material property that must be stored in a particular place: "The 'placelessness' of virtual possessions stored online rather than on a computer often enhanced their value because they were always available" [V17] (Virtual possessions..., 2011). Second, virtual property can be more easily shared through cloud computing and the like. Third, they are far more malleable—personalized, tagged, morphed etc., to suit the owner's present tastes [V15].

It is important to understand that this will be a major element in future social virtuality for central areas of life in which we will interact deeply with a virtual world of our making. The reasons for this are demographic and economic.

Demographically, the world will reach a population plateau sometime in the mid-twenty-first century (at around 9–10 billion people), because of the historically unprecedented decline in birth rates almost everywhere around the world. Most post-industrial countries already have a birth rate significantly under the replacement value of 2.1 children per female. Thus, their populations are aging rapidly. The result within the coming decades: a very large elderly cohort that cannot be supported financially, nor cared for physically by a greatly reduced young, working-age generation.[8] The civilian AI revolution, today merely in its initial stages, will receive huge injections of R&D money and will undoubtedly evolve to take over the massive "loss" of workers in almost all areas of the economy: robots as elderly caretakers; avatars teaching the very young; agricultural and industrial automatons; AI-based driverless transportation; sales "people" (whether physical or virtual); as well as AI expert systems in information, entertainment, culture, and so on.

This is an additional, and important, sense in which virtuality and reality have become two sides of the same coin, as discussed in Chap. 12. True, as seen throughout this book we have always lived in a world surrounded by virtuality of our own making, but in most cases (except for some types of profession, that interacted with the virtuality of their own field on a constant basis e.g., moneylenders, artists, priests) our interaction with such virtuality was discrete and sporadic: praying to God at certain times of the day or week; indulging in art during our limited leisure hours; using symbolic money when shopping. The AI revolution, on the other hand, entails many types of virtuality on a constant basis: the home robot; the android, geriatric caregiver; the AI teacher spending many hours a day with the child. All this without mentioning the even greater interface with AI as noted earlier: ubiquitous computing, with the entire external environment aiding us in "intelligent" fashion, unseen by the naked eye and without material manifestation. Which leads to...

---

[8] Given the cultural and political tensions inherent in massive migration from South and "East" into the North and "West" (with accompanying increased xenophobia), it is questionable whether the migratory solution is politically sustainable over the long run. In addition, even the most populous country in the world, China, has begun to suffer from a labor shortage (Cheng, 2021).

## 13.5  Artificial Intelligence (Macro)

If everything and everyone will be informationally connected, aren't we possibly creating some form of macro-intelligence that is the sum of human minds and even more numerous AI artifacts? This question has now advanced beyond dystopian science fiction.

There are two forms of "macro-intelligence." The first is like an ant colony. Every individual ant by itself is a rather stupid creature. However, when several thousand ants self-aggregate and communicate with each other through chemical pheromones, what emerges is a remarkably "intelligent" colony behaving purposefully to achieve complex goals: building a nest, foraging for food, fighting wars, etc. Can this agglomeration of rationally behaving ants be considered intelligent? It would seem so; we can call this a virtually intelligent society—not "virtually" in the sense of "almost" intelligent but rather in that we have trouble delineating the exact "entity" harboring that intelligence [V6; V15].

The second example is different in scope, but with a similar underlying principle: the human brain. Like the individual ant, each human neuron (brain cell) has little "intelligence"—basically it can "fire" when stimulated, or not. However, when we connect 80 billion brain cells by way of trillions of axons and synapses, once again intelligence emerges: a human being. Here we can more easily see the "entity" housing that intelligence: the brain within the human body.

How, then, does intelligence come into being? To provide a full picture of countless philosophers, psychologists, neuroscientists, biologists, theologians, and others, would digress too much from our core discussion. Suffice it to say that notwithstanding *how* human intelligence evolved over time, it "emerges" out of complexity. We intuitively understand that a simple amoeba is less "intelligent" than the more biologically complex brain of a dog, that's less intelligent than the far more complex human brain.

However, intelligence is not merely a function of the *number* of neurons involved; it is highly dependent on the number and strength of the *interconnections between* the neurons. This is why an adult is far more intelligent than a baby, even though both have almost the same number of brain cells. The developing brain after birth adds synapse connections between brain cells (based on external stimuli), prunes away underutilized synapses, and then strengthens the more useful ones (a result of repetitive external stimuli). Therefore, although the brain is purely organic ("real"), its intelligence can be viewed as somewhat "virtual" because the whole is far more than the sum of its parts—not in a metaphysically "spiritual" sense, but rather in the meta-physical sense of $2 + 2 = 8$.

Returning to the initial question, if we are presently taking intelligent humans and AI objects and interconnecting them in trillions of potential links, couldn't a macro-intelligence emerge—as occurs with the ant colony and/or the human brain? Indeed, several thinkers have suggested that we are moving towards a world in which we are inadvertently creating a Global Mind (Stock, 1993; Levy, 1998, pp. 85–88).[9] Obviously, whatever the "meaning" or manifestation of this concept, it would be a virtual mind working on a physical "brain" platform in much the same way that the human brain works, with the difference being that the human mind's platform is biological material (neurons) whereas the Global Mind's platform would be a combination of organic (human brain; biological computer) and inorganic entities (silicon and quantum AI).

Would—could—a Global Mind also possess consciousness? Teilhard de Chardin, who wrote extensively on the issue of intelligence and complexity offered a possible answer: "The more complex a being is... the more it is centered upon itself and therefore the more *aware* does it becomes" (1964, p. 111). The key word is "being": a discrete biological being, or could it include a social grouping such as an ant colony, a human organization (e.g., the "collective intelligence" of a company), and ultimately Earth's entire biosphere and infosphere? If the latter is possible, then we as individual humans might find ourselves in the future living on, and within, at least two planes: as discrete, individual beings in the immediate physical plane of home, spouse, artifacts; and as a small part of a much larger, virtual plane of information, providing bits of information to each other, thereby forging a "higher" collective intelligence.

The next two sub-chapters describe a different technological approach to the issue, explaining why and how such a possibility is less fantastical (and many would add: less fantastic) than it might seem at first thought.

## 13.6 Brain-to-Machine-to-Brain Communication

Human communication has always required some sort of physical action on the part of the communicator—from speech (moving our vocal cords and using body language) to writing (moving our hand) to reading (holding a book, turning pages) to manipulating another medium (turning on a radio, typing on a computer).

There is at least one diverse population group that heretofore was unable to communicate naturally but now can greatly benefit from AI: paralyzed (or otherwise physically challenged) people. Scientists are studying ways of enabling close-to-normal communication for such individuals. The most promising technology can best be described as "Brain-to-Machine" (B2M) communication (Gunkel, 2020)—communicating with the external world *through pure thought alone!*

This technology already exists, enabling paraplegics (or anyone for that matter) to manipulate a cursor or digital images on the computer screen (Edinger et al., 2009; Cerf et al., 2010), through pure thought alone. The individual undergoes trial and error "training" with electrodes attached to the head. The cursor/computer itself is

---

[9] Political philosophers John of Salisbury and Thomas Hobbes viewed the polity as a collective of people, as a "creature" – "Policraticus" and "Leviathan," respectively. The idea that many humans could form an intelligent entity greater than the individual is not a modern one.

connected to other machines, enabling the person to manipulate objects (wheelchair, room lights, radio/TV) solely by thinking. This is the exact opposite of virtual communication until now. Whereas we have heretofore been doing physical activity to communicate/manipulate a virtual product (e.g., writing a blog on our computer), in B2M we mentalize virtually [V25] to produce a physically real outcome (brain signaling to move a wheelchair).

More recent advances—neuroprosthesis—enable verbal communication through "mentalizing" to produce a virtual product e.g., cerebrating a message directly to the computer screen that types it out (currently 15–18 words a minute), by producing comprehensible words and sentences through mere thinking them (Belluck, 2021).

If one runs this process backwards, we arrive at a machine communicating with our human brain *without us having to use any of our sensory organs* (bypassing eyes, ears, touch)! Technically, this reverse direction of Machine-to-Brain (M2B) communication is not as "easy" to do, but neuroscience is making great progress in "deciphering" exactly what mini-sections of the brain are in charge of what mental or sensory process, so that the concept of M2B is definitely possible: "By using machine learning to analyze complex patterns of activity in a person's brain when they think of a specific number or object, read a sentence, experience a particular emotion or learn a new type of information, the researchers can read minds and know the person's specific thoughts and emotions" (Fields, 2020). Moreover, the U.S. National Institute of Health established an immensely ambitious project: the Connectome (Van Essen & Glasser, 2016)—ultimately mapping all 100–150 *trillion* synapse connections in the human brain, furthering our understanding of emergent intelligence.

Understanding how tiny brain sections work separately and in conjunction with others, as opposed to making them do our bidding, are quite different things. Yet here too, brain experiments have successfully activated very specific areas and even single neurons, through optogenetics. This combines optical (light) stimulation with genetic engineering by injecting a photo-sensitive protein into neuronal genes to be light sensitive (as are plants) and then switching them on or off with light (Joshi et al., 2020).

When we put the two processes together—output and input—we arrive at full Brain-to-Brain (B2B) communication, otherwise called "Synthetic Telepathy": "The bidirectional interface between the brain and the computer would ultimately lead to the development of a 'Brain-to-Brain Interface' (BBI), in which neural activities from individual brains are linked and mediated by computers" (Yoo et al., 2013). Such "computers" could ultimately be mere nano-sensors/transmitters embedded in the person's brain.

In short, communication in the future might very well be totally virtual in that there will be no mediation whatsoever—not even human voice (if one wishes to call our organs a "medium") or body language [V25]. Theoretically, this could be dyadic communication (one-to-one), group (few-to-few), mass (one-to-many), or plural (any and all of the previous three). However, it is not at all far-fetched to foresee the day when most (or perhaps all) of our traditional, *physical* means of communication become obsolete,[10] and instead humans will communicate with each other—as well as with our compunicated environment—in silent, virtual fashion with no discernible utilization of any physical body part: invisible [V19]; and perhaps incorporeal [V25].[11]

## 13.7  Augmented Reality

As described in Chap. 12, virtuality and reality can co-exist in many natural ways as well as humanly produced artificial ones. One can add to this augmented reality (AR) (Chen, 2019). The idea and practice of AR is not new at all. Anyone who wears makeup, puts on a toupee, wears a girdle or padded bra, undergoes plastic surgery, and through other forms of bodily manipulation changes the look and/or form of the body, is in essence "augmenting" their own reality [V5]. The same principle can be applied to painting murals on the side of a rundown building and other forms of environmental prettification.

These examples "add" (or subtract) from reality to *hide* or *distort* the real situation. Modern technology has added ammunition: digital technology now enables us to easily "morph" real pictures, video, and voices into something close to but not exactly like the original [V1; V13]—deepfakes, mentioned in Chap. 10.

AR on the other hand, has a different purpose: to *enhance* our understanding of the world around us by *adding* text, pictures, and other visuals, sounds, haptic (touch) feedback, and (eventually) smell within the context of our natural environment. These augmentations are sensed through "visual displays": simple, personal projection gadgets that can add information when we arrive at a particular site; AR screen/eyeglasses with audio-visual capabilities. As we move around the natural world, our AR system adds data to what we encounter e.g., touring an historical site and seeing a holographic video of what occurred there years earlier; walking down the street, seeing a familiar "stranger" coming at you, and immediately reading who they are and where you last met that person (Taub, 2010).

The augmented information is not permanently "there" i.e., eye-tracking technology can ensure that only when a specific person looks at a discrete object will the added information become visible. Alternatively, AR can be offered to the individual

---

[10] Ironically, this type of non-mediated communication brings us back to the original state of human intercourse that was frontal (face-to-face), using only natural body organs (larynx, hands etc.). However, I deem B2B communication "virtual" and not "natural-real" because there is no discernible utilization of any physical body part. At the least, it is invisible [V19]; beyond that, it can also be considered incorporeal [V25].

[11] There is at least one serious problem with this scenario, also hinging on virtuality: brain implants make the person vulnerable to infection by a computer virus! By enabling direct digital communication with the outside world, we also become vulnerable to external digital attack. Thus, this type of cyborg development cannot advance without developing a virtual immune system ("firewall") to protect the person from being bitten by bytes (Could humans be infected by 'computer viruses'? 2010).

with context-aware computing (Gay, 2011, p. 288) that detects a person's location, identity, date/time, and even activity—and then based on these specific parameters provides personalized information (textual, aural and/or visual) regarding the immediate external environment. Augmented reality, then, can even be considered a "sixth sense" (Hood, 2010), or at the least a significant expansion of the other five senses [V21].

Essentially, this is not "masking" or "distorting" the real world for purposes of making it more palatable [V5]—although AR could do that too—but rather it *complements* the real world with useful information not normally attached to common objects and places. This is another type of "internet of things"—not *between* objects, but instead a virtual layer *on top of* things. It transfers the noösphere of knowledge from the purely virtual realm of computers and computerized artifacts into the rest of our everyday world of places and things. This does not change the world from being real to being virtual, but rather *combines* the two into an integrated whole—returning us to the Global Mind.

The technologies of AR and B2M2B will lead us still further in the direction of a Global Mind, although each in a different way. AR is a straightforward adjunct of AI. As noted earlier, through artificially intelligent compunication systems with which we interface and exchange information, a Global Mind begins to take shape. AR adds a spatial dimension to such information as it appears in discrete physical spaces within the real world, supplementing the natural information of that specific territorial site.

Conversely, B2M2B communication removes spatial territory from the equation, with communication taking place directly between human minds or between human minds and AI entities. This type of interconnection is potentially far more revolutionary for (at least in principle) it would enable the "hookup" of all intelligences in the world—human and artificial—without the need of a physical infrastructure, a "World Wide Mind" as Chorost's book title puts it (2011). Right now, this is still more science fiction than fact, but as we have repeatedly seen, yesterday's magic is today's science is tomorrow's working technology. Anything approaching this would move humanity much deeper into the realm of social virtuality [V17; V25].

## 13.8  Invisibility and Opaque Transparency

In principle, invisibility is one of the definitions and characteristics of virtuality [V19]. Until very recently, there was only one way in which something could be invisible: when our visual sensory organs cannot see—either because they are defective (sight blindness) or more generally, when the objects did not emit light waves that could be picked up by our sensory apparatus (e.g., x-rays). However, modern technology has added a third category by successfully creating materials that bend light, thereby rendering invisible a formerly visible object. This is the mirror image of augmented reality; instead of adding information to something real, invisibility cloaking subtracts (visual) information from a real object [V4].

Stealth technology was first successfully employed in aircraft, starting in World War II. However, the planes using this technology were not really "invisible"; rather they significantly reduced their radar, infra-red, visual, radio frequency (audio) "footprint," to make it much harder for the enemy to locate the plane. More recent technologies have provided advances, although none yet offer complete invisibility (Scharping, 2020).

Conversely, technology is reducing the number of objects that cannot be seen because they are behind other opaque objects. The most advanced technology, non–line-of-sight (NLOS) imaging, captures and produces pictures using reflected light, computer processing and other tools, to see objects that are behind opaque barriers, around corners, or through open passages [V21]—even from a mile away (Wu et al., 2021).

Objects temporarily out of sight can be called "temporary virtuality"—elements in the world that we know (or believe) to be present but for physical reasons cannot be perceived at the moment [V19]. However, it is also possible to argue the reverse. If and when these technologies become standard (and improved)—the problems are less technological than legal and ethical (given privacy issues)—it would significantly expand our ability to perceive all elements in our immediate surroundings, thereby *reducing* the number of objects that are usually unseen based on fundamental laws of optics and physics.

## 13.9 Historical Memory: Micro

As noted in Chap. 3, human memory is malleable. In general, we "suffer" from an inability to remember every detail of our past,[12] as well as selecting those past details that make the most sense in our personal "historical narrative." Thus, memory can be said to be a distorted "construct" of our mind [V5], or to be more precise: a *reconstruction by* our mind. Can this be overcome? Can we reduce the amount of memory distortion virtuality?

The book *Total Recall* (Bell & Gemmell, 2009) described how Bell obsessively scanned, logged, and otherwise captured every bit (and byte) of personal data that he consciously produced and also indirectly generated, every day of his life. Nor is Bell alone—a growing number of people have taken to "self-tracking" by quantifying and recording all events, ideas, and activities in their life (Lupton, 2016). By collecting greater amounts of quantitative information about themselves, people can crunch the data to achieve a level of self-understanding otherwise impossible. Moreover, such "Terabyters" could make a lifetime income doing this, selling such "personal life"

---

[12] Some unlucky few have the opposite problem, suffering from a traumatic event or experience that they cannot forget [V11; V21]. For such unfortunates, the past that for others has become virtual, remains all too real; they need to "virtualize" the past i.e., to forget or at least moderate the effect of that memory. Recent research shows that this is possible (Clem & Huganir, 2010).

information to marketers, health professionals, even government agencies who need comprehensive but also very detailed data on the micro-level (Frey, 2011).

In principle, one can envision a time in the not too far off future when any person can be equipped with a nano-camera, its lens stretched around a fashionable head-band—to film, orally record and wirelessly transmit to a server—everything in 360 degrees around the person for the entire duration of that individual's lifetime (with the ability to turn it off at intimate [sex] or otherwise private [restroom] moments). We already have miniaturized bodycams capable of many storage hours (depending on recording resolution) and wireless transmission to sites of the person's choosing.

In the Internet of Things future, this won't be limited to "equipment" *on* our person; it will include stuff *around* us and also *within* us:

> …an assortment of tiny, unobtrusive cameras, microphones, location trackers, and other sensing devices that can be worn in shirt buttons, pendants, tie clips, lapel pins, brooches, watchbands…. Even more radical sensors will be available to implant inside your body, quantifying your health. Together with various other sensors embedded in gadgets and tools you use and peppered throughout your environment, your personal sensor network (Bell & Gemmell, 2009, p. 4)

This could all climax with what Peter Cochrane (British Telecom's Chief Technologist in 1999) forecast: "a computer chip so small and powerful that it could be implanted behind the eye and used to record every sight, thought and sensation in a person's life, from cradle to grave" (Jonscher, 1999, p. 19).

Conceptually, contemporary society is already there with people recording 24/7 in their home, Twittering incessantly about what they're doing, Instagramming their activities, Facebooking and TikToking their latest antics and social relationship developments. Facebook's "Timeline" is an excellent example: a chronological index of *every* item the person ever put on their homepage: pictures, messages, links, notes, etc. Life lived—*and* virtualized in its entirety!

Who would have the time to "re-view" their own (or another's) complete life? Visual search engine technology will enable us or future generations to ask the system: "give me two minutes interesting footage of every birthday in my life, especially focusing on family and friends"—and the 60-year-old adult would receive a professionally edited, two-hour movie of "Sam's birthdays." Add to this all the data collected constantly and consistently throughout Sam's sixty years, and one can produce a fascinating profile of a lived life. At the least, this would overcome many of the brain's memory deficiencies.

There are some serious problems inherent in such heightened and perfected virtual memory. First, significant privacy issues; as some/most/all of this material is accessible to others too, we lose a huge amount of informational privacy through greater *access* to others' life events and their greater *accuracy*. With all sorts of "Total Recall" systems in the future, anything each of us does can be dredged up by anyone else in the future—an exponentially larger audience. Put another way, in the past we had a large measure of "privacy" from most of the world even when we were "living our life" in public; that is no longer the case. However, such a situation might not be viewed negatively in the future. In an age of growing outdoor and public site surveillance cameras, internet cookies, data mining, location devices, and social network sites where many people bare all (figuratively and literally), fewer of us seem to have much expectation of "privacy" in the public sphere, nor do many (mostly younger) people seem to care about such privacy loss.

A second, related problematic consequence is paradoxical. On the internet people can change their identity i.e., "reinvent themselves"—either starting fresh with new friends or at least temporarily (e.g., within an artificial world, or self-help community forum) through a self-chosen avatar stand-in, the use of a pseudonym, or simple anonymity. In other words, on the internet one can live a "virtual" life divorced from one's "physical" persona. Moreover, even beyond the virtual internet, in the past one could always "reboot" one's life, whether by moving elsewhere or by repenting one's past deeds and "starting over" behaviorally or professionally—what Lifton (1993) called the "protean self."

"Total Recall" systems will not enable anyone to escape their ugly or mundane past.[13] Although the virtual world in principle enables change of identity, and whereas in today's highly individualized society one can always try to start anew, such possibilities are undercut by these same online (and offline) technologies.[14]Consequently, there have been analytical attempts (Mayer-Schönberger, 2009) to turn back the clock technologically to preserve our ability to start anew, or at least not be perpetually in fear that past indiscretions will come back in virtual guise to haunt us. Europe's landmark GDPR "Right To Be Forgotten" law is a significant step in that direction, although it cannot apply to search engines and servers outside the EU (Kelion, 2019).

The result of all this might be that in the future everyone could have several lives: 1—the original real that we lived in; 2—the virtual that we live through (e.g. artificial worlds; virtual reality); 3—the virtual presentation of the real (or virtual) past; 4—the virtual reconstruction of that same past, with significant parts filtered/edited (Zhai, 1998, p. 180); 5—the newly generated/transformed, mid-life/mid-career, real life. Obviously, these will not be clearly separated or differentiated chronologically or functionally.

Will all this lead to less virtuality or more? Both. As noted, memory distortion would decrease significantly so that the past would be "remembered" (perhaps "reviewed" would be more fitting) in more accurate fashion—unless filtering/editing was permitted or became standard operating procedure. The past, therefore, would exist in its original "real" state and so be less virtual for one doing the recalling or re-viewing[15]—whether that person herself, or others interested in that individual's life. Conversely, it could be virtually "refilmed" with automated editing technologies, thus rendering one's past even more virtual than is faulty memory today. In addition, the ability to bring up any part of our past (and others') might mean that in the future

---

[13] This has already occurred many times, especially regarding public figures who might have sent out a reprehensible tweet ten years ago as a young adult – and is now "called out" for insensitivity and worse. In the past few years, many candidates for office (political elections or civil service nominations) have been "taken out" because of a textual indiscretion many years earlier.

[14] See Rosen (2010) for a useful survey and serious discussion of this overall issue, including a list of several academic researchers dealing with the general subject.

[15] Such re-viewing would not only be *seeing* past events but also *hearing*, and perhaps even *smelling*, *tasting*, and *touching* (through virtual reality technology).

people will spend much more time than they do today re-viewing personal histories (literally *YOU* Tube!); from that standpoint our lives here too would have greater virtuality [V11]. Conversely and finally, what was previously "out of sight, behind closed doors" or "ephemerally fleeting public activity" will be forever recallable, rendering the formerly "unseen" part of our present lives now clearly visible in the future—a significant reduction in virtuality.

## 13.10   Historical Memory: Macro

As with any individual person, one of the biggest areas of virtuality for humanity is history. For those living today, history is doubly virtual. First, historical records and other evidence are by nature highly incomplete—"history" becomes a pale, distorted, descriptive image [V5; V9] of what actually happened. Moreover, selective memory is compounded when such historical memory becomes a product of the collectivity. This is the Rashomon syndrome: several people can participate in, or view from the side, the same event and afterwards perceive/interpret it differently [V11; V20]. All the more, then, when historians look at events that they haven't seen with their own eyes.

All this is changing rather drastically—regarding the present, and the future as past. From "today" onwards, our lives increasingly will be captured by several means of media storage. The first of these was just discussed at the individual, micro level. Assuming that these individual records will be made public, the totality of such micro-life-recording would significantly enrich the historical record.

Even if micro-recording (Lifelogging) never becomes a general phenomenon, we have already started leaving a vast digital trail of our actions, thoughts and relationships, a record that will only expand as the digital age deepens and widens. For example, the Library of Congress decided to store all Twitter messages (Lohr, 2010a) for future scholarship and others who might wish to see how we communicated back in the early twenty-first century. However, the amount of digital data produced by all of humanity is humongous—2.5 quintillion bytes of data is being created every day in 2021, close to one sextillion (a thousand trillion) bytes a year![16] As societal activity (economics, politics, entertainment, interpersonal relationships) gradually migrates to the internet-based, digital sphere, it becomes easier—almost automatic—to retain an "historical record" far more detailed than was ever possible.

Beyond all these forms of micro and macro historical recording, there is one technological system that forms a bridge between the public and the individual: "the internet of things." In "augmented reality" we are able to leave a historical "record" at the physical site of a place or on a specific object by enabling anyone to record comments, insights, descriptions of the specific thing, with ensuing visitors adding

---

[16] Storage capacity is keeping up and should have little problem in the future continuing to do so, given relatively new technologies such as holographic data storage, not to mention DNA storage, noted earlier.

their own comments—building up over time a "living record" of the object/place. This is similar to Wikipedia; every encyclopedic term and accompanying information also maintains the "revision record" i.e., the various drafts, emendations, and discussions/arguments between the various editors of that Wiki-item.

The significance of all this digital memorization of both byte (computer) and atom (real world things) activity is somewhat paradoxical: we are using greater amounts of virtual information to *lessen* the extent to which history becomes informationally virtual! If one aspect of history's virtuality is the lack of data regarding what happened [V4], by increasing the amount of original-source, first-hand information, we can significantly fill in many of the inevitable, huge holes from the past. Moreover, until the mid-twentieth century almost all history was written by, or related about, the elites, generally the only ones who left a record (certainly anything written) and were also the ones considered to have made "real history." Both these things began to change in the latter half of the twentieth century, but the historical record left behind by the masses was still very skimpy.

The saving and storing of macro-history, the comings and goings of the general public, as well as the "history" of places and things, will significantly increase our understanding of the contemporary present and all new future "pasts." This will render historicizing less virtual, at least from a data perspective. Indeed, the relatively new sub-discipline of History—Cliometrics, studying the past through statistical analyses of broad-based data—will obviously receive a huge boost in the future,[17] as will the related Annales School (Braudel, 1996) studying large-scale socioeconomic factors of the average citizen and not only of the elites.

However, whether more data will make the future's past less or more virtual from an *interpretational*, "Rashomon" perspective, is a different question altogether. For one, more evidence makes previously indeterminate situations less ambiguous; conversely, more data can lend fodder for each side of differing schools of historical thought. Finally, more information will not "bring back" the past to the present; for that, we will have to wait (probably forever) for the invention of time machines.

## 13.11 Simulated Reality

Biological bodies are complex. To study how they work (internally-biologically or externally-behaviorally), life scientists and social scientists face several obstacles. First, there is the ethical problem: ensuring that the study causes no harm. Second, given the human body's "slow" development, and long-time psychological-behavioral effects of environmental stimuli, human experiments can take an inordinately long time. Third and finally, biological and behavioral effects of environmental

---

[17] An even more recently developed sub-discipline is Culturomics – the computational analysis of large text archives (news, fiction etc.) through the Google Books Project, going back hundreds of years, to discern cultural trends e.g., changing social opinion regarding women's rights, homosexuality, individualism etc. (Reading by numbers…, 2010).

stimuli are complex, necessitating running many experiments, each one incorporating different independent variables, to fully understand all the possible outcomes and effects of the sundry factors. And in many cases an independent variable can also be a dependent variable: other humans impacting and being influenced by each other.

As a result, several new, related research methodologies are being developed, as a group called "virtual simulations" [V10] (Casti, 1997). In one type, a group of virtual people is created inside a computer, where each virtual person respectively is given different real-life parameters e.g., some rich, some economically comfortable, others poor; some extroverted, others introverted; educational level variance; and so on [V1]. The number of people with each sub-variable is proportional to what exists in the real-life society that is being simulated—based on social scientific, biological, and other types of data and knowledge that modern science (and government statistics) has accumulated [V6].

These virtual people are put through Monte Carlo simulations, performed in a vast number of iterations, each varying slightly (Ball, 2010). Traditional simulations manipulated concrete empirical data regarding external variables: weather, macro-educational levels, political regime, GNP, interest rates—almost ad infinitum. With virtual people the variables also have a psychological component. Such virtual simulations can test the *interaction* between each of the virtual people, and not just the interaction between inanimate variables as in traditional simulations. Obviously, this demands a relatively high level of precise knowledge regarding personality traits and concomitant human actions and reactions—something that we are gradually succeeding in developing, in large part a result of data mining of internet activity, as well as what we learn from discrete socio-psychological and neuropsychological experiments in the field and lab.

The potential consequences of such a third-stage Social Simulation approach, at present in its infancy, are revolutionary for social science research, and only time will tell the extent to which it can carry out the promise of comprehensive (and perhaps universal) understanding of the social world—far beyond the first stage of pre-modern philosophizing and second-stage, Industrial Age field experimentation and statistical analysis. In principle, however, Social Simulation is promising, notwithstanding the huge difficulties and complexities of the endeavor.

That's because Social Simulation decreases variable and data incompleteness, vastly increasing the number of units that substitute for their real-world counterparts [V6]. Such full-blown simulation [V10] offers a level of verisimilitude heretofore unapproachable by social scientists. Of course, final exactitude depends on how well we understand the underlying independent variables: human traits, thought processes, etc. This constitutes the second advantage of Social Simulation methodology: it is based in large part (but not exclusively) on the Life Sciences—neuropsychology, physiology, environmental ecology, etc. Humans are creatures of their body, of their organic brain, and their immediate physical surroundings—subject areas that now have been researched in-depth with a decent level of understanding—so that Social Simulation can better predict human behavior than previous Social Science.

Finally, Social Simulation does a far better job of dealing with the feedback loop. Humans will change their behavior based on the results of prior behavior. We learn from our mistakes, from others' errors, and in general will react to whatever is happening in our environment. Social Science equations cannot adequately deal with this feedback; Social Simulation, based on "computer agents" that mimic human behavior in its reactive mode, does take this into account—indeed, it might be the main strength of the methodology. As a result, agent-based simulations are dynamic and ongoing; we can see how social behavior plays out over a long period of time, as each human "agent" acts, reacts, and then acts again in a continual process that we call "real world life."

As more human activity gravitates to the internet and other digital media that can store and analyze such behavior, it becomes easier to data mine and macro-analyze such group behavior. In other words, Social Simulation could not have worked very well in the non-internet, non-digitally-networked past, even if we had powerful computers back then. Much of our understanding of human behavior in the future will not come from social science lab experiments but rather from real human activity in the various virtual worlds we inhabit, whether the internet of work, social communication, data accumulation, information augmentation, or artificial worlds of play.

The Life Sciences are part of this simulated reality revolution. For instance, virtual Stage 2 and 3 pharmaceutical trials replicate on the computer the physiology of various human organs [V7] (Lee, 2020). Virtual simulations, of course, are far faster than real-life clinical trials, meaning that successful drugs get to market years earlier.

A third type of virtual simulation involves the accelerated evolution of life forms within the computer [V25]. Most biological creatures have a "lengthy" lifespan; to study their evolutionary development (either natural or with mutations induced by human intervention) would take a very long time—even years for creatures with relatively short lifespans. By virtually simulating and artificially accelerating such evolution, we can discover evolutionary effects and outcomes in hours or days instead of decades or longer (Brahic, 2010).

Such simulations can be carried out based on our knowledge of the creature's biology (like the two previous simulations above) or even more intriguingly can be done on virtual creatures that do not exist, but that we "invent." This could take the form of adding (in the computer) a heretofore non-existent trait to a real creature [V22] or through *creatio ex nihilo* (virtually) of a totally new type of creature [V13]. In the former case, scientists better understand the effects of "mild" genetic engineering; in the latter, a better idea about such issues as how alien creatures would look and react in a physical environment radically different than Earth, which creatures would be worth creating for salutary purposes (e.g., oil-slick-eating bacteria), and by combining all three types of simulation we could assess the potential environmental impact of new creatures on other, existing life forms and on the environment as a whole.

In sum, virtuality is being further developed to represent, copy, and even "create" reality [V1; V13], where and when the latter is too unwieldy for in-depth study. As such, the overall paradoxical goal of all these virtual simulations is to *reduce* virtuality (knowledge lacunae or error) [V4; V5]—to improve our reality, whether through a better understanding of how to manipulate our physical surroundings, control our social environment, or create new (f)actors within both.

However, if so, a far more intriguing and profound issue arises: perhaps we ourselves are already living within such a simulation produced by a vastly advanced future civilization? The question has been given serious, philosophical-scientific thought (Bostrom, 2003). Among other things, there is almost no way to determine whether this is the state of our "existence," but Bostrom argued that there is an approximately one-third chance of this being the case.[18] It should be noted that this is a statistically or quasi-empirically oriented variation of Descartes "dream simulation" argument mentioned in Chap. 5.

Clearly, the blurring and intermingling of reality and virtuality in our life has so permeated our culture that we have now reached the point where serious scientists are contemplating the possibility that *all* (our) reality is nothing but a virtual chimera [V31]! A disturbing thought, even if it is the ultimate, logical extension of virtuality's expansion through human history. Indeed, it is possible to offer a wild speculation that humanity's steady expansion of its own virtuality is somehow an unconscious, mimetic expression of the (un)reality in which we exist and of which we are somehow dimly aware.

## 13.12  Virtual Reality

The technology of Virtual Reality (VR) is commonly considered today to be the "core" of virtuality. VR is a technological *system* (using more than one specific technology) that gives the user a strong, three-dimensional impression [V7] of being a part of some "reality"—instead of the two-dimensional ones [V8] that we were used to in the past (paintings, mirrors, photographs, TV screens). This is accomplished through a VR set comprising a helmet (sight and sound; in the future also smell and perhaps even taste [Gatto, 2011]), and gloves or full body suit (touch/haptics). The VR set is connected (by wire or wirelessly) to a computer system with strong computing speed and power enabling the presentation of any type of un/reality in full 3D with other sensory stimuli *in ongoing, dynamic, interactive fashion*. Put simply, VR offers the user not only a realistic "world" but does so in dynamic fashion, reacting "naturally" to every movement that the user might make. Considering Moore's Law and the development of the new types of computers discussed earlier in this chapter, it is merely a matter of time before VR becomes a household system of entertainment and education, regarding at least the senses of sight, sound, and touch—and in the more distant future, perhaps smell and taste as well.

Thus, VR is "virtual" in several ways. First, and most trivially, it uses computer bytes [V17] and not material atoms. Second, and far less trivially, it can present a

---

[18] His work has engendered a lively debate in scholarly and other circles. See: www.simulation-argument.com

realistic version of completely (or partly) imaginary worlds/scenes/characters/objects [V28]. These can range from true historical events limited only by our knowledge of what actually occurred [V4], to completely fantastical scenarios limited only by our imagination [V35]. Third and most important of all, VR gives us the subjective impression of the situation being "real" in a *material* sense [V7; V10], while in fact being entirely incorporeal [V25] (aside from the haptic pressure placed on our hand or body to give the impression of mass).

All this sounds like great fun, but in fact VR contains the seed for possibly the greatest virtual revolution of all, practically and existentially—especially if it becomes an integral part of the "metaverse": the internet enabling virtual reality experiences across the globe (obviously something necessitating a huge increase in bandwidth)—already a central future project of Facebook ('Metaverse': the next internet revolution? 2021). Here are a few illustrative practical examples of VR (within or even without the metaverse), keeping in mind that VR can potentially "invade" *every* field of human endeavor.

## 13.12.1 Education: Biology (Martisiute, 2020)

The ninth-grade biology teacher begins a lesson on "The Heart" with her pupils wearing their VR sets. The pupils find themselves walking as a group *inside* the left ventricle. The blood cells—red and white corpuscles—jostle the pupils' bodies as the blood flows by. In the distance, they hear the whooshing of lungs breathing. Some pupils scrape off some fatty deposits from the heart's wall, whole others check the mitral valve opening and shutting like a dam lock. With their teacher explaining various elements, the class walks through the four heart chambers, finishing the lesson trying to withstand the massive pumping action of blood streaming into the aorta.

## 13.12.2 Education: History

After putting on their VR sets, the twelfth-grade pupils find themselves on a small knoll overlooking the Waterloo battlefield. The sounds and sights are horrendous—soldiers crying out in pain, muskets firing at random (with an occasional "ball" glancing off a pupil's helmet). Throughout, the teacher explains Napoleon's battle strategy and Wellington's counter-tactics. A few pupils go down to the battlefield, but quickly return to the knoll, overwhelmed by the stench of death.

### 13.12.3  Training

You've bought that $2000 ticket (how did they afford $450,000 back in 2021?) to tour outer space but on condition that you practice sleeping, eating, "walking" and doing your bathroom stuff in zero gravity. No problem: the VR set puts you through your paces daily—literally a weight off your shoulders…

It is hard to think of *any* important field of human endeavor that will not be significantly affected by VR: sex, sports, "churchgoing" (Hardwick, 2021), earth-bound tourism (why "shlep" to Hawaii when you can enjoy the exact same surf and sun at home?), medical care, shopping, entertainment (Disneyland without long lines!), and so on. Digital VR is theoretically capable of producing any scene, object or situation—just as can movies today: imaginary or "real."

## 13.13  The Ultimate Existential Question

From here it is but a small step to the ultimate existential question: will future generations wish to live in the "real" world when they can go through life with far more "satisfying" virtual reality experiences? The quotation marks around "real" and "satisfying" signify the subjectivity of each experience—the external and the internal. Obviously, this existential question has several aspects. First, can the economy survive if everyone is living in a world of virtual reality? The answer: yes, if we transfer all forms of production to artificially intelligent robots and ubiquitous expert systems built into all economic institutions and activity. Second, can society survive if everyone is in virtual reality? The answer to this depends on one's definition of "society." If defined as a large agglomeration of people *physically* interacting with each other, then society cannot exist when each person is physically separated from all others. However, as should be clear by now, humans have an almost unbounded ability and propensity to indulge in social activities without physical contact or even proximity. No one feels that they are not indulging in "real" social-human activity when they are reading a pen-to-paper letter from a friend, talking on the phone, chatting by computer, video-conferencing across the ocean, and so on. If that is true when we get to employ only one or two senses, then communicating with others through VR should feel even closer to the "real thing" because most/all senses will be employed by such systems.

However, this is only half the answer. The main "social problem" begins if and when people in the future will use VR not only to interact with other people but also (or rather) to experience unreal worlds or even our own "world" in non-realistic (VR) fashion. This is not far-fetched, for as we saw in Chap. 11, one of the main reasons we "do" virtuality is to escape from the real world's social restrictions, physical constrictions, and experiential boredom. If future humans choose to spend most of their time in virtual interaction with imaginary creatures (or even imaginary "people"), each person might not feel any loss of "social interaction" *individually*, but there will no longer be a human *society*.

We will return to the "limits to virtuality" question in the next, final chapter (14). For now, two points should be noted. First, there is a "technical" reason why it would be a bad idea for humanity to move lock, stock, and barrel into a VR existence: the safety net of redundancy. Humanity has always lived in two life realms: the real and the virtual. We shift from one to another not only to enrich our life but also when one of them becomes unbearable or dysfunctional. A complete shift to VR in human life would weaken the "real" world—not physically, of course (Nature would remain), but in its usefulness for modern-life—through desuetude. That would leave us highly vulnerable to cyber-collapse, or what Zhai calls "the Reality Engine collapse": "we could have an accident of virtual reality annihilation or cyberspace collapse, comparable to a collision of the earth with another planet in the actual world" (1998, p. 155). This could happen for reasons of the cyber-system's inherent program fragility, or resulting from a massive, malicious, and deadly attack on the system by neo-Luddites, or even by traditional enemies who try to cybernetically decapitate the enemy and instead bring down the entire edifice on everyone's heads (the Samson Syndrome). Therefore, it would be highly prudent to maintain the real world in its modern, socially constructed incarnation as well as sustaining a working virtuality.[19]

Second, it is possible that VR might lead to a radical shift in human society, a revolution whose intensity we have not witnessed since at least the invention of agriculture and the advent of human civilization many millennia ago. On the other hand, perhaps VR is not much more than going "back to the future," providing "an experience of comfort with multiple realities for a lot of people in western civilization, an experience which is otherwise rejected. Most societies on earth have some method by which people experience life through radically different realities at different times, through ritual, through different things" (Zhai, 1998, p. 184). Virtual reality, then, might well return us to an age of wondrous, subjectively unmediated, experiential multiplicity.

## 13.14  Virtual Immortality

I will end this chapter with one final "wild speculation." Earlier, I broached the possibility of our life being merely a massive computer simulation. However, there's an altogether different (and more probable) sort of potential, ultimate virtuality:

---

[19] Zhai's admonishment indirectly raises the converse problem: *real*-world destruction. This is a potential danger, albeit one that rarely occurs in terms of human time. Still, we are almost "guaranteed" to be physically wiped out sometime in the far-off future, as happened to the dinosaurs and *most* other species that have roamed the Earth at one time or another. Other than space colonization (entailing several extremely serious obstacles), the fallback solution is VR embedded in space (colonization) ships – entire societies, generation after generation, living and dying within a virtual reality, as their spaceships travel eons to reach their goal: another humanly-habitable planet. Today this sounds "crazy," but is it that much crazier than how our forebearers would have viewed contemporary technology and modern life?

"virtual immortality" (Graziano, 2013). This is not an altogether new idea (leaving aside the religious belief in life after death); Arthur C. Clarke's 1956 novel *The City and the Stars* described people living forever by being "archived" in a computer.

The concept is relatively simple; carrying it out technologically would be fiendishly difficult, if at all possible. Let's start with the concept noted earlier in this chapter of total recall i.e., capturing and storing the entire life memory of a person. Add to this physiological and neurological data (heart rate and ongoing brain wave patterns). Now imagine that after death we could "upload" all this memory data and other related information onto some sort of computer "platform" within an android body. Wouldn't this entity—a near (or complete?) clone [V1]—continue to live the life of the deceased, believing that it was the deceased person? Or maybe the deceased person continues to "live" in and through the android?

It is not even clear that these related questions make sense. However, if one concludes—a la the discussion in Chap. 3 on the brain, and in the present chapter noting that the brain is merely organic matter and chemical processes with consciousness "emerging" out of huge, interconnected brain "activity"—then a virtual "immortal" cannot be dismissed out of hand. That is certainly the case if the android is created out of organic matter!

Even if one does not accept the possibility of "consciousness transfer" i.e., the deceased person is not "living out" a continuation of life, the memories within that android (or any type of non-human entity) are still a virtual copy of the original [V1]. We could view this as an immortal, quasi-virtual meme [V6]—the original person's consciousness does not continue to exist but rather the virtual memory of a life's experience goes on forever, continuing to influence another sentient being's behavior (whether artificial or human).

In fact, this might be a workable way to retain the brilliance, genius and otherwise phenomenal mental powers of extraordinary people who could continue to significantly contribute to humanity even after the inevitable demise of their biological bodies. Extending the mental life spans of future giants on the same plane as Plato, da Vinci, Shakespeare, Kant, Einstein (and especially those who died too early e.g., Mozart), would be a great boon to future generations. The issue, of course, is fraught with dilemmas, mainly who decides who gets to be virtually "memed," based on what criteria, and for how long? Most humans would probably be very interested in having their experiences and wisdom continue to influence the world (however circumscribed—even at the level of immediate family members, as a great-grandfather-like avatar dispensing advice to one and all of his progeny).

Indeed, Blascovich and Bailenson report a very interesting (and for them, surprising) story that emerged out of an avatar experiment they called "Alive Forever." I bring their narrative here at some length because of what it says (albeit impressionistically) about people's attitude towards virtual immortality:

> Our idea was to build each person's avatar and to allow them to keep digital versions of themselves stored forever, perhaps to interact with distant progeny. Our plan, as psychologists, was to study the reactions and the ways people invested themselves in their digital counterparts…

For the first time since creating a virtual-reality lab, we experienced outrage from some of our participants. Some actually became so irate after learning that it was just a psychology experiment (and that we weren't actually going to store their digital selves "for centuries") that we had to end the study… a number of them proceeded to contact us long after the study was over in order to find out if we still had their avatars. (2011, pp. 145–146)

## 13.15  Conclusion: Thinking About the Future

Realistically speaking, not all forms of future virtuality outlined in this chapter will come to pass. As von Braun is reputed to have said: "I have learned to use the word impossible with the greatest caution" (unsourced). Yet even if only a few of the technologies and phenomena surveyed here become a significant part of our future world, most would immeasurably intensify and broaden the scope of virtuality in our lives. Indeed, for the critics of virtuality that were quoted in the book's introductory Chap. 1, the present chapter reads as a nightmare, for the phenomena discussed here make contemporary virtuality concerns look like a picnic. Certainly, some of these scenarios will make many readers uncomfortable—understandably so, given our all-too-natural predilection to favor the familiar over the strange. Moreover, a few of the above technologies are not merely "strange" but rather have the potential to radically change the way humans have lived from time immemorial.

In addition, this chapter's all too schematic survey only lightly touched on an important point that will increase the amount, and deepen the substance, of virtualizing technologies in the future. Twenty-first century science and concomitantly its technologies are breaking down the disciplinary walls, finding ways to merge and mesh their knowledge into newly *combined* technologies and systems. For instance, we now have a new scientific subfield: "bio-nanotechnology" (Bionanotechnology has new face…, 2010) that marries semiconductors and even hard metals to biological products. I have already mentioned earlier biological computers—a combination of computer science and biology—not to mention quantum physics and cybernetics, to create quantum computers. Such combinatorial science exponentially increases the possibilities of breakthrough technologies, many (but not all) also increasing virtuality in our future world.

Before reacting negatively, one should consider the much smaller amount of virtuality that existed thousands of years ago and how much virtuality has been added since then. Has life improved as a result? Any cold-eyed, unromantic assessment of "then" and "now" ineluctably leads to a positive answer. Thus, while we ought to move forward carefully regarding some of these virtualizing technologies and systems, we must also control our instinctive reaction that "not done before" = "shouldn't do it." Just as we have benefited greatly from all the virtualities that our ancestors have bestowed upon us, so too we owe it to future generations not to short-circuit the continuity of ever-expanding virtuality, without very good reasons for doing so.

Does this mean that the circle of virtuality should/will expand forever? It is to this ultimate question that we now turn in the final chapter of this book.

# Chapter 14
# Conclusions: Should There Be Limits to Human Virtualizing?

## 14.1 Introduction

Is there an upper limit to virtualizing beyond which humans start to suffer more harm than benefit?

This final chapter will not provide a definitive answer but rather note some relevant philosophical issues and raise methodological obstacles towards a future "research program." From there, society can deal with future virtuality challenges: Will we relate to "money" differently when it becomes completely ethereal? Will the legal concept of personal responsibility change when neuroscience finds that we are beholden to our virtual unconscious, itself exclusively a product of (occasionally flawed) organic matter? Will religion lose adherents if we have incontrovertible evidence that spiritual uplift ("transcendence" etc.) is a purely brain-organic phenomenon? and so on.

---

The *Appendix: A Taxonomy of Virtuality*—listing by numbers 1-35, e.g. [V13] or [V34], some of the various types of Virtuality mentioned throughout this chapter—is freely available to the public online at https://link.springer.com/book/10.1007/978-981-16-6526-4

## 14.2  Virtuality in Light of Value Mutability and Lifeworld Adaptability

To the empirical question "*are* there limits to human virtualizing?" the answer is probably "no." However, just because we *can* do something does not mean that we *should*.

As already mentioned, a "central value" for one generation might not be cherished the same way (if at all) by another. *Freedom?* Slavery was once considered to be an immutable fact of social life. *Democracy?* Divine Right of Kings and other undemocratic forms of regime were considered the most moral and "true" systems of government. *Individualism?* Most societies past and present view(ed) the collective as taking precedence over individual needs and rights. *Privacy?* The concept (and the actual word) doesn't appear anywhere in the U.S. Constitution or in France's Declaration of the Rights of Man. Given technological advances—data mining; panoptic surveillance; location finders; biometric ID cards—informational privacy will most probably not be considered as important in the future.

Similarly, "hyper-virtuality." Most contemporary youth are multitasking their eyeballs, earlobes and fingers with/on/through screens of various sizes and functions, but hardly exercising their bodies as in the past. Beyond body health issues from less physical exercise, are there mental health issues involved in lack of face-to-face interaction? If so, perhaps on the other hand our Twittering, friending, Instagramming, YouTubing, children are also gaining something no less valuable? Even more fundamentally, perhaps this is not a "secondary" social change but rather a far more radical and basic *primary* shift in how humanity experiences life?

Two historical matters should be considered. First, as this book has shown, the virtuality "shift" in human lifestyle has been evolving over time, parallel to human civilization's development. Second, human beings have already undergone radical changes in our mode of living—no less revolutionary than the shift from "reality" to "virtuality": the agricultural revolution changed human life forever as we became sedentary and developed large scale social institutions and varied cultural artifacts. Similarly, the Industrial Revolution's radical changes. In neither case can we say that the transformation was only for the better, but in totality (especially over the past two hundred years) life has improved for much of the planet's population (Pinker, 2018)—no longer a Hobbesian world. Thus, one can't presently determine that a further radical change—mainly involved with digital bytes and relating to the world's (re)presentation from afar—must necessarily be to our detriment.

This brings us to an important existential concept: *lifeworld*. As Husserl rhetorically asks: "Is not the life-world as such what we know best, what is always taken for granted in all human life, always familiar to us in its typology through experience?" (1970, pp. 123–124). If one took a child and enveloped her in an intellectually stimulating and emotionally satisfying world of virtual artifice, wouldn't her lifeworld—as defined by Husserl—be a "real"ly satisfying one based on the "typology of her experience"? Indeed, given the incredible variety of economic, cultural, environmental "lifeworlds" in which humans have grown up and flourished over the millennia, each with a significant albeit different measure of virtuality, why assume that only the "virtual lifeworld" is something to which humans cannot adapt and thrive?

Not everyone is sanguine about living life more virtually. One can discern four general, normative/philosophical approaches to VR (as a stand-in for virtuality writ large), listed here in order from negative to positive: 1—a false or degraded approximation of reality (Doel and Clarke, 1999, p. 261) through which "immediate experience is compromised" (Heim, 1998, p. 37); 2—the mirror image of reality—just as virtuality is all surface, to too is our perception of reality: "'true' reality itself is posited as a semblance of itself, as a pure symbolic edifice" (Zizek, 1994, p. 44); 3—a supplement to impoverished reality: "suppletion" (Doel and Clarke, 1999, p. 268); 4—a complete resolution of the real, i.e. escape from the world, as in uploading our entire mind into a cybernetic receptacle (Moravec, 1988).

Notwithstanding Moravec's certainly far-into-the-future scenario (discussed in Chap. 13), very few of virtuality's strongest proponents suggest a lifeworld devoid of corporeal contact. We must eat; we need to be clothed and housed against the elements; it is even questionable whether full VR (including haptics) could take the place of some critical functions involved in body contact, between parent and baby, or between people in general (the "hug"). Indeed, as explained in the previous chapter, some newer technologies might bring the virtualizing technologies *closer* to our body. Just as the traditional lifeworld of yesteryear was largely based on concrete matter with a few doses of virtuality here and there, a future lifeworld largely based on virtual experience would still maintain a modicum of corporeality.

Of course, it is not inevitable that we move toward "the transformation of life itself, of everyday life, into virtual reality" (Baudrillard, 1995, p. 107). We instinctively feel that face-to-face is "good" (Covid-19 Zoom fatigue seems to prove that) or at least crucial to a healthy body and mind. Thus, increasing virtuality is not historically "ineluctable"; it is eminently understandable why many people would try to stop the virtuality bulldozer from increasingly taking over our lives. Indeed, just because virtuality has expanded over time, its continued expansion is not "inevitable" or necessarily highly beneficial.

## 14.3 Limits to Virtuality: Philosophical and Methodological Issues

Whether one is pro, con, or indifferent to virtuality, the potential impact of hyper-virtuality leads to the question: Where do we draw the line, if at all? That question entails four interrelated, subsidiary ones:

1.  Should a person's involvement with virtuality take up a *maximum number of hours per day*? If the average contemporary individual spends most wakeful hours in virtual activity (watching TV, reading a book, tele-communicating), is there a point when this begins to be significantly harmful (even if they are physically active)?
2.  Should our virtual behavior be *restricted to a set and finite number of areas of life*? Spending many hours daily in virtual play might be largely positive; hours and years of education in a completely virtual learning experience, could be mostly negative. What about virtual religious services? Political town hall meetings? In short, by what criteria do we differentiate between beneficial and deleterious virtuality?
3.  Should any *human activities be off limits to virtualization*? Sex immediately springs to mind—but we already fantasize (or view erotic/pornographic material) while masturbating or making love—a hybrid virtual/corporeal activity. Would a baby really suffer harm if provided with all needs in the first year of life—tactilely, aurally, visually, nutritionally—by an android, parental stand-in?

Most people recoil at the thought. It is certainly unnatural. However, "unnatural" is not necessarily "bad." We fly airplanes, we perform plastic surgery, we talk to each other across great distances—all "unnatural."

4.  *What type of "virtuality" are we referring to*? In each of the above first three subsidiary questions, one must first define which types of virtuality are under discussion, among the 35 described in Chap. 2. These virtualities could be considered "good" and/or "bad." "Metaphysical" or "fantastical" things display the imaginative powers of human beings; "simulacra" and "synthetic" things tend to fool people into thinking that they are dealing with something natural or real. Conversely, "metaphysical" and "fantastical" virtuality is completely divorced from material reality, whereas "simulacra" and "synthetic" things could be functional tools to enable humans to conduct life more efficiently.

Complicating matters, specific activities have different effects within the same general "genre" of virtuality, depending on the specific medium and/or whether the activity is performed alone or with many others i.e., one must consider the social context as well as the specific technological vehicle of that virtual activity. And of course, the effect can be harmful and also positive at one and the same time. An example among many: a comprehensive review of recent research studies found numerous positive *and* negative effects of virtual gaming (Prot et al., 2014). This also assumes that effects are objective, but "positive" or "negative" assessments also depend on one's subjective life goals.

Complicating matters even more: are we referring to negative effects on an *individual* level or on *society*, or both? A man could spend all his waking hours (except for necessary bodily functions) playing intellectually stimulating and emotionally satisfying video games. Such virtuality is personal Nirvana; for family, friends and general society, a total loss. Conversely, someone who devotes most of her waking time to wiki activities for society's betterment but is quite miserable in her loneliness.

Societally, what happens when numerous people take their "positive" virtuality to extremes? Collective hyper-virtualizing could be very harmful to society at large: tax revenues plummet; virtual cliques form, cut off from wider, socially necessary activities e.g., millions of immigrant citizens leading "full" lives in virtual contact with their homeland, without integrating into their new country.

A counterargument: the highly virtualized future offers humanity greater social choice than ever before. Because of the interface between reality and virtuality, individuals in the future could choose to live within a very wide spectrum of almost total Physicality to almost total Virtuality—and all points in between—without undercutting the basic foundations of human society.

In short, "injurious" activity must be defined on the micro as well as the macro planes. Moreover, if society concluded that something Virtual can be harmful on either plane, we would still have to decide what qualitative *degree* and quantitative *number* of people crosses some danger threshold before deciding "to do something" about the problem, assuming intervention is feasible.[1]

Methodologically, ascertaining a virtuality "danger threshold" is fraught with difficulties. There are so many types of virtual activities that it will be extremely hard for researchers to separately test all the platform and functional independent variables (e.g., computer games, TV/video, social networking, texting, etc.) and correlate them with the myriad dependent variables: sleep, social intelligence, memory, creativity, physical functioning, health, concentration, emotional maturity, self-control, and so on. Note, this takes into account only virtuality *platforms*; how would we measure other significant virtuality activities surveyed in Part II, like religion, philosophy, literature, virtual economics, and so on?

The second research obstacle relates to the great variety of individuals (plural) and the disparate needs and makeup of different people. Virtuality is like food: we all have to eat, but each of us has different taste preferences and a different level of necessary caloric consumption. Moreover, foodstuffs answer other needs (emotional, athletic, social). Similarly, any virtual experience could be attractive and useful to one individual and harmful to another. Therefore, we can try to provide some general answers for the macro-social questions regarding "limits to virtuality" but these are hardly the answer for every person.

The third methodological issue: virtuality and reality are often complementary—existing within the same activity. If a good part of our future life will involve doses of augmented reality, how will we "measure" this: as virtuality? reality? half and half?

---

[1] Such "upper limit" considerations are found everywhere in real life. Most societies tolerate a certain low number of people who choose not to work, but if the proportions became too large the government would incentivize work and even require work of every able-bodied person.

Such complementarity has always existed: books are virtual symbols on a material "platform" (paper, ink); television pictures appear on a glass screen embedded in a plastic or composite frame.

In short, abolishing virtuality or existing without it returns us to the cave—and even there, we face virtuality if Plato has anything to say about shadows on the wall, or as the ancient Lascaux wall paintings suggest. Thus, the question is not "virtuality: yes or no?"; rather: "virtuality—how much (quantitatively), what type (of activity), for whom (society or the individual, and which individual), and at what cost/benefit for each?"

These are *empirical* questions, not amenable to philosophizing but still based on "normative values." For example, if we decide as parents or teachers that the prime goals of education are intellectual growth and emotional balance, then we would measure different levels of virtual versus face-to-face education to see which better produced the desired result. Or if our society insists on a certain minimal level of social glue (and not a completely atomized/individualistic situation), then again, we can test the number and types of mediated communication that push us below that minimal threshold (however defined empirically). But such a "good/bad" outcome must be *preceded* by a social consensus regarding the goal of human activity, individual or social—a philosophical/normative/value judgment.

## 14.4   Virtuality: Facing the Unknown Future

If and when society or an individual make(s) such a normative, goal-related decision, we have to be careful not to shut off other forms of thought, values, and lifestyles that are antithetical to what we believe in the present but that might have significant future utility. For instance, an early fifteenth century visionary who proposed widespread use of a "printing press" would probably have faced the following argument: "If we allow mass, printed material, we will lose our oral culture; the literate culture that replaces it will undercut the mysterium of our world, as science undercuts the verities of ancient religious beliefs. This would weaken the Church and ancient Authority in other areas of life. Such an outcome is patently bad; society would irrevocably change, and life as we know it disappear forever."

The possibility that hyper-virtuality in the future might significantly change our way of life does not render that eventuality "harmful"; what takes the place of contemporary "classical" social life could be a marked "improvement"—or it could be "worse." We must honestly try to understand and assess the "new life" that future hyper-virtuality might bring—and not merely worry about the "collapse of civilization as we know it." Indeed, there's a third possibility: increased virtuality might not cause much change, especially as much of human life has always been imbued with types and degrees of virtuality, even if we didn't recognize them as such.

In short, virtuality cannot be ethically or naturally good or bad intrinsically; indeed, as this book has shown, virtuality has always been part of human society, and thus *ipso facto* a "natural" aspect of the human condition. Not "natural" in any biological sense but rather socially "natural," constituting an important part of being human. This doesn't "prove" that it is good (murder too has been around from the start: Cain and Abel…), but it does place the onus on those who wish to argue against virtuality or warn of its depredations.

The evidence accumulated in this book goes further than that. The historical record indicates that virtuality in its variegated expressions is almost universal across cultures and through the ages. Unlike other social constants like warfare and murder, however, virtuality usually seems to serve some socially useful or psychologically necessary purpose. Therefore, the question of virtuality's harms shifts focus—from whether it is *intrinsically* problematic, to the conditions rendering it *situationally* harmful.

Moreover, as noted throughout this book (and especially in Chap. 12), reality and virtuality are not necessarily antithetical to each other; they constitute "two sides of the same coin." In most cases we are not faced with a zero-sum game where one's expansion must come at the other's expense. In any case, virtuality and reality are increasingly merging in our social and physical environment.

Nevertheless, there are situations where the relationship *is* zero-sum. It is at this point—and perhaps *only* at this point—that we have to assess the "damage" that triumphant virtuality might cause our "real" life, fully understanding that we must free ourselves from the "comfort of the habitual." As a commentator notes regarding the virtuality of cyberspace (similarly for any other type): "If we lose something by turning to the cyber realms, well, so we lose. Life is a game of lose-and-gain, and by losing something (even a constitutive part of our being) we at the same time gain something. New spaces are often opened up by the shrinking of the old ones" (Sikora, 2000). Economics provides a classic example: without symbolic money, our economy would still be stuck in its barter stage—a case of virtuality not merely (re)presenting reality but even creating a new (and better?) one for us. Thus, it is only in hindsight that we can perceive whether the advance is worth the loss: no pain, no gain.

The bottom line: not every new type of virtuality nor every expansion of long-standing virtuality is a "threat" to our well-being, even if at the start of the specific virtual phenomenon we may believe this. Moreover, often without virtuality we have no chance to experience—or even comprehend—elements of the real world. Without telescopes and advanced spectrometers capturing starlight from eons earlier [V20], we could not see distant nebulae or neutron stars. Similarly, the nano-world around us and within our bodies [V19].

True, occasionally real-world activity is lost in the face of virtuality's onslaught. Yet as we can see from experience, in most cases human society seems not only to have adapted well to such a shift but even benefited immensely. There is no guarantee that this will continue in the coming decades, but humanity's very long and wide-ranging experience with virtuality and its inexorable quantitative and qualitative expansion over time, suggest that there is far less to be worried about than many pundits fear. On the contrary, if the past is anything to go by, our approach to future virtuality should be one of cautious optimism that we can, and will, shape it to the benefit of humanity.

## 14.5   Coda: The Reality of Virtuality

The virtual is not opposed to the real; it possesses a full reality by itself. (Deleuze, 1994, p. 211)

The title of this book is *Virtuality and Humanity*. As noted at the start, it is *not Virtuality is Humanity*. Virtuality in its multiple forms has been inextricably connected to the human condition from the start, in relatively increasing measure. Hopefully, this can undo the misimpression—and perhaps allay fears—that our present massive use of virtuality threatens our mental health, social solidarity, or even our "humanity."

However, it is hugely important to emphasize that *by itself virtuality does not "define" what it is to be human*. We lived in the corpo*real* world in the past, we still do today, and almost certainly will continue to do so into the far future. Humans are born out of mother's womb and ultimately return to material dust or ashes; people ingest organically-based food (plant or animal); individuals make carnal love to, and with, flesh-and-bone bodies; we reside in materially-constructed structures; we are physically transported in, and by, material vehicles (or use our sinewy legs for locomotion); all people are affected by environmental dynamics—precipitation, earthquakes, pollution, even sunspot activity.

With so much diverse, almost limitless, physicality in our world, why then would we need or wish to indulge in virtuality? The several answers in Chap. 11 can be summed up thus: *the human race became too big for its own (cognitive) britches*. To raise the chances of survival in this physical world, our brain's intellectual capacity increased to the extent that it began to transcend the physical world's challenges, asking meta-questions and seeking non-material ways of dealing with our increasingly complex social environment.

As already explained in previous chapters, this is called evolutionary "pre-adaptation," "exaptation," "cooption," or "co-opted adaptation"[2]—entailing a change in a specific trait's function over time, from something necessary for survival to something less critical or even unnecessary, but still "useful." For instance, bird feathers evolved to regulate body heat, but then subsequently birds co-opted their feathers for use in mating display and still later they were exapted for flight! Several other examples were noted in Chaps. 4 (religion), and 6 (the arts).

Human intelligence(s) can be viewed as exaptation(s). Each organism evolves traits to ensure its survival. One early hominid example: we started developing "tools" to break hard nuts more easily or cut animal carcasses (and later to kill the latter without getting too close). At some point, such toolmaking became more sophisticated, enabling the creation of jewelry and crafts; from an initial functional purpose our brains exapted to create (virtual) art! Thus, human virtuality constitutes clever, productive, useful—and harmful—exaptations of human brain development. However, and this is a crucial point, *exaptations are no less "natural" than are the*

---

[2] Darwin used the term "pre-adaptation" but this suggested some sort of teleology i.e., a "pre-planned," evolutionary goal. Thus, Gould and Vrba invented the term "exaptation" with two forms of "cooption"; Buss et al. (1998) then introduced "co-opted adaptation."

*original traits.* Feathers are not (or no longer) a "secondary" feature of birds that fly. Similarly, humanity's ability and penchant to "virtualize" is not subsidiary; perhaps only "auxiliary," but the preponderance of evidence surveyed in this book more than suggests that it is deeply "supplementary" to human cognition.

This exaptative, supplementary character of virtuality also explains the difficulty we have in perceiving so many "normal" human activities as containing significant elements of virtuality. They come so naturally to us that we don't question their nature, except when a new form of virtual expression or activity appears. The novelty makes us stop and ask: "what is this?" Two opposing answers are possible.

On the one hand, late modern society is quite open to invention, novelty, paradigm breaking. When something new arrives on the scene we're open to accepting its "irregular" nature. For example, because we don't have much historical experience with an artificial entity such as the computer, we describe several of its components as "virtual." Conversely, people in the past were wary of anything new ("tradition" was a preeminent social value), predisposed to interpreting new expressions of what we today call "virtual" as constituting a continuation of what always existed. To take an already noted example: the late medieval concept of "incorporation" from which we derive the modern word "corporation." The corporation is a patently legal "fiction" or piece of "social artifice" [V13; V23; V25] designed to enable large scale economic activity. But the term itself contains the familiar, for the term "corporation" is derived from the Latin "corpus"—*body!* Such a lexical framing device masked the "non-corporeal" nature of this beast; consequently, society didn't view it as something "virtual."[3]

This is not to say that every new phenomenon is virtual: modern inventions such as the automobile, airplane, skyscraper, are certainly not virtual—although they *result* from our ability to mentally virtualize (combinatorial creativity, inquisitive discovery) and they also *influence* other fields of social virtuality. Furthermore, not every novel type of general phenomenon or specific example mentioned in this study is *ipso facto* virtual. By its very nature, virtuality is a somewhat amorphous concept—but "amorphous" does not mean "unreal," "nonexistent," "narrow," or lacking a common denominator. As I have argued, a common thread does exist—actually, two.

What holds "virtuality" together as a phenomenological idea are the warp of *immateriality* and the woof of *deficient verisimilitude* (compared to the real world). The first emanates from our internal world: what we mentally create, envision, imagine. On the list of 35 virtuality definitions, this general category appears more in the higher numbers (i.e., towards the bottom of the list): abstract, alien, imaginative, illusory, metaphysical, fantastical. The second is connected more to the external world, although here too occasionally a product of mental construction. These tend to be found near the top of the list: duplicate, incomplete, substitute, representative,

---

[3] Gunkel (2018) spends an entire book explicating (actually "deconstructing") several terms related to video games, avatars, virtual worlds, and the like. His analyses show the power inherent in the use of a specific terminology i.e., how it leads us to think in specific ways about these virtual phenomena: "different words reveal different things or different aspects of things" (p. 17).

description, simulation. Most of the definitions in the middle of the list incorporate elements of immateriality and of low-level verisimilitude: multi-dimensional,
incommensurable, deterritorialized, invisible, misperceived.

By looking not at specific definitions but at these two underlying categorizations,
it becomes possible to see why one can "lump together" such seemingly disparate
things as angels, literature, quarks, hallucinations, credit cards, nationhood, television, paintings etc., into the catch-all term "virtuality." *None of them existed before
humanity came on the scene*; they are constructs of our individual mind and/or social
interaction. In most instances, they are intrinsic products of our humanity; in a few
cases, they were tools that helped us evolve into *homo sapiens*.

Virtuality, then, is not merely an "ethereal" phenomenon somehow existing astride
our real life, but rather is part and parcel of human affairs. How? One must differentiate between the object or force that generates virtuality and its effects. For instance,
virtual pictures or text on a computer screen might consist of immaterial bytes, but
their influence on our behavior is as real as a punch in the face (e.g., cyber-bullying);
the completely abstract nature of a modern painting doesn't represent anything in the
world, but when we view such art, it forces us to think and at times even react in some
physical fashion (speech). To repeat Chap. 12's title: "Virtuality and Reality—Two
Sides of the Same Coin."

Of course, not all virtual phenomena have an equal *impact* on human affairs. In
addition, one can argue about the relative, *intrinsic* virtuality "strength" of each
type—as well as finding counterarguments to my sub-thesis that virtuality has
increased in quantity and quality over the course of human history.

Nevertheless, the overall bottom line seems beyond dispute. Given the mountain
of evidence in this book,[4] it is hard to deny the centrality of virtuality in the human
condition, past and present—heading towards the future. And if so, virtuality should
not be a major cause of worry, even if one reluctantly agrees with the extrapolation that
virtuality will continue to increase in quantity and quality into the foreseeable future.
Metaphorically and perhaps literally as well, humanity-and-virtuality constitute a
psycho-cultural twin of our bio-genetic double helix: inextricably intertwined and
constantly "progressing" in complexity.

Such has it always been and (most probably) so will it continue to be. You can be
virtually certain of that.

---

[4] Once again, I note that despite the numerous areas of life and fields of endeavor covered in this
book, the treatment is far from complete. Other areas worth considering (some briefly mentioned
in the book) and expanding: 1—Language: reflection of virtuality in the way we speak e.g., "out
of sight, out of mind"; "absence makes the heart grow fonder"; "mind over matter." 2—Ethics:
social mores and norms that internally guide our real-world behavior. 3—Advertising: image and
information manipulation of consumers—what you buy is not really what's promised. 4—Sports:
(a) As virtual war e.g., chess (military pieces); football (conquering territory); javelin throw. (b) As
sublimated inter-personal violence e.g., judo, karate, fencing.

# Appendix
# A Taxonomy of Virtuality

1. **"duplicate/clone"**: a near/perfect imitation of a real object e.g., android, hologram, and biological clone (genetically almost indistinguishable from the "parent").
2. **"action-at-a-distance"/"telepresence"**: modern communication technologies enable us to be physically located in one place and yet directly act upon a subject in another place.
3. **"distanced"/"media(ted)"/"presence"**: as opposed to face-to-face, any communication between people (or from a non-human source to a human) that has to go through some form of technology, whether in close proximity (presence) or from afar (distanced).
4. **"incomplete"**: not having all the qualities or elements of the original.
5. **"distorted"**: appearing in different form than in its original incarnation.
6. **"substitution"/"similarity"**: something or someone that takes the place of, or acts as a stand-in for, the original. This can also be a "perfect" forgery.
7. **"three-dimensional, representational object"**: a (relatively) precise rendition of an original object or human interaction.
8. **"two-dimensional, representational image"**: accurately portraying reality but missing a dimension.
9. **"description"**: oral or written depiction of a real situation, event, character etc., e.g., a history book; a storyteller.
10. **"simulation"**: replication of a dynamic scenario, a real-life interaction, and/or processes in the physical and social spheres.
11. **"remembrance"/"associative memory"**: the "virtual" can be anything with an inherent potentiality to bring up memories through a sensory immersion in the real.
12. **"simulacrum"**: substituting signs-of-the-real for the real—neutralizing the essential meaning of the original.
13. **"artificial"/"synthetic"**: something made by human hands (or technology) in a way significantly different from the way it naturally exists. "Artificial" = ersatz (e.g., plastic "diamond"); "Synthetic" = a meaningful version of an original (e.g., a diamond, lab-manufactured under high pressure).

14. **"multi-dimensional"**: something existing in more than three dimensions.
15. **"indeterminate"/"ambiguous"**: while these two meanings are not identical, they are similar enough to constitute one definition. "Indeterminate" is something that we know exists but find difficult to decide what it is qualitatively or how much it is quantitatively. "Ambiguous" means unclear—something that could be one or the other "thing" (answer, interpretation, meaning). Both can be a function of constant flux in the world.
16. **"incommensurable"**: that which cannot be counted or measured *in principle*.
17. **"deterritorialized"/"cyberspace"**: "deterritorialized" = not taking up space, not occupying a specific point in space; "cyberspace" = a specific type of "nonspace" that in fact actually does occupy a space (servers, computers).
18. **"atemporal"**: an indeterminate amount of *time*.
19. **"not sensed"/"invisible"/"null"**: real and physical but too small, too large, too fast to be perceived by our unaided senses—or simply invisible. This can include "nothing" (the number zero, musical pause etc.).
20. **"mis-perceived"/"illusion"**: perceived as real and physical but no longer in existence—or existing in different form from the way it is sensed.
21. **"over-sensed"**: having perceptions that are beyond the norm.
22. **"quasi-representational"**: similar to something real but in changed form or essence.
23. **"abstract"**: non-representational but vaguely suggesting something familiar and real.
24. **"symbolic"/"signs"**: having no intrinsic significance other than what humans have arbitrarily assigned to it.
25. **"incorporeal"/"spectral"**: non-physical "substance" but within a physical framework and impacting the real world.
26. **"potential"/"latent"**: In Aristotle's scheme of things, entities are either actual or potential i.e., every type of existence is an actualization of a potentiality.
27. **"non-actualized"**: something that *could* have happened but didn't.
28. **"imaginative"**: significantly different from what exists but still theoretically possible.
29. **"alien"**: a foreign being that might exist but who/which is so different from any form of life that we know that we might not be able to even recognize that it exists.
30. **"ideal"**: a Platonic conception of the best form of any object, as in "ideal-type."
31. **"mistaken"/"illusory"**: something that people understand to be true (in the case of "illusory"—attainable) but in fact it is not.
32. **"false"/"fictitious"/"fake"**: anything that attempts to give the impression of truth but is known or should be known (by the communicator) to be untrue.
33. **"metaphysical"**: beyond physical or corporeal matter; might—or might not—be real, but it's impossible to find direct physical evidence of the actual existence of anything metaphysical.
34. **"other-worldly"**: physically (not spiritually, as in "metaphysical") beyond our world e.g., Multiverse.
35. **"fantastical"**: impossible in principle/theory, at least insofar as present-day science and knowledge understands the world.

# References

50 ideas to change science: Neuroscience (Top-down processing) (2010, October 6). *NewScientist.com*. http://www.newscientist.com/article/mg20827815.700-50-ideas-to-change-science-neuroscience.html?full=true

100 best novels (2000). *The modern library*. http://www.modernlibrary.com/top-100/100-best-novels/

Aaronovitch, D. (2009). *Voodoo histories: The role of the conspiracy theory in shaping modern history*. Jonathan Cape.

Abbott, E. A. (1884). *Flatland: A romance of many dimensions*. Seeley & Co.

Ackerman, D. (1991). *A natural history of the senses* (4th ed.). Vintage Books.

AI can invent—Australia is first to recognise non-human inventorship (2021, August 3). *The National Law Review, XI*(215). https://www.natlawreview.com/article/ai-can-invent-australia-first-to-recognise-non-human-inventorship

Allen-Robertson, J. (2017). The materiality of digital media: The hard disk drive, phonograph, magnetic tape and optical media in technical close-up. *New Media & Society, 19*(3), 455–470. http://doi.org/10.1177/1461444815606368

Alzheimer's disease facts and figures 2021 (2021). Alzheimer's Association. https://www.alz.org/media/documents/alzheimers-facts-and-figures.pdf

Ananthaswamy, A. (2010a, March 17). Firing on all neurons: Where consciousness comes from. *New Scientist*. http://www.newscientist.com/article/mg20527520.400-firing-on-all-neurons-where-consciousness-comes-from.html?page=1

Ananthaswamy, A. (2010b, March 24). Mind over matter? How your body does your thinking. *New Scientist*. http://www.newscientist.com/article/mg20527535.100-mind-over-matter-how-your-body-does-your-thinking.html?

Anderson, B. (1991). *Imagined communities: Reflections on the origin and spread of nationalism* (rev. and extended ed.). Verso.

Anderson, C. A., Shibuya, A., Ihori, N., Swing, E. L., Bushman, B. J., Sakamoto, A., Rothstein, H. R., & Saleem, M. (2010). Violent video game effects on aggression, empathy, and prosocial behavior in eastern and western countries: A meta-analytic review. *Psychological Bulletin, 136*(2), 151–173. https://pubmed.ncbi.nlm.nih.gov/20192553/

Anderson, P. W. (2005). What do you believe is true even though you cannot prove it? https://www.edge.org/response-detail/11811

Angier, N. (2008, April 1). Blind to change, even as it stares us in the face. *The New York Times*. http://www.nytimes.com/2008/04/01/science/01angi.html?scp=1&sq=Natalie%20Angier,%20Blind%20to%20Change&st=cse

Anitei, S. (2007). 10 amazing facts about human speech. *Softpedia*. http://news.softpedia.com/news/10-Amazing-Facts-About-Human-Speech-70397.shtml

Apollinaire, G. (1971). *Selected writings of Guillaume Apollinaire* (R. Shattuck, Trans.). New Directions.

Applebaum, W. (1996). Keplerian astronomy after Kepler: Researches and problems. *History of Science, 34*, 451–504.

Aristotle (1984). *The politics* (C. Lord, Trans.). Chicago University Press.

Arnoldi, J. (2004). Derivatives: Virtual values and real risks. *Theory, Culture & Society, 21*(6), 23–42.

Art forgery (2021, July 22). *Wikipedia.* http://en.wikipedia.org/wiki/Art_forgery

Art in the 'fourth dimension' (2004). *The telematic society.* https://www.artpool.hu/2004/4D-links.html

Artificial life forged in a lab? Scientists create synthetic cell that grows and divides normally (2021, March 31). *SciTechDaily.* https://scitechdaily.com/artificial-life-forged-in-a-lab-scientists-create-synthetic-cell-that-grows-and-divides-normally/

Atran, S. (2002). *In gods we trust: The evolutionary landscape of religion.* Oxford University Press.

Automating programming: AI is transforming the coding of computer programs (2021, July 10). *The Economist.* https://www.economist.com/science-and-technology/2021/07/07/ai-is-transforming-the-coding-of-computer-programs

Axe, D. (2008, August 20). Army eyes invisibility cloak. *WIRED.com.* http://www.wired.com/dangerroom/2008/08/army-eyes-invis/

Bainbridge, W. S. (2010). *Online worlds: Convergence of the real and the virtual.* Springer.

Ball, P. (2010, November 4). Model citizens: Building SimEarth. *NewScientist.com.* http://www.newscientist.com/article/mg20827841.500-model-citizens-building-simearth.html?

Ball, P. (2021, June 7). Replotting the human: The thorny ethics of growing babies outside the womb. *Prospect Magazine.* https://www.prospectmagazine.co.uk/magazine/replotting-the-human-the-thorny-ethics-of-growing-babies-outside-the-womb

Barrett, J. L. (2004). *Why would anyone believe in god?* AltaMira Press.

Barth, J. (1994, September). Virtuality. *Johns Hopkins Magazine.* http://www.jhu.edu/~jhumag/994web/culture1.html

Battersby, S. (2008, November 20). It's confirmed: Matter is merely vacuum fluctuations. *New Scientist.* http://www.newscientist.com/article/dn16095-its-confirmed-matter-is-merely-vacuum-fluctuations.html

Baudrillard, J. (1994). *Simulacra and simulation* (S. F. Glaser, Trans.). University of Michigan Press.

Baudrillard, J. (1995). The virtual illusion: Or the automatic writing of the world (Interview by R. Boyne & S. Lash). *Theory, Culture & Society, 12*(4), 79–95, 97–107.

Bechtel, W., & Richardson, R. C. (1992). Emergent phenomena and complex systems. In A. Beckermann, H. Flohr & J. Kim (Eds.), *Emergence or reduction? Essays on the prospects of nonreductive physicalism* (pp. 257–288). Walter de Gruyter Verlag.

Begault, D. R. (1994). *3D sound for virtual reality and multimedia.* Academic Press Limited.

Behrent, M. C. (2013). Foucault and technology. *History and Technology, 29*(1), 54–104. https://doi.org/10.1080/07341512.2013.780351

Bell, G., & Gemmell, J. (2009). *Total recall: How the e-memory revolution will change everything.* Dutton Adult.

Belluck, P. (2021, July 14). Tapping into the brain to help a paralyzed man speak. *The New York Times.* https://www.nytimes.com/2021/07/14/health/speech-brain-implant-computer.html

Bendel, O. (2021). Love dolls and sex robots in unproven and unexplored fields of application. *Paladyn, Journal of Behavioral Robotics, 12*, 1–12. https://doi.org/10.1515/pjbr-2021-0004/html

Benjamin, W. (1985). *One-way street and other writings* (E. Jephcott & K. Shorter, Trans.). Verso.

Benjamin, W. (1996). In M. Bullock & M. W. Jennings (Eds.), *Selected writings.* Belknap Press.

Berger, K. (2019, February 7). Gustav Klimt in the brain lab. *Nautilus.* http://nautil.us/issue/69/patterns/gustav-klimt-in-the-brain-lab

Berger, P. L. (1969). *A rumor of angels: Modern society and the rediscovery of the supernatural.* Doubleday.

Bergson, H. (1991). *Matter and memory* (N. M. Paul & W. S. Palmer, Trans.). Zone (Original work published 1896).

Bering, J. (2010). *The god instinct: The psychology of souls, destiny and the meaning of life.* Nicholas Brealey Publishing.

Berkeley, G. (2003). *A treatise concerning the principles of human knowledge* (Gutenberg Project e-text edition) (Original work published 1710). https://www.gutenberg.org/files/4723/4723-h/4723-h.htm

Berns, L. (1976). Rational animal-political animal: Nature and convention in human speech and politics. *The Review of Politics, 38*(2), 177–189. http://www.jstor.org/stable/1405935

Bhattacharjee, Y., & Winters, D. (2017, May 18). Why we lie: The science behind our deceptive ways. *nationalgeographic.com.* www.nationalgeographic.com/magazine/2017/06/lying-hoax-false-fibs-science

Biocca, F. (1997). The cyborg's dilemma: Progressive embodiment in virtual environments. *Journal of Computer-Mediated Communication, 3*(2). http://doi.org/10.1111/j.1083-6101.1997.tb00070.x

Biological joints could replace artificial joints soon (2011, January 5). *PhysOrg.com.* http://www.physorg.com/news/2011-01-biological-joints-artificial.html

Bionanotechnology has new face, world-class future (2010, April 19). *PhysOrg.com.* http://www.physorg.com/news190899969.html

Birds' eye view is far more colorful than our own (2011, June 23). *PhysOrg.com.* http://www.physorg.com/news/2011-06-birds-eye-view.html

Birkerts, S. (2010, Spring). Reading in a digital age. *The American Scholar.org.* http://www.theamericanscholar.org/reading-in-a-digital-age

Blackmore, S. (1999). *The meme machine.* Oxford University Press.

Blake, W. (1906). *The marriage of heaven and hell.* John W. Luce & Company (Original work published 1793). https://www.gutenberg.org/files/45315/45315-h/45315-h.htm

Blascovich, J., & Bailenson, J. (2011). *Infinite reality: Avatars, eternal life, new worlds, and the dawn of the virtual revolution.* William Morrow.

Bloom, H. (1998). *Shakespeare: The invention of the human.* Riverhead Books.

Bloom, P. (2010). *How pleasure works: The new science of why we like what we like.* The Bodley Head.

Blu Buhs, J. (2009). *Bigfoot: The life and times of a legend.* University of Chicago Press.

Blum, D. (2006, December 30). Ghosts in the machine. *The New York Times.* http://www.nytimes.com/2006/12/30/opinion/30blum.html

Blume, M. (2009). The reproductive benefits of religious affiliation. In E. Voland & W. Schuefenhovel (Eds.), *The biological evolution of religious mind and behavior* (pp. 117–126). Springer.

Boellstorff, T. (2008). *Coming of age in second life: An anthropologist explores the virtually human.* Princeton University Press.

Booth, W. C. (1991). *The rhetoric of fiction* (2nd ed.). Penguin.

Bostrom, N. (2003). Are you living in a computer simulation? *Philosophical Quarterly, 53*(211), 243–255.

Bowley, G. (2011, January 2). The new speed of money, reshaping markets. *The New York Times.* http://www.nytimes.com/2011/01/02/business/02speed.html?_r=1&nl=todaysheadlines&emc=globaleua24

Boyd, B. (2009). *On the origin of stories: Evolution, cognition, and fiction.* Belknap Press of Harvard University Press.

Boyer, P. (2001). *Religion explained: The evolutionary origins of religious thought.* Basic Books.

Bracken, C. C., & Skalski, P. (Eds.). (2010). *Immersed in media: Telepresence in everyday life.* Routledge.

Brady, T., & Schachner, A. (2008, December 16). Blurring the boundary between perception and memory. *Scientific American.* http://www.scientificamerican.com/article.cfm?id=perception-and-memory

Brahic, C. (2010, August 4). Artificial life forms evolve basic intelligence. *NewScientist.com*. http://www.newscientist.com/article/mg20727723.700-artificial-life-forms-evolve-basic-intell igence.html

Braudel, F. (1996). *The Mediterranean and the Mediterranean world in the age of Philip II*. University of California Press.

Brooks, M. (2010, September 8). Laws of physics may change across the universe. *New Scientist, 2777*. http://www.newscientist.com/article/dn19429-laws-of-physics-may-change-acr oss-the-universe.html

Brooks, M. (2011, October 6). Quantum life: The weirdness inside us. *New Scientist, 2832*. http:// www.newscientist.com/article/mg21128321.500-quantum-life-the-weirdness-inside-us.html

Brooks, M. (2015, December 2). Is quantum physics behind your brain's ability to think? *New Scientist, 3050*. https://www.newscientist.com/article/mg22830500-300-is-quantum-physics-behind-your-brains-ability-to-think/

Bryson, N. (1983). *Vision and painting: The logic of the gaze*. Yale University Press.

Buchanan, M. (2011). Quantum minds: Why we think like quarks. *New Scientist, 2828*. http:// www.newscientist.com/article/mg21128285.900-quantum-minds-why-we-think-like-quarks. html?DCMP=NLC-nletter&nsref=mg21128285.900

Buhrman, H., Cleve, R., & Van Dam, W. (2001). Quantum entanglement and communication complexity. *SIAM Journal of Computing, 30*(6), 1829–1841. http://epubs.siam.org/sicomp/res ource/1/smjcat/v30/i6/p1829_s1

Burns, H. (2020, November 19). Sr. Helena Burns: Virtual reality and theology of the body. *The Catholic Register*. https://www.catholicregister.org/opinion/columnists/item/32396-sr-hel ena-burns-virtual-reality-and-theology-of-the-body

Buss, D. M., Haselton, M. G., Shackelford, T. K., Bleske, A. L., & Wakefield, J. C. (1998). Adaptations, exaptations, and spandrels. *American Psychologist, 53*(5), 533–548. http://www.sscnet. ucla.edu/comm/haselton/webdocs/spandrels.html

Cain, F. (2009, January 28). How many stars? *Universe Today*. http://www.universetoday.com/ 24328/how-many-stars/

Calleja, G. (2010). Digital games and escapism. *Games and Culture, 5*(4), 335–353.

Callisen, C. T., & Adkins, B. (2011). Pre digital virtuality: Early modern scholars and the republic of letters. In D. W. Park, N. W. Jankowski & S. Jones (Eds.), *The long history of new media: Technology, historiography, and contextualizing newness* (pp. 55–72). Peter Lang.

Camouflage. (2021, August 9). *Wikipedia*. https://en.wikipedia.org/wiki/Camouflage#Non-vision_ camouflage

Campbell, C. (1987). *The romantic ethic and the spirit of modern consumerism*. Blackwell-IDEAS.

Campbell, M. (2011, January 5). Game on: When work becomes play. *NewScientist*. https://www. newscientist.com/article/mg20927940-300-game-on-when-work-becomes-play/

Campion, N. (2008). *The dawn of astrology—A cultural history of western astrology* (Vol. 1: The Ancient and Classical Worlds). Continuum.

Cantor, G. (1891). Ueber eine elementare Frage derMannigfaltigkeitslehre [On an elementary question of Manifold Theory]. *Jahresbericht der Deutschen Mathematiker-Vereinigung, 1*, 75–78. https://gdz.sub.uni-goettingen.de/download/pdf/PPN37721857X_0001/LOG_0029.pdf

Carmeli, O. (2017, May 18). The physicist who denies dark matter. *Nautilus*. http://nautil.us/issue/ 48/chaos/the-physicist-who-denies-dark-matter?

Carpenter, S. (2010, December 23). Body of thought: How trivial sensations can influence reasoning, social judgment and perception. *Scientific American*. http://www.scientificamerican.com/article. cfm?id=body-of-thought&WT.mc_id=SA_CAT_physics_20101224

Carr, N. (2010). *The shallows: What the internet is doing to our brains*. W.W. Norton & Co.

Carruthers, P. (2017). The illusion of conscious thought. *Journal of Consciousness Studies, 24*(9–10), 228–252.

Casey, J. (2009). *After lives: A guide to heaven, hell, and purgatory*. Oxford University Press.

Castells, M. (1996). *The rise of networked societies* (Vol. 1). Blackwell Publishing.

Casti, J. L. (1997). *Would-be worlds: How simulation is changing the frontiers of science*. Wiley.

Castronova, E. (2003). On virtual economies. *Game Studies, 3*(2). http://www.gamestudies.org/0302/castronova

*Catechism of the Catholic Church* (n.d.). https://www.vatican.va/archive/ENG0015/_INDEX.HTM

Cavell, S. (1971). *The world viewed*. Viking.

Cepelewicz, J. (2019, September 30). Your brain chooses what to let you see. *QuantaMagazine*. https://www.quantamagazine.org/your-brain-chooses-what-to-let-you-see-20190930/

Cerf, M., Thiruvengadam, N., Mormann, F., Kraskov, A., Quiroga, R.Q., Koch, C., & Fried, Y. (2010). On-line, voluntary control of human temporal lobe neurons. *Nature, 467*, 1104–1108. http://www.nature.com/nature/journal/v467/n7319/full/nature09510.html

Chabris, C., & Simons, D. (2010). *The invisible gorilla and other ways our intuitions deceive us*. Crown Archetype.

Chang, L. (2006). *Wisdom for the soul: Five millennia of prescriptions for spiritual healing*. Gnosophia Publishers.

Changizi, M. (2011). *Harnessed: How language and music mimicked nature and transformed ape to man*. BenBella Books.

Chayka, K. (2021, July–August). The very online novel: Who's afraid of internet writing? *Colloquy*, 58–60.

Chen, Y., Wang, Q., Chen, H., Song, X., Tang, H., & Tian, M. (2019). An overview of augmented reality technology. *Journal of Physics: Conference Series, 1237*. http://doi.org/10.1088/1742-6596/1237/2/022082

Cheng, E. (2021, January 27). China reports a growing shortage of factory workers. *CNBC*. https://www.cnbc.com/2021/01/27/china-reports-a-growing-shortage-of-factory-workers.html

Chirimuuta, M. (2019, May). The reality of color is perception: An argument for a new definition of color. *Aeon*. http://wise.nautil.us/feature/384/the-reality-of-color-is-perception?mc_cid=46d3631354&mc_eid=222ed6e6fb

Choi, C. Q. (2010, April 29). Seeing the closest aliens will take centuries. *PhyOrg.com*. http://www.physorg.com/news191776050.html

Chomsky, N. (1957). *Syntactic structures*. Mouton & Co.

Chorost, M. (2011). *World wide mind: The coming integration of humanity, machines, and the internet*. Free Press.

Chudoba, K. M., Wynn, E., Lu, M., & Watson-Manheim, M. B. (2005). How virtual are we? Measuring virtuality and understanding its impact in a global organization. *Information Systems Journal, 15*, 279–306. https://web.archive.org/web/20060511074429/http://www.ariadne.bz.it/documenti/pdf/how_virtual.pdf

*Citizens United v. Federal Election Commission* (2010). 558 U.S. 50.

Clark, A. (2008). *Supersizing the mind: Embodiment, action, and cognitive extension*. Oxford University Press.

Clem, R. L., & Huganir, R. L. (2010). Calcium-permeable AMPA receptor dynamics mediate fear memory erasure. *Science, 330*(6007), 1108–1112. http://www.sciencemag.org/content/330/6007/1108.short

Coeckelbergh, M. (2010a). Moral appearances: Emotions, robots, and human morality. *Ethics & Information Technology, 12*, 235–241. https://doi.org/10.1007/s10676-010-9221-y

Coeckelbergh, M. (2010b). Robot rights? Towards a social-relational justification of moral consideration. *Ethics & Information Technology, 12*, 209–221. https://doi.org/10.1007/s10676-010-9235-5

Coghlan, A. (2017, March 9). Sneaky beetles evolved disguise to look like ants, then eat them. *New Scientist*. https://www.newscientist.com/article/2124050-sneaky-beetles-evolved-disguise-to-look-like-ants-then-eat-them

Cohen, N. (2020, January 4). Research continues showing gains in DNA computing. *TechXplore*. https://techxplore.com/news/2020-01-gains-dna.html

Cohen, P. (2010, March 31). Next big thing in English: Knowing they know you know. *The New York Times*. http://www.nytimes.com/2010/04/01/books/01lit.html

Cohen, S., & Taylor, L. (2007). *Escape attempts: The theory and practice of resistance to everyday life* (Rev. ed.). Taylor & Francis.

Conard, N. J., Malina, M., & Münzel, S. C. (2009). New flutes document the earliest musical tradition in southwestern Germany. *Nature, 460*, 737–740. http://www.nature.com/nature/journal/v460/n7256/full/nature08169.html

Conti, R., & Schmidt, J. (2021, May 14). What you need to know about non-fungible tokens (NFTs). *Forbes.* https://www.forbes.com/advisor/investing/nft-non-fungible-token/

Coogan, P. (2006). *Superhero: The secret origin of a genre.* MonkeyBrain Books.

Cooper, D. (1970). *The Cubist epoch.* Phaidon.

Copenhaver, B. (2015). *The book of magic: From antiquity to the enlightenment.* Penguin.

Corballis, M. C. (2011). *The recursive mind: The origins of human language, thought, and civilization.* Princeton University Press.

Cottier, C. (2021, July/August). Where did music come from? *DISCOVER Magazine*, 16–17.

Could humans be infected by 'computer viruses'? (2010, May 26). *ScienceDaily.* https://www.sciencedaily.com/releases/2010/05/100526095830.htm

Cunliffe, R. (2021, July 28). Timothy Gowers: The man who changed Dominic Cummings's mind on Covid-19. *NewStatesman.* https://www.newstatesman.com/politics/uk/2021/07/timothy-gowers-man-who-changed-dominic-cummings-s-mind-covid-19

Dali, S. (1942). *The secret life of Salvador Dali.* Dial Press.

Damasio, A. R. (1994). *Descartes' error: Emotion, reason and the human brain.* Putnam.

Darwin, C. (1883). *The descent of man and selection in relation to sex.* Appleton and Co.

Davies, P. (2010). Universe from bit. In P. Davies & N. H. Gregersen (Eds.), *Information and the nature of reality: From physics to metaphysics* (pp. 65–91). Cambridge University Press.

Davies, S. (1991). *Definitions of art.* Cornell University Press.

Davies, W. (2017a, January 19). How statistics lost their power—And why we should fear what comes next. *The Guardian.* https://www.theguardian.com/politics/2017/jan/19/crisis-of-statistics-big-data-democracy

Davis, W. (2017b). *Visuality and virtuality: Images and pictures from prehistory to perspective.* Princeton University Press.

Dawkins, R. (1998). *Unweaving the rainbow: Science, delusion and the appetite for wonder.* Houghton Mifflin Harcourt.

Dawson, L. L., & Cowan, D. E. (Eds.). (2004). *Religion online: Finding faith on the internet.* Routledge.

De Brigand, F. (2020, July 20). Nostalgia reimagined. *Aeon.* https://aeon.co/essays/nostalgia-doesnt-need-real-memories-an-imagined-past-works-as-well

de Chardin, P. T. (1964). *The future of man* (N. Denny, Trans.). Harper & Row.

de Jesus, A. (2018, December 12). Augmented reality shopping and artificial intelligence—Near-term applications. *Emerj.* https://emerj.com/ai-sector-overviews/augmented-reality-shopping-and-artificial-intelligence/

de Souza e Silva, A., & Sutko, D. M. (2011). Placing location-aware media in a history of the virtual. In D. W. Park, N. W. Jankowski & S. Jones (Eds.), *The long history of new media: Technology, historiography, and contextualizing newness* (pp. 299–315). Peter Lang.

De Vynck, G. (2021, July 7). The U.S. says humans will always be in control of AI weapons. But the age of autonomous war is already here. *The Washington Post.* https://www.washingtonpost.com/technology/2021/07/07/ai-weapons-us-military/

de Waal, F. (2009, November 19). Monkey see, monkey do, monkey connect. *Discover.* https://www.discovermagazine.com/mind/monkey-see-monkey-do-monkey-connect

Dean, T. (2007). A review of the Drake equation. *Cosmos, 14.* https://web.archive.org/web/20121208043158/http://www.cosmosmagazine.com:80/features/print/2915/are-we-alone

Deep sea hydrothermal vents (2010). *Extreme Science.* https://web.archive.org/web/20110828072106/http://www.extremescience.com:80/zoom/index.php/life-in-the-deep-ocean/42-deep-sea-hydrothermal-vents

Deleuze, G. (1994). *Difference and repetition* (P. Patton, Trans.). Columbia University Press.

Dertouzos, M. (1997). *What will be: How the new world of information will change our lives.* HarperOne.

Descartes, R. (1911). Meditations on the first philosophy. In *The philosophical works of Descartes* (Vol. 1; E. S. Haldane & G. R. T. Ross, Trans.). Cambridge University Press (Reprinted with corrections, 1931).

Descartes, R. (2008). *Discourse on method.* Pomona Press (Original work published 1637).

d'Espagnat, B. (1979, November). The quantum theory and reality. *Scientific American, 241*(5), 158–181. http://www.scientificamerican.com/media/pdf/197911_0158.pdf

Deutsch, D. (1998). *The fabric of reality: The science of parallel universes and its implications.* Penguin.

Deutsch, D. (2017). What scientific term or concept ought to be more widely known? Illusory conjunction. *Edge.* https://www.edge.org/response-detail/27027

Diamond, M., Jozifkova, E., & Weiss, P. (2011, October). Pornography and sex crimes in the Czech Republic. *Archives of Sexual Behavior, 40*(5), 1037–1043. https://pubmed.ncbi.nlm.nih.gov/211 16701/

Dibbell, J. (2006). *Play money: Or, how I quit my day job and made millions trading virtual loot.* Basic Books.

Digital payments worldwide (2020). *statista.* https://www.statista.com/outlook/296/100/digital-pay ments/worldwide

Dijkstra, N. (2021, July 28). The fine line between reality and imaginary. *Nautilus.* https://nautil. us/issue/104/harmony/the-fine-line-between-reality-and-imaginary

Doel, M. A., & Clarke, D. B. (1999). Virtual worlds: Simulation, suppletion, s(ed)uction and simulacra. In M. Crang, P. Crang & J. May (Eds.), *Virtual geographies: Bodies, space and relations* (pp. 261–283). Routledge.

Doidge, N. (2015). *The brain's way of healing: Remarkable discoveries and recoveries from the frontiers of neuroplasticity.* Viking Press.

Doty, R. L. (2007). Smelling and tasting problems. *The Dana guide to brain health.* https://web.arc hive.org/web/20101214045839/http://dana.org/news/brainhealth/detail.aspx?id=9876

Drachmann, A. B. (1922). *Atheism in pagan antiquity.* Christiana.

Dutta-Bergman, M. J. (2004). Complementarity in consumption of news types across traditional and new media. *Journal of Broadcasting and Electronic Media, 48*, 41–60.

Dutton, D. (2009). *The art instinct: Beauty, pleasure & human evolution.* Bloomsbury Press.

Earleywine, M. (Ed.). (2005). *Mind altering drugs: The science of subjective experience.* Oxford University Press.

Ecklund, E. H., & Long, E. (2011). Scientists and spirituality. *Sociology of Religion, 72*(3), 253–274.

Eco, U. (1986). *Travels in hyper reality* (W. Weaver, Trans.). Harcourt, Inc.

Edinger, G., Krausz, G., Groengrass, C., Holzner, C., & Slater, M. (2009). Brain-computer interfaces for virtual environmental control. In *13th International Conference on Biomedical Engineering, IFMBE Proceedings* (Vol. 23, No. 1, pp. 366–369). https://doi.org/10.1007/978-3-540-92841-6_90

Edmonson, M. E. (1971). *LORE: An introduction to the science of folklore.* Holt, Rinehart & Winston.

Edson, E. (2001). Bibliographic essay: History of cartography. *CHOICE: Current Reviews for Academic Libraries, 38*(11/12), 1899–1909.

Edwards, E. D. (2005). *Metaphysical media: The occult experience in popular culture.* Southern Illinois University Press.

Einstein, A. (1956). *Out of my later years.* Philosophical Library.

Einstein, A. (1971). *The Born-Einstein letters; Correspondence between Albert Einstein and Max and Hedwig Born from 1916 to 1955* (Letter from Einstein to Max Born, 1947, March 3). Walker.

Eisenberg, A. (2010, November 7). When a camcorder becomes a life partner. *The New York Times.* http://www.nytimes.com/2010/11/07/business/07novel.html?_r=1&nl=technology& emc=techupdateema3

Eisenstein, E. L. (1983). *The printing revolution in early modern Europe*. Cambridge University Press.

Eliot, T. S. (1943). *Four quartets*. Harcourt. http://www.davidgorman.com/4quartets/

Ellis, E. G. (2019, August 23). You are already having sex with robots. *Wired*. https://www.wired.com/story/you-are-already-having-sex-with-robots/

Engelking, C. (2015, August 19). Your color perception changes with the seasons. *Discover*. https://www.discovermagazine.com/mind/your-color-perception-changes-with-the-seasons

Engler, B., & Müller, K. (Eds.). (1994). *Historiographic metafiction in modern American and Canadian literature*. Schöningh.

Erickson, P. (2003). Help or hindrance? The history of the book and electronic media. In D. Thornburn & H. Jenkins (Eds.), *Rethinking media change: The aesthetics of transition* (pp. 95–116). MIT Press.

Esterow, M. (2021, September 1). Joining plastic, glass and metal on the recycle list: Fake art. *The New York Times*. https://www.nytimes.com/2021/09/01/arts/design/fake-art-recycled.html

Eucharistic theology (2021, July 21). *Wikipedia*. https://en.wikipedia.org/wiki/Eucharistic_theology

Farman, J. (2010). Mapping the digital empire: Google Earth and the process of postmodern cartography. *New Media & Society, 12*(6), 869–888.

Feeney, S. M., Johnson, M. C., Mortlock, D. J., & Peiris, H. V. (2011). First observational tests of eternal inflation: Analysis methods and WMAP 7-year results. *Physical Review D*. http://arxiv.org/abs/1012.3667v2?

Felton, D. (1998). *Haunted Greece and Rome: Ghost stories from classical antiquity*. University of Texas Press.

Fields, R. D. (2020, March 10). Mind reading and mind control technologies are coming. *Scientific American*. https://blogs.scientificamerican.com/observations/mind-reading-and-mind-control-technologies-are-coming/

Fingeroth, D. (2007). *Disguised as Clark Kent: Jews, comics, and the creation of the superhero*. Continuum.

Fisher, M. (2021, July 25). Disinformation for hire, a shadow industry, is quietly booming. *The New York Times*. https://www.nytimes.com/2021/07/25/world/europe/disinformation-social-media

Fiske, A. P. (1992). The four elementary forms of sociality: Framework for a unified theory of social relations. *Psychology Review, 99*, 689–723.

Fleischer, J. (2006, August). Superman's other secret identity. *World Jewish Digest*. http://www.jeffleischer.com/up-up-and-oy-vey

Floridi, L. (2007). A look into the future impact of ICT on our lives. *The Information Society, 23*(1), 59–64.

Flyn, C. (2021, July 20). What it's like to be an ant. *Prospect Magazine*. https://www.prospectmagazine.co.uk/magazine/what-its-like-to-be-an-ant-cal-flyn

Folger, T. (2018, November). Your daily dose of quantum. *DISCOVER Magazine*, 31–37.

Fox, D. (2008, November 5). The secret life of the brain. *NewScientist*. http://www.newscientist.com/article/mg20026811.500-the-secret-life-of-the-brain.html

Frank, A. (2010, April). Who wrote the book of physics? *Discover*, 32–37.

Frank, A. (2017). Minding matter. *Aeon*. https://aeon.co/essays/materialism-alone-cannot-explain-the-riddle-of-consciousness

Freud, S. (1928). *The future of an illusion* (W. D. Robson-Scott, Trans.). Hogarth Press.

Frey, T. (2011). The coming of terabytes: Lifeblogging for a living. *The Futurist, 45*(1), 35–36.

Fry, C. L. (2008). *Cinema of the occult: New age, satanism, wicca, and spiritualism in film*. Rosemont Publishing & Associated University Presses.

Gallagher, B. (2021, July 21). Reports of a baleful internet are greatly exaggerated. *Nautilus*. https://nautil.us/issue/104/harmony/reports-of-a-baleful-internet-are-greatly-exaggerated

Gallagher, J. (2018, April 10). More than half your body is not human. *BBC News*. https://www.bbc.com/news/health-43674270

Gallagher, W. (2009). *Rapt: Attention and the focused life*. Penguin Press.

Gardels, N. (2021, September 3). Planetary politics when the nation-state falters. *NOEMA*. https://www.noemamag.com/planetary-politics-when-the-nation-state-falters/

Gardner, H. (1983). *Frames of mind: The theory of multiple intelligences*. Basic Books.

Garvey, J. (2010, October 25). Hacker's challenge. *TPM: The Philosopher's Magazine, 51*. https://integral-options.blogspot.com/2010/11/peter-hacker-tells-james-garvey-that.html

Gatto, K. (2011, April 19). A virtual reality scent system that fools human taste. *PhysOrg.com*. http://www.physorg.com/news/2011-04-virtual-reality-scent-human.html

Gaudiosi, J. (2017, September 13). Virtual touch: Inside technology that makes VR feel real. *Rolling Stone*. http://www.rollingstone.com/culture/features/virtual-touch-inside-technology-that-makes-vr-feel-real-w438867

Gay, G. (2006). The role of social navigation and context in ubiquitous computing. In P. Messaris & L. Humphreys (Eds.), *Digital media: Transformations in human communication* (pp. 287–297). Peter Lang.

Gazzaniga, M. (2017, January 14). The schnitt. *Edge: What scientific term of concept ought to be more widely known?* https://www.edge.org/response-detail/27049

Geary, D. C., & Bjorklund, D. F. (2000). Evolutionary development psychology. *Child Development, 71*(1), 57–65. http://web.missouri.edu/~gearyd/EvoDevPsy.pdf

Gellner, E. (1965). *Thought and change*. Weidenfeld and Nicolson.

Gellner, E. (1983a). *Nationalism and the two forms of cohesion in complex societies*. British Academy.

Gellner, E. (1983b). *Nations and nationalism*. Cornell University Press.

Geraci, R. M. (2010). *Apocalyptic AI: Visions of heaven in robotics, artificial intelligence, and virtual reality*. Oxford University Press.

Gerrig, R. J. (1993). *Experiencing narrative worlds: On the psychological activities of reading*. Yale University Press.

Gershenfeld, N. (2005). *FAB: The coming revolution on your desktop—From personal computers to personal fabrication*. Basic Books.

Gerson, E. S., & Woolsey, B. (2009). The history of credit cards. *CreditCards.com*. http://www.creditcards.com/credit-card-news/credit-cards-history-1264.php

Getlen, L. (2017, January 7). How scientists actually could bring dinosaurs back to life. *New York Post*. http://nypost.com/2017/01/07/how-scientists-actually-could-bring-dinosaurs-back-to-life

Golding, D. (2021, July). The memory of perfection: Digital faces and nostalgic franchise cinema. *Convergence*. http://doi.org/10.1177/13548565211029406

Goldstein, J. (2020). *Money: The true story of a made-up thing*. Hachette Books.

Goldstein, J. S. (2011). *Winning the war on war: The decline of armed conflict worldwide*. Dutton Adult/Penguin Group.

Gomelsky, V. (2010, May 20). Illusions vs. reality in a world of gems. *The New York Times*. http://www.nytimes.com/2010/05/21/arts/21iht-acajpaste.html

Goodman, J. D. (2010, September 26). Learning to share, thanks to the web. *The New York Times*. http://www.nytimes.com/2010/09/26/weekinreview/26goodman.html?emc=tnt&tntemail1=y

Gopnik, A. (2008). Imagination is real. *Edge: World question center—"What have you changed your mind about?"*. https://web.archive.org/web/20190712130212/https://www.edge.org/q2008/q08_14.html

Gould, S. J., & Eldredge, N. (1977). Punctuated equilibria: The tempo and mode of evolution reconsidered. *Paleobiology, 3*(2), 115–151.

Gould, S. J., & Vrba, E. S. (1982). Exaptation—A missing term in the science of form. *Paleobiology, 8*(1), 4–15.

Grabiner, J. V. (1988). The centrality of mathematics in the history of western thought. *Mathematics Magazine, 61*(4), 220–229. http://www.jstor.org/pss/2689357

Graziano, M. (2013, December 18). Endless fun. *Aeon*. https://aeon.co/essays/the-virtual-afterlife-will-transform-humanity

Greenblatt, S. (2011, August 8). The answer man. *The New Yorker*, 28–33.

Greene, B. (2011). *The hidden reality: Parallel universes and the deep laws of the cosmos.* Alfred Knopf.

Grellard, C., & Robert, A. (Eds.). (2009). *Atomism in late medieval philosophy and theology.* Brill.

Grimes, W. (2011, April 23). Max Mathews, pioneer in making computer music, dies at 84. *The New York Times.* http://www.nytimes.com/2011/04/24/arts/music/max-mathews-father-of-computer-music-dies-at-84.html

Grimshaw, M., & Garner, T. (2015). *Sonic virtuality: Sound as emergent perception.* Oxford University Press.

Grosvenor, E. S., & Wesson, M. (1997). *Alexander Graham Bell: The life and times of the man who invented the telephone.* Abrams.

Grotius, H. (1814). *On the law of war and peace* (A. C. Campbell, Trans.) (Original work published 1625). https://books.google.co.il/books/about/The_Rights_of_War_and_Peace.html?id=MtM8AAAAYAAJ&printsec=frontcover&source=kp_read_button&hl=en&redir_esc=y#v=onepage&q&f=false

Grotstein, J. S. (2000). *Who is the dreamer who dreams the dream?* The Analytic Press.

Gunkel, D. J. (2010). The real problem: Avatars, metaphysics and online social interaction. *New Media & Society, 12*(1), 127–141.

Gunkel, D. J. (2018). *Gaming the system: Deconstructing video games, games studies, and virtual worlds.* Indiana University Press.

Gunkel, D. J. (2020). Brain–computer interface. In M. Filimowicz & V. Tzankova (Eds.), *Reimagining communication: Mediation* (Chap. 19). Routledge.

Guthrie, S. E. (1995). *Faces in the clouds: A new theory of religion.* Oxford University Press.

Hackett, R. I. J. (2006). Religion and the internet. *Diogenes, 211*, 67–76. https://www.academia.edu/37442959/Religion_and_the_Internet

Hairopoulos, K. (2009, July 20). Sports fans, say hello to huge-screen TV. *The Dallas Morning News.* https://web.archive.org/web/20100107024857/http://www.dallasnews.com/sharedcontent/dws/news/localnews/cowboysstadium/stories/072009dnspogoldcuptv.42bc674.html

Haldane, J. (2011, January). Philosophy lives. *First Things.* https://www.firstthings.com/article/2011/01/philosophy-lives

Haldane, J. B. S. (1928). *Possible worlds and other essays.* Chatto & Windus.

Han, B.-C. (2018, March 8). The copy is the original. *Aeon.* https://aeon.co/essays/why-in-china-and-japan-a-copy-is-just-as-good-as-an-original?

Hansell, S. (2006, March 12). As internet TV aims at niche audiences, the slivercast is born. *The New York Times.* http://www.nytimes.com/2006/03/12/business/yourmoney/12sliver.html?pagewanted=all

Hansen, M. B. N. (2006). *Bodies in code: Interfaces with digital media.* Routledge.

Hardwick, C. (2021, July 13). The post-pandemic church in the Great Virtuality. *The Presbyterian Outlook.* https://pres-outlook.org/2021/07/the-post-pandemic-church-in-the-great-virtuality/

Harris, I. (2011, Fall). Edmund Burke. In E. N. Zalta (Ed.), *The Stanford Encyclopedia of Philosophy.* http://plato.stanford.edu/archives/fall2011/entries/burke/

Harris, S., Kaplan, J. T., Curiel, A., Bookheimer, S. Y., Iacoboni, M., & Cohen, M. S. (2009). The neural correlates of religious and nonreligious belief. *PLoS One, 4*(10), e7272.

Hartley, J., & McWilliam, K. (Eds.). (2009). *Story circle: Digital storytelling around the world.* Wiley-Blackwell.

Hauser, M. D. (1996). *The evolution of communication.* MIT Press/Bradford Books.

Hawking, S. (1988). *A brief history of time: From the big bang to black holes.* Bantam Dell Publishing.

Hayles, N. K. (1999a). The condition of virtuality. In P. Lunenfeld (Ed.), *The digital dialectic: New essays on new media* (pp. 68–94). MIT Press.

Hayles, N. K. (1999b). *How we became posthuman: Virtual bodies in cybernetics, literature, and informatics.* University of Chicago Press.

Heck, Z. S., War, J., Yi, P., Borowick, J., & MacArthur, A. P. (2021). Actual legal questions now being asked in virtual worlds. *Corporate Law Advisory.* https://www.lexisnexis.com/communities/corporatecounselnewsletter/b/newsletter/archive/2017/03/07/actual-legal-questions-raised-by-virtual-reality-from-assault-to-virtual-property-rights-to-intellectual-property-concerns.aspx

Heeks, R. (2010, January 4). Gaming for profits: Real money from virtual worlds. *Scientific American.* http://www.scientificamerican.com/article.cfm?id=real-money-from-virtual-worlds

Heim, M. (1993). *The metaphysics of virtual reality.* Oxford University Press.

Heim, M. (1998). *Virtual realism.* Oxford University Press.

Helland, C. (2000). Religion online/online religion and virtual communitas. In J. K. Hadden & D. E. Cowan (Eds.), *Religion on the internet: Research prospects and promises* (pp. 205–224). JAI Press/Elsevier Service.

Henig, R. M. (2008, February 17). Taking play seriously. *The New York Times.* http://www.nytimes.com/2008/02/17/magazine/17play.html?pagewanted=all

Henshilwood, C. S., d'Errico, F., van Niekerk, K. L., Coquinot, Y., Jacobs, Z., Lauritzen, S.-E., Menu, M., & Garcia-Moreno, R. (2011). A 100,000-year-old ochre-processing workshop at Blombos Cave, South Africa. *Science, 334*(6053), 219–222. http://www.sciencemag.org/content/334/6053/219

Heraclitus. (2001). *Fragments—The collected wisdom of Heraclitus* (B. Haxton, Trans.). Viking.

Herbert, N. (1993). *Elemental mind: Human consciousness and the new physics.* Dutton Publishing.

Higham, J. P. (2018, February 6). The red and green specialists: Why human colour vision is so odd. *Aeon.* https://aeon.co/ideas/the-red-and-green-specialists-why-human-colour-vision-is-so-odd

Hill, A. (2011). *Paranormal media: Audiences, spirits and magic in popular culture.* Routledge.

Hill-Smith, C. (2009). Cyberpilgrimage: A study of authenticity, presence and meaning in online pilgrimage experiences. *Journal of Religion and Popular Culture, 21*(2). https://web.archive.org/web/20110922093541/http://www.usask.ca:80/relst/jrpc/art21(2)-Cyberpilgrimage.html

Hinton, C. H. (1980). What is the fourth dimension? In *Speculations on the fourth dimension: Selected writings of Charles H. Hinton.* Dover Publications, Inc. (Original work published in 1884). http://www.ibiblio.org/eldritch/chh/h1.html

History of printing in East Asia (2021, July 27). *Wikipedia.* http://en.wikipedia.org/wiki/History_of_printing_in_East_Asia

Hobart, M. E. (2018, July 6). When numeracy superseded literacy—And created the modern world. *Zocalo.* https://www.zocalopublicsquare.org/2018/07/06/numeracy-superseded-literacy-created-modern-world/ideas/essay/

Hobbes, T. (1651). *Leviathan.* http://www.gutenberg.org/etext/3207

Hoffman, H. G. (2004, July 26). Virtual-reality therapy. *Scientific American.* http://www.scientificamerican.com/article.cfm?id=virtual-reality-therapy

Hogan, J., & Fox, B. (2005, April 7). Sony patent takes first step towards real-life matrix. *NewScientist.com.* https://web.archive.org/web/20080906172244/http://www.newscientist.com:80/article.ns?id=mg18624944.600

Holograms you can touch and feel (2021, May 6). *Science Connected Magazine.* https://magazine.scienceconnected.org/2021/05/holograms-can-touch-feel/

Holographic TV market—Growth, trends, COVID-19 impact, and forecasts (2021–2026) (2021). *Research and Markets.* https://www.researchandmarkets.com/reports/4591900/holographic-tv-market-growth-trends-covid-19

Holt, F. L. (2021). *When money talks: A history of coins and numismatics.* Oxford University Press.

Hood, M. (2010, April 6). Computer-enhanced vision adds a 'sixth sense'. *PhysOrg.com.* http://www.physorg.com/news189747401.html

Horgan, J. (2010, December 21). Science 'faction': Is theoretical physics becoming 'softer' than anthropology? *Scientific American.* http://blogs.scientificamerican.com/cross-check/2010/12/21/science-faction-is-theoretical-physics-becoming-softer-than-anthropology

Horner, J., & Gorman, J. (2009). *How to build a dinosaur: Extinction doesn't have to be forever.* Dutton.

Howard, R. G. (2010). Enacting a virtual 'ekklesia': Online Christian fundamentalism as vernacular religion. *New Media & Society, 12*(5), 729–744.

Humans 'predisposed' to believe in gods and the afterlife (2011, July 14). *Science Daily.* http://www.sciencedaily.com/releases/2011/07/110714103828.htm

Hunt, T. (2021, August 3). Electrons may very well be conscious. *Nautilus*. https://nautil.us/blog/-electrons-may-very-well-be-conscious

Husserl, E. (1970). *The crisis of European sciences and transcendental phenomenology* (D. Carr, Trans.). Northwestern University Press.

Huxley, A. (1954). *The doors of perception*. Harper & Row. https://maps.org/images/pdf/books/HuxleyA1954TheDoorsOfPerception.pdf

Hybrid network (2020). *SYNCH*. http://synch.eucoord2020.com/concept/

Ignatieff, M. (2000). *Virtual war: Kosovo and beyond*. Picador-Henry Holt & Co.

Ingemard, D., & Ingemark, C. A. (2007). Teaching ancient folklore. *The Classical Journal, 102*(3), 279–289. https://www.jstor.org/stable/30037990

Innis, H. (1950). *Empire and communications*. Oxford University Press.

*Internet seen as positive influence on education but negative on morality in emerging and developing nations* (2015, March 19). PEW Research Center. http://www.pewglobal.org/2015/03/19/internet-seen-as-positive-influence-on-education-but-negative-influence-on-morality-in-emerging-and-developing-nations/

Intraub, H., & Dickinson, C. A. (2008). False memory 1/20th of a second later: What the early onset of boundary extension reveals about perception. *Psychological Science, 19*(10), 1007–1014.

Itzkoff, D. (2007, June 24). A brave new world for TV? Virtually. *The New York Times*. https://www.nytimes.com/2007/06/24/arts/television/24itzk.html

Jacques, L. L. (2010, December). The currency wars. *Le Monde Diplomatique* (English ed.). http://mondediplo.com/2010/12/03currency

James, W. (1890). *The principles of psychology* (2 Vols.). Henry Holt.

James, W. (2002). *The varieties of religious experience* (New introductions by E. Taylor & J. Carrette). Routledge, Taylor & Francis (Original work published 1902).

Jayanti, A. (Ed.). (2020). *Synthetic biology*. Belfer Center for Science and International Affairs. Harvard Kennedy School. https://www.belfercenter.org/sites/default/files/2020-10/tappfactsheets/SyntheticBiology.pdf

Johanson, D. (2001, May). *Origins of modern humans: Multiregional or out of Africa?* American Institute of Biological Sciences. http://thorpsci.weebly.com/uploads/1/7/0/0/17001640/origins_of_modern_humans.pdf

Johnson, N. F. (2009). *The multiplicities of internet addiction*. Ashgate.

Johnson, R. (2011, August 31). This new stealth ship is about to revolutionize the US navy. *Business Insider.com*. http://www.businessinsider.com/this-new-stealth-ship-is-set-to-revolutionize-2011-8

Johnson, S. (2005). *Everything bad is good for you: How today's popular culture is actually making us smarter*. Riverhead-Penguin.

Johnson, S. (2010). *Where good ideas come from: The natural history of innovation*. Riverhead-Penguin.

Jonscher, C. (1999). *The evolution of wired life*. Wiley.

Joshi, J., Rubart, R., & Zhu, W. (2020, January 29). Optogenetics: Background, methodological advances and potential applications for cardiovascular research and medicine. *Frontiers in Bioengineering and Biotechnology*. http://doi.org/10.3389/fbioe.2019.00466

Jung, C., & Jaffe, A. (1961). *Memories, dreams, reflections*. Fontana.

Juvenal. (1998). *The sixteen satires* (3rd rev. ed.; P. Green, Trans.). Penguin Books.

Kahlenberg, S. M., & Wrangham, R. W. (2010). Sex differences in chimpanzees' use of sticks as play objects resemble those of children. *Current Biology, 20*(24), R1067–R1068. http://www.cell.com/current-biology/fulltext/S0960-9822(10)01449-1

Kaiser, W. L., & Wood, D. (2001). *Seeing through maps: The power of images to shape our world view*. ODT Inc.

Kant, I. (1784). *Idea for a universal history from a cosmopolitan point of view. On history* (L. W. Beck, Trans.). The Bobbs-Merrill Co. http://www.marxists.org/reference/subject/ethics/kant/universal-history.htm

Kant, I. (1910). In Akademie der Wissenschaften (Eds.), *Gesammelte schriften* [Collected writings]. Reimer.

Kant, I. (1965). *Critique of pure reason* (N. K. Smith, Trans.). St. Martin's Press.

Kaplan, M. (2013). *The science of monsters: The origins of the creatures we love to fear*. Scribner's.

Kapogiannis, D., Barbey, A. K., Su, M., Zamboni, G., Krueger, F., & Grafman, J. (2009). Cognitive and neural foundations of religious belief. *Proceedings of the National Academy of Sciences USA, 106*(12), 4876–4881.

Kastrenakes, J. (2021, March 11). Beeple sold an NFT for $69 million. *The Verge*. https://www.the verge.com/2021/3/11/22325054/beeple-christies-nft-sale-cost-everydays-69-million

Katz, V. J., & Barton, B. (2007). Stages in the history of algebra with implications for teaching. *Educational Studies in Mathematics, 66*(2), 185–201.

Kaufman, E. P. (2010). *Shall the religious inherit the earth? Demography and politics in the twenty-first century*. Profile Books.

Kelemen, D. (2004). Are children 'intuitive theists'? *Psychological Science, 15*(5), 295–301.

Kelion, L. (2019, September 24). Google wins landmark right to be forgotten case. *BBC News*. https://www.bbc.com/news/technology-49808208

Kelly, J. (2007). Pop music, multimedia, and live performance. In J. Sexton (Ed.), *Music, sound and multimedia: From the live to the virtual* (pp. 105–120). Edinburgh University Press.

Kessler, R. C., Chiu, W. T., Demler, O., & Walters, E. E. (2005). Prevalence, severity, and comorbidity of twelve-month DSM-IV disorders in the National Comorbidity Survey Replication (NCS-R). *Archives of General Psychiatry, 62*(6), 617–627.

Killingsworth, M. A., & Gilbert, D. T. (2010, November 12). A wandering mind is an unhappy mind. *Science, 330*(6006), 932. http://www.sciencemag.org/cgi/content/full/330/6006/932

Kinstler, L. (2021, July 16). Can Silicon Valley find god? *The New York Times*. https://www.nyt imes.com/interactive/2021/07/16/opinion/ai-ethics-religion.html

Kivy, P. (2009). *Antithetical arts: On the ancient quarrel between literature and music*. Clarendon Press.

Klein, R. G. (2000). Archaeology and the evolution of human behavior. *Evolutionary Anthropology, 9*, 17–36.

Klinger, E. (2009). Daydreaming and fantasizing: Thought flow and motivation. In K. D. Markman, W. M. P. Klein & J. A. Suhr (Eds.), *Handbook of imagination and mental stimulation* (pp. 225–240). Psychology Press.

Klopfer, E., Osterweil, S., Groff, J., & Haas, J. (2009). *The instructional power of digital games, social networking, simulations, and how teachers can leverage them*. MIT Education Arcade—Creative Commons License. https://web.archive.org/web/20150216015609/http://education.mit.edu/papers/GamesSimsSocNets_EdArcade.pdf

Klosterman, C. (2010, December 5). My zombie, myself: Why modern life feels rather undead. *The New York Times*. http://www.nytimes.com/2010/12/05/arts/television/05zombies.html?_r=1&nl=todaysheadlines&emc=a28

Koerner, B. I. (2011, April). In praise of ugly robots. *Wired*, 21–22. https://web.archive.org/web/201 61222060411/https://www.wired.com/2011/03/st_essay_ugly_robots/ (Online title: The trouble with humanoid droids).

Kormendy, J., & Richstone, D. O. (1995). Inward bound—The search for supermassive black holes in galactic nuclei. *Annual Reviews of Astronomy and Astrophysics, 33*(1), 581–624.

Koslofsky, C. (2011). *Evening's empire: A history of the night in early modern Europe*. Cambridge University Press.

Kotenidis, E., & Veglis, A. (2021). Algorithmic journalism—Current applications and future perspectives. *Journalism and Media, 2*, 244–257. https://www.mdpi.com/2673-5172/2/2/14

Krackow, E., Lynn, S. J., & Payne, D. G. (2005–06). The death of Princess Diana: The effects of memory enhancement procedures on flashbulb memories. *Imagination, Cognition and Personality, 25*(3), 197–219.

Krahenbuhl, L. (2006). A theater before the world: Performance history at the intersection of Hebrew, Greek, and Roman processional. *The Journal of Religion and Theatre, 5*(1), 61–72.

Kramer, M., Pita, M., Zhou, J., Ornatska, M., Poghossian, A., Schoning, M., & Katz, E. (2008). Coupling of biocomputing systems with electronic chips: Electronic interface for transduction of biochemical information. *Journal of Physical Chemistry, 113*, 2573–2579.

Kramer, S. N. (1959). *History begins at Sumer*. Doubleday Anchor Books.

Krauss, L. M. (1995). *The physics of Star Trek*. HarperPerennial.

Krauss, L. M. (2005). *Hiding in the mirror: The mysterious allure of extra dimensions, from Plato to string theory and beyond*. Viking Penguin.

Kunstler, J. H. (2010). Virtual is no refuge from the real. https://web.archive.org/web/201603040 13059/http://www.kunstler.com/mags_virtual.html

Lam, L. T., & Peng, Z.-W. (2010, August 2). Effect of pathological use of the internet on adolescent mental health: A prospective study. *Archives of Pediatrics & Adolescent Medicine*. https://pub med.ncbi.nlm.nih.gov/20679157/

Langer, S. K. (1953). *Feeling and form*. Routledge & Kegan Paul.

Lanier, J. (2011). *You are not a gadget: A manifesto*. Knopf.

Lanier, J. (2017). *Dawn of the new everything: Encounters with reality and virtual reality*. Henry Holt and Co.

Lanier, J. (2018). *Ten arguments for deleting your social media accounts right now*. Picador/Macmillan.

Lant, K. (2017, March 28). These high-tech glasses will give you superhuman vision. *Futurism*. https://futurism.com/these-high-tech-glasses-will-give-you-superhuman-vision

Lanza, R. (2007, Spring). A new theory of the universe: Biocentrism builds on quantum physics by putting life into the equation. *The American Scholar*. https://web.archive.org/web/200812171 44353/http://www.theamericanscholar.org:80/archives/sp07/newtheory-lanza.html

Lastovicka, J. L., & Sirianni, N. J. (2011, January 25). Truly, madly, deeply: Consumers in the throes of material possession love. *The Journal of Consumer Research, 38*(2), 323–342.

Lastowka, G., & Hunter, D. (2006). Virtual crime. In J. M. Balkin & B. S. Noveck (Eds.), *The state of play: Law and virtual worlds* (pp. 31–54). New York University Press.

Lavers, C. (2009). *The natural history of unicorns*. Granta/Harper Collins.

Leake, J. (2010, April 25). Don't talk to aliens, warns Stephen Hawking. *The Sunday Times*. https://www.thetimes.co.uk/article/dont-talk-to-aliens-warns-stephen-hawking-rftn7kpjgbx

Leder, D., Hermann, R., Hüls, M., et al. (2021). A 51,000-year-old engraved bone reveals Neanderthals' capacity for symbolic behaviour. *Nature Ecology & Evolution*. https://doi.org/10.1038/s41559-021-01487-z

LeDoux, J. E. (1996). *The emotional brain*. Simon & Schuster.

Lee, B. Y. (2020, January 13). The future of clinical trials? Here is a simulation model of the heart. *Forbes*. https://www.forbes.com/sites/brucelee/2020/01/13/the-future-of-clinical-tri als-here-is-a-simulation-model-of-the-heart/?sh=4605c61f3aa4

Lee, K. M. (2009). Presence theory. In *Encyclopedia of communication theory*. SAGE Publications. https://sk.sagepub.com/reference/communicationtheory/n301.xml

Lehdonvirta, V. (2014). Virtuality in the sphere of economics. In M. Grimshaw (Ed.), *The Oxford handbook of virtuality*. Oxford University Press. http://doi.org/10.1093/oxfordhb/978019982 6162.001.0001

Lehdonvirta, V., & Castronova, E. (2014). *Virtual economies: Design and analysis*. MIT Press.

Lehman-Wilzig, S. (1981). Frankenstein unbound: Toward a legal definition of artificial intelligence. *FUTURES: The Journal of Forecasting and Planning, 13*(6), 442–457. https://www.sciencedirect. com/science/article/abs/pii/0016328781901002

Lem, S. (1974). *The cyberiad stories*. Seabury Press.

Leo, B., et al. (1919). *The life and work of Alan Leo*. "Modern Astrology" Office. http://www.arc hive.org/stream/lifeworkofalanle00leob#page/n7/mode/2up

Leslie, A. M. (1987). Pretense and representation: The origins of 'theory of mind.' *Psychology Review, 94*, 412–426.

Levack, B. (2013). *The devil within: Possession and exorcism in the Christian west*. Yale University Press.

Levinson, M. (2021, August 12). A fourth globalisation. *Aeon.* https://aeon.co/essays/the-globalisation-of-ideas-will-be-different-than-that-of-goods

Levitin, D. J. (2008). *The world in six songs: How the musical brain created human nature.* Dutton.

Levy, D. (2007). *Love + sex with robots: The evolution of human-robot relationships.* HarperCollins Publishers.

Lewis, C. (2007, December 23). The trouble with Mary. *nationalpost.com.* https://nationalpost.com/news/the-trouble-with-mary

Lieberman, D. (2021). *Exercised: Why something we never evolved to do is healthy and rewarding.* Pantheon.

Lifton, R. J. (1993). *The protean self: Human resilience in an age of fragmentation.* The University of Chicago Press.

Lim, D. (2010, April 4). A bold vision, still ahead of its time. *The New York Times.* https://www.nytimes.com/2010/04/04/movies/04wire.html?scp=1&sq=april%204,%2010%20-%20dennis%20lim&st=cse

Linden, D. J. (2007). *The accidental mind: How brain evolution has given us love, memory, dreams, and god.* The Belknap Press of Harvard University Press.

Lloyd, S. (2006). *Programming the universe.* Knopf.

Locke, J. (1689). *Second treatise of government.* http://www.gutenberg.org/etext/7370

Loftus, E. F. (1997). Creating false memories. *Scientific American, 277*(3), 70–75.

Logan, R. K. (2004). *The alphabet effect: A media ecology understanding of the making of western civilization.* Hampton Press.

Lohr, S. (2010a, April 14). Library of Congress will save tweets. *The New York Times.* http://www.nytimes.com/2010/04/15/technology/15twitter.html?th&emc=th

Lohr, S. (2010b, April 19). Global strategy stabilized I.B.M. during downturn. *The New York Times.* http://www.nytimes.com/2010/04/20/technology/20blue.html?scp=2&sq=IBM&st=cse

Lupton, D. (2016). *The quantified self.* Wiley.

MacIntyre, A. (2004). *The unconscious: A conceptual analysis.* Routledge and Kegan Paul (Original work published 1958).

Makin, T. R., Holmes, N. P., & Zohary, E. (2007). Is that near my hand? Multisensory representation of peripersonal space in human intraparietal sulcus. *The Journal of Neuroscience, 27*(4), 731–740.

Malpas, J. (2009). On the non-autonomy of the virtual. *Convergence: The International Journal of Research into New Media Technologies, 15*(2), 135–139. http://con.sagepub.com/content/15/2/135.full.pdf+html

Mammone, A. (2011, September 20). The future of Europe's radical right. *Foreign Affairs.* http://www.foreignaffairs.com/articles/68286/andrea-mammone/the-future-of-europes-radical-right

Mandelbrot, B. (1977). *Fractals: Form, chance and dimension.* W.H. Freeman and Co.

Manjoo, F. (2008). *True enough: Learning to live in a post-fact society.* Wiley.

Marcus, G. (2008). *Kluge: The haphazard construction of the human mind.* Houghton Mifflin Co.

Marescaux, J., Leroy, J., Gagner, M., Rubino, F., Mutter, D., Vix, M., Butner, S. E., & Smith, M. K. (2001, September 27). Transatlantic robotic assisted remote telesurgery. *Nature, 413*(6854), 379–80. http://doi.org/10.1038/35096636

Marin, M.-F., Hupbach, A., Maheu, F. S., Nader, K., & Lupien, S. J. (2011). Metyrapone administration reduces the strength of an emotional memory trace in a long-lasting manner. *Journal of Clinical Endocrinology & Metabolism, 96*(8), E1221–E1227. https://academic.oup.com/jcem/article/96/8/E1221/2833629

Markley, R. (1996). *Virtual realities and their discontents.* The Johns Hopkins Press.

Marquit, M. (2010, April 19). Could printers produce human tissue? *PhysOrg.com.* http://www.physorg.com/news190911197.html

Martin, J.-C. (2014). Emotions and altered states of awareness: The virtuality of reality and the reality of virtuality. In M. Grimshaw (Ed.), *The Oxford handbook of virtuality.* Oxford University Press. https://doi.org/10.1093/oxfordhb/9780199826162.001.0001

Martinez-Conde, S. (2007, September 4). Blindsight: When the brain sees what you do not. *Scientific American*. http://www.scientificamerican.com/blog/post.cfm?id=blindsight-when-the-brain-sees-what

Martinez-Conde, S., & Macknik, S. L. (2008a, May 28). The neuroscience of illusion: How tricking the eye reveals the inner workings of the brain. *Scientific American*. http://www.scientificameri can.com/article.cfm?id=the-neuroscience-of-illusion

Martinez-Conde, S., & Macknik, S. L. (2008b, November 24). Magic and the brain: How magicians 'trick' the mind. *Scientific American*. http://www.scientificamerican.com/article.cfm?id=magic-and-the-brain

Martisiute, L. (2020, December 20). Virtual reality (VR) in education: A complete guide. *E-student.org*. https://e-student.org/virtual-reality-in-education/

Maruta, T., Colligan, R. C., Malinchic, M., & Offord, K. P. (2000). Optimists vs pessimists: Survival rate among medical patients over a 30-year period. *Mayo Clinic Proceedings, 75*(2), 140–143.

Marx, K. (1938). *Critique of the Gotha programme*. International Publishers Co. (Original; work published 1875).

Marx, P. (2010, March 29). Four eyes. *The New Yorker*, 46–50. http://www.newyorker.com/report ing/2010/03/29/100329fa_fact_marx

Maslow, A. H. (1943). A theory of human motivation. *Psychological Review, 50*(4), 370–396.

Maugham, W. S. (1977). *Books and you*. Arno Press.

Mayer-Schönberger, V. (2009). *Delete: The virtue of forgetting in the digital age*. Princeton University Press.

McBain, S. (2021, June 30). The plants that change our consciousness. *NewStatesman*. https://www. newstatesman.com/culture/books/2021/06/plants-change-our-consciousness

McLuhan, M. (1962). *The Gutenberg galaxy: The making of typographic man*. University of Toronto Press.

McLuhan, M. (1964). *Understanding media: The extensions of man*. McGraw-Hill.

Mendelsohn, D. (2010, April 5). Epic endeavors. *The New Yorker*, 74–78. http://www.newyorker. com/arts/critics/books/2010/04/05/100405crbo_books_mendelsohn

Menninger, K. (1992). *Number words and number symbols: A cultural history of numbers*. MIT Press; Dover Publications (Original work published 1969).

Merali, Z. (2010, April). Back from the future. *Discover*, 38–44. https://web.archive.org/web/201 91117104740/http://discovermagazine.com:80/2010/apr/01-back-from-the-future

Merin, G. (2021, June 28). The medieval nuns' guide to virtual travel. *1843 Magazine*. https://www. economist.com/1843/2021/06/28/the-medieval-nuns-guide-to-virtual-travel

Merlin, M. D. (2003). Archaeological evidence for the tradition of psychoactive plant use in the old world. *Economic Botany, 57*(3), 295–323.

Mesch, G. S., & Talmud, I. (2010). *Wired youth: The social world of adolescence in the information age*. Routledge.

'Metaverse': The next internet revolution? (2021, July 28). *TechXplore*. https://techxplore.com/ news/2021-07-metaverse-internet-revolution.html

Metz, C. (2017, May 15). Google's AI invents sounds humans have never heard before. *WIRED.com*. https://www.wired.com/2017/05/google-uses-ai-create-1000s-new-musical-instru ments/?mbid=nl_51517_p8&CNDID=27736618

Meyrowitz, J. (1986). *No sense of place: The impact of electronic media on social behavior*. Oxford University Press.

Mihhailova, G. (2006). From ordinary to virtual teams: A model for measuring the virtuality of a teamwork. In *ICEB 2006 Proceedings*. https://aisel.aisnet.org/iceb2006/58/

Miller, G. (2001). Aesthetic fitness: How sexual selection shaped artistic virtuosity as a fitness indicator and aesthetic preferences as mate choice criteria. *Bulletin of Psychology and the Arts, 2*(1), 20–25.

Milton, J. (1667). *Paradise lost*. https://en.wikisource.org/wiki/Paradise_Lost_(1667)/Book_I

Mind-body dualism (2021, July 16). *Wikipedia*. https://en.wikipedia.org/wiki/Mind%E2%80%93b ody_dualism

Mironov, V. (2011). The future of medicine: Are custom-printed organs in the horizon? *The Futurist*, *45*(1), 21–24.

Mithen, S. (1999). *The prehistory of the mind: The cognitive origins of art, religion and science*. Thames & Hudson.

Monmonier, M. (1996). *How to lie with maps* (2nd ed.). University of Chicago Press.

Monmonier, M. (2005). Lying with maps. *Statistical Science, 20*(3), 215–222.

Morales, J., Bax, A., & Firestone, C. (2020). Sustained representation of perspectival shape. *PubMed.gov*. https://pubmed.ncbi.nlm.nih.gov/32532920/

Moravec, H. (1988). *Mind children: The future of robot and human intelligence*. Harvard University Press.

More, H. (1712). *The immortality of the soul* (4th ed.).

Morgan, S. (2020, November 13). Cybercrime to cost the world $10.5 trillion annually by 2025. *Cybercrime Magazine*. https://cybersecurityventures.com/hackerpocalypse-cybercrime-report-2016/?utm_source=angellist

Morowitz, H. J. (2002). *The emergence of everything: How the world became complex*. Oxford University Press.

Morris, S. (2008, November 13). 'Second Life' ends first marriage for British man. *The Guardian*. https://www.theguardian.com/technology/2008/nov/13/second-life-divorce

Morse, M. (1998). *Virtualities: Television, media art, and cyberculture*. Indiana University Press.

Mulhall, D. (2002). *Our molecular future: How nanotechnology, robotics, genetics, and artificial intelligence will transform our world*. Prometheus Books.

Mullen, W. (2010, September 24). Before Twitter, tweets were made in stone and clay. *Chicago Tribune*. https://www.chicagotribune.com/news/ct-xpm-2010-09-24-ct-met-writing-invention-20100924-story.html

Multiple personality disorder (2010). *The free dictionary: Medical dictionary*. http://medical-dictionary.thefreedictionary.com/multiple+personality+disorder

Nagel, T. (1974). What is it like to be a bat? *The Philosophical Review, LXXXIII*(4), 435–450. https://warwick.ac.uk/fac/cross_fac/iatl/study/ugmodules/humananimalstudies/lectures/32/nagel_bat.pdf

Narula, H. (2019, December 29). A billion new players are set to transform the gaming industry. *WIRED*. https://www.wired.co.uk/article/worldwide-gamers-billion-players

Nass, C., & Brave, S. (2005). *Wired for speech: How voice activates and advances the human-computer relationship*. The MIT Press.

Neiger, M. (2007). Media oracles: The cultural significance and political import of news referring to future events. *Journalism, 8*(3), 309–321.

Nell, V. (1988). *Lost in a book: The psychology of reading for pleasure*. Yale University Press.

Newberg, A., & Waldman, R. (2010). *How god changes your brain: Breakthrough findings from a leading neuroscientist*. Random House.

Newberg, A. B., D'Aquili, E. G., & Rause, V. (2002). *Why god won't go away: Brain science and the biology of belief*. Ballantine Books.

Newport, F., & Strausberg, M. (2001, June 8). Americans' belief in psychic and paranormal phenomena is up over the last decade. *Gallup*. http://www.gallup.com/poll/4483/americans-belief-psychic-paranormal-phenomena-over-last-decade.aspx

Newton, J. H. (2006). Influences of digital imaging on the concept of photographic truth. In P. Messaris & L. Humphreys (Eds.), *Digital media: Transformations in human communication* (pp. 3–14). Peter Lang.

Nguyen, T. H., & Preston, D. S. (2006). *Virtuality and education: A reader*. Brill.

Nicholls, H. (2008, October 29). Can animals escape the present? *NewScientist.com*. http://www.newscientist.com/article/mg20026801.600-can-animals-escape-the-present.html

Niebuhr, R. (1932). *Moral man and immoral society: A study in ethics and politics*. Charles Scribner's Sons.

Nohe, P. (2018, September 27). 2018 cybercrime statistics: A closer look at the 'web of profit'. *HashedOut*. https://www.thesslstore.com/blog/2018-cybercrime-statistics/

O'Brien, M., & Kellan, A. (2011, January 24). Virtual self can affect reality self. *PhysOrg.com.* http://www.physorg.com/news/2011-01-virtual-affect-reality.html

O'Connor, A. R., & Moulin, C. J. A. (2006). Normal patterns of deja experience in a healthy, blind male: Challenging optical pathway delay theory. *Brain and Cognition, 62*(3), 246–249.

O'Leary, B. (1997). On the nature of nationalism: An appraisal of Ernest Gellner's writings on nationalism. *British Journal of Political Science, 27,* 191–222.

O'Leary, S. D. (1996). Cyberspace as sacred space: Communicating religion on computer networks. *Journal of the American Academy of Religion, 64*(4), 781–808.

Oliver, M. B., & Raney, A. A. (2011). Entertainment as pleasurable and meaningful: Identifying hedonic and eudaimonic motivations for entertainment consumption. *Journal of Communication, 61*(5), 984–1004.

Ong, W. (1982). *Orality and literacy: The technologizing of the word.* Methuen & Co.

Opie, I., & Opie, P. (1969). *Children's games in street and playground: Chasing, catching, seeking, hunting, racing, dueling, exerting, daring, guessing, acting, and pretending.* Oxford University Press.

Osborne, P. (2002). *Conceptual art (Themes and movements).* Phaidon.

Otto, P. (2011). *Multiplying worlds: Romanticism, modernity, and the emergence of virtual reality.* Oxford University Press.

Overbye, D. (2007, December 18). Laws of nature, source unknown. *The New York Times.* http://www.nytimes.com/2007/12/18/science/18law.html

Palmer, A. (2020, August 31). Amazon wins FAA approval for prime air drone delivery fleet. *CNBC.* https://www.cnbc.com/2020/08/31/amazon-prime-now-drone-delivery-fleet-gets-faa-approval.html

Panero, J. (2009, December). The art market explained. *The New Criterion, 28,* 26. http://www.newcriterion.com/articles.cfm/The-art-market-explained-4337

Papagiannis, H. (2020, October 7). How AR is redefining retail in the pandemic. *Harvard Business Review.* https://hbr.org/2020/10/how-ar-is-redefining-retail-in-the-pandemic

Pearson, J., Clifford, C. W. G., & Tong, F. (2008). The functional impact of mental imagery on conscious perception. *Current Biology, 18*(13), 982–986.

Penrose, R. (2016). *The emperor's new mind: Concerning computers, minds and the laws of physics* (Illustrated ed.). Oxford University Press.

Perez, O., Bar-Ilan, J., Gazit, T., Aharony, N., Amichai-Hamburger, Y., & Bronstein, J. (2018). The prospects of e-democracy: An experimental study of collaborative e-rulemaking. *Journal of Information Technology & Politics, 15*(3): 278–299. https://doi.org/10.1080/19331681.2018.1485605

Pergams, O. R. W., & Zaradi, P. A. (2008, February 4). Evidence for a fundamental and pervasive shift away from nature-based recreation. *Proceedings of the National Academy of Sciences of the U.S.A.* http://www.pnas.org/content/105/7/2295.full

Perkowitz, S. (2021, June 23). The math of living things. *Nautilus.* https://nautil.us/issue/102/hidden-truths/the-math-of-living-things

Pfaff, W. (2008, January 4). Pfaff: Paying for virtual oil. *The New York Times.* http://www.nytimes.com/2008/01/04/opinion/04iht-edpfaff.1.9024521.html

Phillips, H. (2009, March 25). Déjà vu: Where fact meets fantasy. *New Scientist.* http://www.newscientist.com/article/mg20127011.400-deja-vu-where-fact-meets-fantasy.html?full=true

Pinker, S. (1997). *How the mind works.* Norton.

Pinker, S. (2007, January 19). The brain: The mystery of consciousness. *TIME Magazine.* http://content.time.com/time/magazine/article/0,9171,1580394,00.html

Pinker, S. (2011). *The better angels of our nature: Why violence has declined.* Viking.

Pinker, S. (2018). *Enlightenment now: The case for reason, science, humanism, and progress.* Penguin Books.

Plato. (1950). *Republic* (A. D. Lindsay, Trans.). A. P. Dutton.

Platt, C. (1999, November). You've got smell! *Wired, 7*(11). http://www.wired.com/wired/archive/7.11/digiscent.html

Polonska-Kimunguyi, E., & Kimunguyi, P. (2011). The making of the Europeans: Media in the construction of pan-national identity. *International Communication Gazette, 73*(6), 507–523.

Popper, K. (1965). *Of clouds and clocks: An approach to the problem of rationality and the freedom of man* (The Arthur Holly Compton memorial lecture). Washington University Press.

Popper, K. (1977, November 8). *Natural selection and the emergence of mind.* First Darwin Lecture, Delivered at Darwin College, Cambridge. http://www.informationphilosopher.com/solutions/phi losophers/popper/natural_selection_and_the_emergence_of_mind.html

Postman, N. (1992). *Technopoly: The surrender of culture to technology.* Random House.

Press release (2010, October 4). Nobel Assembly at Karolinska Institutet. https://www.nobelprize. org/uploads/2018/06/press-8.pdf

Price, R. (1978). *A palpable god.* Atheneum.

Prot, S., Anderson, C. A., Gentile, D. A., Brown, S. C., & Swing, E. L. (2014). The positive and negative effects of video game play. In A. Jordan & D. Romer (Eds.), *Media and the well-being of children and adolescents* (pp. 109–128). Oxford University Press.

Quiggin, A. H. (1949). *A survey of primitive money: The beginnings of currency.* Methuen & Co.

Qvortrup, L. (2002). Cyberspace as representation of space experience: In defense of a phenomeno-logical approach. In L. Qvortrup (Ed.), *Virtual space: Spatiality in virtual inhabited 3D worlds* (pp. 3–7). Springer.

Raden, A. (2021). *The truth about lies: The illusion of honesty and the evolution of deceit.* Martin's Press.

Rafaeli, S. (1988). Interactivity: From new media to communication. In R. P. Hawkins, J. M. Wiemann & S. Pingree (Eds.), *Sage annual review of communication research: Advancing communication science* (Vol. 16, pp. 110–134). Sage.

Reading by numbers: Science invades the humanities (2010, December 16). *The Economist.* http:// www.economist.com/node/17730198

Reeves, B., & Nass, C. (1996). *The media equation: How people treat computers, television, and new media like real people and places.* Cambridge University Press.

Rescher, N. (2008). Process philosophy (Sect. 8). *Stanford encyclopedia of philosophy.* http://plato. stanford.edu/entries/process-philosophy

Retail e-commerce sales worldwide from 2014 to 2023 (2020). *Statista.* https://www.statista.com/ statistics/379046/worldwide-retail-e-commerce-sales/

Reuben, A., & Schaefer, J. (2017, July 14). Mental illness is far more common than we knew. *Scientific American.* https://blogs.scientificamerican.com/observations/mental-illness-is-far-more-common-than-we-knew

Rheingold, H. (1993). *The virtual community: Homesteading on the electronic frontier.* Harper-Perennial. http://www.rheingold.com/vc/book/intro.html

Roach, M. (2005). *Spook: Science tackles the afterlife.* W.W. Norton & Company.

Roberson, D., & Hanley, J. R. (2007). Color vision: Color categories vary with language after all. *Current Biology, 17*(15), R605–R607.

Robinett, W. (1992). Synthetic experience: A proposed taxonomy. *Presence, 1*(2), 229–247.

Robinson, S. (2011). 'Journalism as process': The organizational implications of participatory online news. *Journalism & Communication Monographs, 13*(3), 137–210.

Robson, D. (2010, September 22). Cosmic accidents: Inventing language, the easy way. *NewScientist.com.* http://www.newscientist.com/article/mg20727796.300-cosmic-accidents-inventing-lan guage-the-easy-way.html

Roman, J. (2008). Whale communication and culture. *The Encyclopedia of Earth.* https://editors. eol.org/eoearth/wiki/Whale_communication_and_culture

Ropolyi, L. (2001). Virtuality and plurality. In A. Riegler, M. F. Peschl, K. Edlinger, G. Fleck & W. Feigl (Eds.), *Virtual reality: Cognitive foundations, technological issues & philosophical implications* (pp. 167–187). Peter Lang. https://www.researchgate.net/publication/220725379_ Virtuality_and_Plurality (Slightly modified version: pp. 1–20).

Rorty, R. (1979). *Philosophy and the mirror of nature.* Princeton University Press.

Rose, F. (2017, January 4). The making of virtually real art with Google's tilt brush. *The New York Times*. https://www.nytimes.com/2017/01/04/arts/design/the-making-of-virtually-real-art-with-googles-tilt-brush.html?_r=0

Rose, N. (Ed.). (1988). *Mathematical maxims and minims*. Rome Press Inc.

Rosen, J. (2010, July 25). The web means the end of forgetting. *The New York Times*. http://www.nytimes.com/2010/07/25/magazine/25privacy-t2.html?pagewanted=1

Rosenberg, A. (2016, July 18). Why you don't know your own mind. *The New York Times*. http://nyti.ms/2a1HtPB

Rosenberg, A. (2018, November 2). Is neuroscience a bigger threat than artificial intelligence? *3:AM Magazine*. https://www.3ammagazine.com/3am/is-neuroscience-a-bigger-threat-than-artificial-intelligence/

Ross, A. (2010, February 8). Close listening. *The New Yorker*, 66–67.

Rothbard, M. N. (2009, October 9). *Karl Marx as religious eschatologist*. Ludwig von Mises Institute. http://mises.org/daily/3769

Roudometof, V. (2005). Translationalism, cosmopolitanism, and glocalization. *Current Sociology*, *53*(1), 113–135.

Rousseau, J. J. (1762). *The social contract*. https://web.archive.org/web/20200706154113/http://constitution.org/jjr/socon.htm

Ryan, M.-L. (1991). *Possible worlds, artificial intelligence and narrative theory*. Indiana University Press.

Ryan, M.-L. (2001). *Narrative as virtual reality: Immersion and interactivity in literature and electronic media*. The Johns Hopkins University.

Sabag, N. N. (2006). Exobiology. *Biology Cabinet*. http://biocab.org/Exobiology.html

Saliba, G. (1994). *A history of Arabic astronomy: Planetary theories during the golden age of Islam*. New York University Press.

Sandywell, B. (2006). Monsters in cyberspace: Cyberphobia and cultural panic in the information age. *Information, Communication & Society*, *9*(1), 39–61.

Sanneh, K. (2011, May 9). The reality principle. *The New Yorker*, 72–77. https://www.newyorker.com/magazine/2011/05/09/the-reality-principle

*Santa Clara County v. Southern Pacific Railroad* (1886). 118 U.S. 394.

Saunders, R. (1999). *Business the Amazon.com way: Secrets of the world's most astonishing web business*. Capstone.

Scalise Sugiyama, M. (1996). On the origins of narrative: Storyteller bias as a fitness-enhancing strategy. *Human Nature*, *7*, 403–425.

Scalise Sugiyama, M. (2003). Cultural relativism in the bush: Towards a theory of narrative universals. *Human Nature*, *14*, 383–396.

Schachter, D. L. (2001). *The seven sins of memory*. Houghton Mifflin Harcourt.

Scharf, C. (2021, June 16). Is the universe open-ended? *Nautilus*. https://nautil.us/issue/102/hidden-truths/is-the-universe-open_ended

Scharping, N. (2020, February 28). The invisible man' isn't real, but this invisibility technology is. *Smithsonian Magazine*. https://www.smithsonianmag.com/science-nature/the-invisible-man-real-invisibility-science-180974301/

Schiesel, S. (2006, May 7). Video games; Welcome to the new dollhouse. *The New York Times*. http://www.nytimes.com/2006/05/07/arts/07schi.html

Schizophrenia. (2020). NIMH (National Institute of Mental Health). http://www.nimh.nih.gov/health/topics/schizophrenia/index.shtml

Schizophrenia: Fact sheet (2021). World Health Organization. http://www.who.int/mediacentre/factsheets/fs397/en/

Schlenoff, D. (2011, September 12). Aliens and Nazis and electric history, Oh My! *Scientific American*. http://blogs.scientificamerican.com/observations/2011/09/12/aliens-get-higher-nielsen-ratings/

Schmandt-Besserat, D. (2004, September–October). Birth of narrative art: How writing led to picture painting. *Odyssey, 36*–43, 55. https://web.archive.org/web/20120325152408/https://web space.utexas.edu/dsbay/Docs/HowWritingLedToPicturePainting.pdf

Schmandt-Besserat, D. (2009, September). Tokens and writing: The cognitive development. *SCRIPTA, 1*, 145–154. https://web.archive.org/web/20120915115315/https://webspace.utexas. edu/dsbay/Docs/Tokens%20and%20Writing%20--%20the%20Cognitive%20Development.pdf

Schmidt, M. S. (2010, July 29). To pack a stadium, provide video better than TV. *The New York Times*. http://www.nytimes.com/2010/07/29/sports/football/29stadium.html?_r=1&th&emc=th

Schofield, A. T. (1888). *Another world, or the fourth dimension*. George Allen & Unwin Ltd.

Schonfeld, M. (2000). *The philosophy of the young Kant: The precritical project*. Oxford University Press.

Schrodinger, E. (1944). *What is life?* Cambridge University Press. https://web.archive.org/web/201 60621003649/http://whatislife.stanford.edu/LoCo_files/What-is-Life.pdf

Schulz, J. (1993). Virtu-real space: Information technologies and the politics of consciousness. In *Computer Graphics: Visual Proceedings (ACM Siggraph, Annual Conference Series)* (pp. 159–163).

Scientist redefines 'Internet addiction' and sets new standards for treatment (2007, August 17). *PhysOrg.com*. http://www.physorg.com/news106574926.html

Scott, A. O. (2010, December 9). How real does it feel? *The New York Times*. http://www.nytimes. com/2010/12/12/magazine/12Reality-t.html?scp=2&sq=documentary&st=nyt

Searle, J. (1980). Minds, brains and programs. *Behavioral and Brain Sciences, 3*(3), 417–457.

Seife, C. (2000). *Zero: The biography of a dangerous idea*. Viking Penguin.

Shakespeare, W. (1600). *A midsummer night's dream*. Thomas Fisher. http://shakespeare.mit.edu/ midsummer/full.html

Shannon, C. E. (1948, July and October). A mathematical theory of communication. *Bell System Technical Journal, 27*, 379–423; 623–656. http://doi.org/10.1002%2Fj.1538-7305.1948. tb01338.x

Sharma, A. (2017, July 6). Japanese men are tying the knot with their favourite Anime characters in VR weddings. *Digit*. https://www.digit.in/features/vr-ar/japanese-men-are-tying-the-knot-with-their-favourite-anime-characters-in-vr-weddings-35911.html

Sheridan, T. B. (1992). Musings on telepresence and virtual reality. *Presence, 1*(1), 120–126.

Shields, R. (2003). *The virtual*. Routledge.

Shklovskii, I. S., & Sagan, C. (1966). *Intelligent life in the universe*. Delta Books.

Shore, D. I. (2021, August 3). Our brains perceive our environment differently when we're lying down. *The Conversation*. https://theconversation.com/our-brains-perceive-our-environment-dif ferently-when-were-lying-down-165234

Siegel, L. (2008). *Against the machine: Being human in the age of the electronic mob*. Spiegel & Grau.

Siegel, R. K. (2005). *Intoxication: The universal drive for mind-altering substances*. Park Street Press.

Sikora, T. (2000). 'Wild as a hawk's dream': Towards an ecology of cyberspace. https://www.aca demia.edu/3358933/Wild_as_a_Hawk_s_Dream_Towards_an_Ecology_of_Cyberspace_

Silva, R. (2021, April 17). 3D TV is dead—What you need to know. *Lifewire*. https://www.lifewire. com/why-3d-tv-died-4126776

Simner, J., Mulvenna, C., Sagiv, N., Tsakanikos, E., Witherby, S. A., Fraser, C., Scott, K., & Ward, J. (2006). Synaesthesia: The prevalence of atypical cross-modal experiences. *Perception, 35*(8), 1024–1033.

Simner, J., & Ward, J. (2008). Synaesthesia, color terms, and color space: Color claims came from color names in Beeli, Esslen, and Jäncke (2007). *Psychological Science, 19*(4), 412–414.

Simon, G., Lambert, C., & Feltz, F. (2006, June 8–9). A network of excellence as a virtual orga-nization: The nanobeams case. In F. Feltz, O. Benoit, A. Oberweis & N. Poussing (Eds.), *AIM 2006—Information Systems and Collaboration: State of the Art and Perspectives, Best Papers of the 11th International Conference of the Association Information and Management*, Luxembourg (pp. 297–317). http://subs.emis.de/LNI/Proceedings/Proceedings92/gi-proc-092-019.pdf

Simons, D. J., & Chabris, C. F. (1999). Gorillas in our midst: Sustained inattentional blindness for dynamic events. *Perception, 28*, 1059–1074.

Simulation. (2021, July 21). *Wikipedia.* http://en.wikipedia.org/wiki/Simulation#cite_note-enviro nment-3

Smith, A. (2003). *The wealth of nations.* Bantam Classics (Original work published 1776).

Smith, A. D. (2000). *Myths and memories of the nation.* Oxford University Press.

Smith, R. (2010). The long history of gaming in military history. *Simulation & Gaming, 41*(1), 6–19. http://sag.sagepub.com/content/41/1/6

Snow, C. P. (1959). *The two cultures and the scientific revolution.* Cambridge University Press.

Specter, M. (2009). *Denialism: How irrational thinking hinders scientific progress, harms the planet, and threatens our lives.* Penguin Books.

Spinney, L. (2020, December 9). What science can learn from religion. *NewStatesman.* https://www. newstatesman.com/science-tech/2020/12/what-science-can-learn-religion

'Spooky survey' gets big response (2006). *ABC Online.* http://www.abc.net.au/science/news/sto ries/2006/1791144.htm

Stahl, J. (2009, April). 20 things you didn't know about money. *Discover Magazine, 30*(4), 80. https://web.archive.org/web/20121017221036/http://discovermagazine.com/2009/apr/20-thi ngs-you-didn.t-know-about-money/

Staiger, J. (2005). *Media reception studies.* New York University Press.

Stapp, H. P. (2009). Quantum reality and mind. *Journal of Cosmology, 3*, 570–579. http://journalof cosmology.com/QuantumConsciousness105.html

Stock, G. (1993). *Metaman: The merging of humans and machines into a global superorganism.* Simon & Schuster.

Stoll, C. (1995). *Silicon snake oil: Second thoughts on the information highway.* Anchor Books.

Strom, D. (2006, October 9). Virtual reality or real virtuality? *TidBITS.* http://db.tidbits.com/article/ 8701

Takahashi, D. (2010, February 3). The Sims celebrates 125 million games sold across 10 years. *GamesBeat.* https://venturebeat.com/2010/02/03/the-sims-celebrates-125-million-games-sold-across-10-years/

Taleb, N. N. (2007). *The black swan.* Random House.

Tangermann, V. (2019, November 7). Hollywood legend James Dean to be reconstructed in CGI for movie. *FUTURISM—The Byte.* https://futurism.com/the-byte/james-dean-cgi-moviehttps:// futurism.com/the-byte/james-dean-cgi-movie

Taub, E. A. (2010, October 4). Augmented reality: From baseball cards to books. *The New York Times.* http://gadgetwise.blogs.nytimes.com/2010/10/04/augmented-reality-from-baseball-cards-to-books/?nl=technology&emc=techupdateemb1

Taylor, M. (1999). *Imaginary companions and the children who create them.* Oxford University Press.

Tegmark, M. (2007, September 25). Shut up and calculate. *arXiv.org.* http://arxiv.org/pdf/0709. 4024v1

Teitel, R. G. (2011). *Humanity's law.* Oxford University Press.

Terranova, T. (2004). *Network culture: Politics for the information age.* Pluto Press.

Thatcher, R. D., North, D. M., & Biver, C. J. (2008). Intelligence and EEG phase reset: A two compartmental model of phase shift and lock. *NeuroImage, 42*(4), 1639–1653.

The history of drugs (2002). *Drug-Rehabs.org.* http://www.drug-rehabs.org/drughistory.php

Thompson, E. (2004). *The soundscape of modernity: Architectural acoustics and the culture of listening in America, 1900–1933.* MIT Press.

Tillema, T. Dijst, M., & Schwanen, T. (2010). Face-to-face and electronic communications in main-taining social networks: The influence of geographical and relational distance and of information content. *New Media & Society, 12*(6), 965–983.

Tilly, C. (1984). *Big structures, large processes, huge comparisons*. Russell Sage Foundation Publications.

Toffler, A. (1970). *Future shock*. Random House.

Toffler, A. (1980). *The third wave*. William Morrow.

Tofts, D., Jonson, A., & Cavallero, A. (Eds.). (2002). *Prefiguring cyberculture: An intellectual history*. The MIT Press.

Tomas, D. (2004). *Beyond the machine: A history of visual technologies*. Continuum.

Tooby, J., & Cosmides, L. (1990). The past explains the present: Emotional adaptation and the structure of ancestral environments. *Ethology and Sociobiology, 11*, 407–424. http://citeseerx.ist. psu.edu/viewdoc/download?doi=10.1.1.115.2770&rep=rep1&type=pdf

Tooby, J., & Cosmides, L. (2001). Does beauty build adapted minds? Toward an evolutionary theory of aesthetics, fiction and the arts. *SubStance, 94/95*, 6–27. https://www.cep.ucsb.edu/papers/bea uty01.pdf

Tooby, J., & Cosmides, L. (2005). Conceptual foundations of evolutionary psychology. In D. M. Buss (Ed.), *The handbook of evolutionary psychology* (pp. 5–67). Wiley.

Trevethan, C. T., Sahraie, A., & Weiskranz, L. (2007). Can blindsight be superior to 'sighted-sight'? *Cognition, 103*(3), 491–501.

Tuan, Y.-F. (1998). *Escapism*. The Johns Hopkins University Press.

Turing, A. (1950). Computing machinery and intelligence. *Mind, LIX*(236), 433–460. https://aca demic.oup.com/mind/article/LIX/236/433/986238

Turkle, S. (2007, May 7). Can you hear me now? *Forbes.com*. https://web.archive.org/web/202007 14215607/https://www.forbes.com/free_forbes/2007/0507/176.html

Turner, J. (2002, January 23). Myron Kreueger live. *CTHEORY, 25*(1&2). https://web.archive.org/ web/20150914082125/http://www.ctheory.net/printer.aspx?id=328

Tversky, A., & Kahneman, D. (1982). Judgment under uncertainty: Heuristics and biases. In D. Kahneman, P. Slovic & A. Tversky (Eds.), *Judgment under uncertainty: Heuristics and biases* (pp. 3–22). Cambridge University Press.

Tyner, J. A. (2018). Maps as language/the language of maps. In S. D. Brunn & R. Kehrein (Eds.), *Handbook of the changing world language map* (pp. 1–6). Springer. https://link.springer.com/ref erenceworkentry/10.1007%2F978-3-319-73400-2_92-1#citeas

Ulmer, G. L. (2002). Reality tables: Virtual furniture. In D. Tofts, A. Jonson & A. Cavallero (Eds.), *Prefiguring cyberculture: An intellectual history* (pp. 110–129). The MIT Press.

University of British Columbia. (2021, August 2). More to pictures than meets the eye: Study. *MedicalXpress*. https://medicalxpress.com/news/2021-08-pictures-eye.html

Valdes Garcia, A. (2017, January 5). IBM 5 in 5: Hyperimaging and AI will give us superhero vision. *IBM Research*. https://www.ibm.com/blogs/research/2017/1/ibm-5-in-5-hyperimaging-and-ai-will-give-us-superhero-vision

van der Veer, P. (2012). Market and money: A critique of rational choice theory. *Social Compass, 59*, 183–192.

Van Essen, D. C., & Glasser, M. F. (2016, September–October). The human connectome project: Progress and prospects. *Cerebrum*. https://www.ncbi.nlm.nih.gov/pmc/articles/PMC5198757/

Vandendorpe, C. (1999). *From papyrus to hypertext: Toward the universal digital library* (Rev. ed. 2009; P. Aronoff & H. Scott, Trans.). University of Illinois Press.

Vasquez, M. A. (2010). *More than belief: A materialist theory of religion*. Oxford University Press.

Veenhoven, R. (2008). Healthy happiness: Effects of happiness on physical health and the consequences for preventive health care. *Journal of Happiness Studies, 9*(3), 449–469.

Vilenkin, A. (2006). *Many worlds in one: The search for other universes*. Hill & Wang.

Vilenkin, A., & Tegmark, M. (2011, July 19). The case for parallel universes. *Scientific American*. http://www.scientificamerican.com/article.cfm?id=multiverse-the-case-for-parallel-universe

Virtual organisations become a reality (2008, December 12). *PhysOrg.com*. http://www.physorg. com/news148318112.html

Virtual possessions have powerful hold on teenagers, researchers say (2011, May 9). *PhysOrg.com*. http://www.physorg.com/news/2011-05-virtual-powerful-teenagers.html

von Däniken, E. (1999). *Chariots of the gods? Unsolved mysteries of the past* (M. Heron & Souvenir Press, Trans.). Berkley Books (Original work published 1968).

Voss, J. L., & Paller, K. A. (2009). An electrophysiological signature of unconscious recognition memory. *Nature Neuroscience, 12*, 349–355. http://www.nature.com/neuro/journal/v12/n3/full/nn.2260.html

Wallin, N. L., Merker, B., & Brown, S. (2000). An introduction to evolutionary musicology. In N. L. Wallin, B. Merker & S. Brown (Eds.), *The origins of music* (pp. 3–24). MIT Press.

Walton, K. (1990). *Mimesis as make-believe: On the foundations of representational art.* Harvard University Press.

Walton, K. (1997). Spelunking, simulation, and slime: On being moved by fiction. In M. Hjort & S. Laver (Eds.), *Emotion and the arts* (pp. 37–49). Oxford University Press.

Ward, K. (2010). God as the ultimate informational principle. In P. Davies & N. H. Gregersen (Eds.), *Information and the nature of reality: From physics to metaphysics* (pp. 282–300). Cambridge University Press.

Warwick, K. (2010). Implications and consequences of robots with biological brains. *Ethics & Information Technology, 12*, 223–234.

Warzel, C. (2018, February 11). Believable: The terrifying future of fake news. *BuzzFeed.News.* https://www.buzzfeednews.com/article/charliewarzel/the-terrifying-future-of-fake-news#.vpv5R5wXV

Watters, E. (2010, January 8). The Americanization of mental illness. *The New York Times.* http://www.nytimes.com/2010/01/10/magazine/10psyche-t.html?pagewanted=all

Waytz, A., Morewedge, C. K., Epley, N., Monteleone, G., Gao, J.-H., & Cacioppo, J. T. (2010). Making sense by making sentient: Effectance motivation increases anthropomorphism. *Journal of Personality and Social Psychology, 99*(3), 410–435.

Wei-Haas, M. (2020, November 4). Prehistoric female hunter discovery upends gender role assumptions. *National Geographic.* https://www.nationalgeographic.com/science/2020/11/prehistoric-female-hunter-discovery-upends-gender-role-assumptions/

Weil, A. (2004). *The natural mind: A revolutionary approach to the drug problem* (Rev. ed.). Houghton Mifflin.

Weintraub, K. (2019, March 25). The adult brain does grow new neurons after all, study says. *Scientific American.* https://www.scientificamerican.com/article/the-adult-brain-does-grow-new-neurons-after-all-study-says/

Weisman, A. (2007). *The world without us.* Thomas Dunne Books.

Wells, H. G. (1895). *The time machine.* William Heinemann.

Wen, T. (2014, September 5). Why do people believe in ghosts? *The Atlantic.* https://www.theatlantic.com/health/archive/2014/09/why-do-people-believe-in-ghosts/379072/

Wertheim, M. (1999). *The pearly gates of cyberspace: A history of space from Dante to the internet.* W.W. Norton & Co.

What is cryptocurrency, how does it work and why do we use it? (2018, August 17). *The Telegraph.* https://www.telegraph.co.uk/technology/0/cryptocurrency/

Whitehead, A. N. (1929). *Process and reality: An essay in cosmology.* Macmillan.

Whitehouse, H. (2000). *Arguments and icons: Divergent modes of religiosity.* Oxford University Press.

Whitely, S., & Rambarran, S. (2016). *The Oxford handbook of music and virtuality.* Oxford University Press.

Williams, C. (2011, August 26). Crittervision: The world as animals see (and sniff) it. *NewScientist.com.* http://www.newscientist.com/special/crittervision

Wohlforth, C. (2010, March). Who are you calling bird brain? *Discover Magazine*, 45–49. https://web.archive.org/web/20191101055048/http://discovermagazine.com:80/2010/mar/01-who-you-callin-bird-brain

Wollstonecraft (Godwin) Shelley, S. (1993). *Frankenstein; or, the modern Prometheus.* Project Gutenberg e-book (Original work published 1817). https://www.gutenberg.org/files/84/84-h/84-h.htm

Wooley, B. (1992). *Virtual worlds*. Blackwell.

Wu, C., Liu, J., Huang, X., Li, Z.-P., Yu, C., Ye, J.-T., Zhang, J., Zhang, Q., Dou, X., Goyal, V. K., Xu, F., & Pan, J.-W. (2021, March 9). Non-line-of-sight imaging over 1.43 km. *PNAS, 118*(10). http://doi.org/10.1073/pnas.2024468118

Yerushalmi, Y. H. (2005). *Zakhor: Jewish history and Jewish memory.* The University of Washington Press.

Yoo, S.-S., Kim, H., Filandrianos, E, Taghados, S. J., & Park, S. (2013, April 3). Non-invasive brain-to-brain interface (BBI): Establishing functional links between two brains. *PLoS One.* http://doi.org/10.1371/journal.pone.0060410

Young, G. (2004). Reading and praying online: The continuity of religion online and online religion in internet Christianity. In L. L. Dawson & D. E. Cowan (Eds.), *Religion online: Finding faith on the internet* (pp. 93–105). Routledge.

Zeltzer, D. (1992). Autonomy, interaction and presence. *Presence, 1*(1), 127–132.

Zhaoping, L., & Jingling, L. (2008, February 15). Filling-in and suppression of visual perception from context: A Bayesian account of perceptual biases by contextual influences. *PLoS Journal of Computational Biology, 4*(2), e14. http://doi.org/10.1371/journal.pcbi.0040014

Zhou, M. (2018, September 25). 3D-printed gun controversy: Everything you need to know. *cnet.* https://www.cnet.com/news/the-3d-printed-gun-controversy-everything-you-need-to-know/

Zimmer, C. (2008, October). Could an inner zombie be controlling your brain? *Discover Magazine.* https://web.archive.org/web/20191022204229/http://discovermagazine.com:80/2008/oct/15-could-an-inner-zombie-be-controlling-your-brain

Zimmer, C. (2010, August 31). Scientists square off on evolutionary value of helping relatives. *The New York Times.* http://www.nytimes.com/2010/08/31/science/31social.html?ref=science

Zizek, S. (1994). *Tarrying with the negative: Kant, Hegel, and the critique of ideology.* Duke University Press.

Zunshine, L. (2006). *Why we read fiction: Theory of mind and the novel.* Ohio State University Press.

# Index

Printed in the United States
by Baker & Taylor Publisher Services

Printed in the United States
by Baker & Taylor Publisher Services